ADVANCES IN CHEMICAL PHYSICS

VOLUME LXXXVI

Advances in
CHEMICAL PHYSICS

Edited by

I. PRIGOGINE

University of Brussels
Brussels, Belgium
and
University of Texas
Austin, Texas

and

STUART A. RICE

Department of Chemistry
and
The James Franck Institute
The University of Chicago
Chicago, Illinois

VOLUME LXXXVI

AN INTERSCIENCE® PUBLICATION
JOHN WILEY & SONS, INC.
NEW YORK • CHICHESTER • BRISBANE • TORONTO • SINGAPORE

CONTRIBUTORS TO VOLUME LXXXVI

M. P. ALLEN, H. H. Wills Physics Laboratory, Bristol, United Kingdom

G. T. EVANS, Department of Chemistry, Oregon State University, Corvallis, Oregon

D. FRENKEL, F.O.M. Institute for Atomic and Molecular Physics, Amsterdam, The Netherlands

F. GRECO, Universita degli Studi di Napoli Federico II, Departimento de Ingegneria Chimica, Napoli, Italy

J. KRZYSTEK, Institute of Physics, Polish Academy of Sciences, Warsaw, Poland

G. MARRUCCI, Universita degli Studi di Napoli Federico II, Departimento di Ingegneria Chimica, Napoli, Italy

B. M. MULDER, F.O.M. Institute for Atomic and Molecular Physics, Amsterdam, The Netherlands

J. U. VON SCHÜTZ, Physikalisches Institut, Universitat Stuttgart, Germany

INTRODUCTION

Few of us can any longer keep up with the flood of scientific literature, even in specialized subfields. Any attempt to do more and be broadly educated with respect to a large domain of science has the appearance of tilting at windmills. Yet the synthesis of ideas drawn from different subjects into new, powerful, general concepts is as valuable as ever, and the desire to remain educated persists in all scientists. This series, *Advances in Chemical Physics*, is devoted to helping the reader obtain general information about a wide variety of topics in chemical physics, a field which we interpret very broadly. Our intent is to have experts present comprehensive analyses of subjects of interest and to encourage the expression of individual points of view. We hope that this approach to the presentation of an overview of a subject will both stimulate new research and serve as a personalized learning text for beginners in a field.

ILYA PRIGOGINE
STUART A. RICE

CONTENTS

ADVANCES IN CHEMICAL PHYSICS

VOLUME LXXXVI

HARD CONVEX BODY FLUIDS

M. P. ALLEN

*H. H. Wills Physics Laboratory,
Tyndall Avenue Bristol BS8 1TL, UK*

G. T. EVANS

*Department of Chemistry, Oregon State University,
Corvallis OR 97331 USA*

D. FRENKEL, B. M. MULDER

*F.O.M. Institute for Atomic and Molecular Physics,
Kruislaan 407, 1098 SJ Amsterdam, The Netherlands*

CONTENTS

Advances in Chemical Physics, *Volume LXXXVI*, Edited by I. Prigogine and Stuart A. Rice.
ISBN 0-471-59845-3 © 1993 John Wiley & Sons, Inc.

I. INTRODUCTION

The basis of our theoretical understanding of statistical mechanics of dilute gases dates back to the nineteenth century. The atomistic description of crystalline solids originated in the first half of the twentieth century. But the theoretical description of simple liquids in the framework of statistical mechanics is of a much more recent origin. A possible reason for the slow development of a molecular theory of liquids is the following. The theory of dilute gases could be developed starting from the *ideal gas*, the properties of which could be evaluated analytically. Similarly, the theory of solids could be constructed starting from the *harmonic crystal*. But, for liquids, there is no corresponding *ideal liquid* model that can be solved exactly. As a consequence, much of the progress in the construction of a molecular theory of liquids had to await the advent of electronic computers that could generate essentially exact results for simple model liquids. During the past four decades, the numerical data on the hard-sphere fluid have provided us with a substitute for exact results on the *ideal liquid*. Much of the subsequent development of the theory of liquids, as described for instance in the book by Hansen and McDonald, [1] relies heavily on the insight gained from the numerical study of hard-sphere fluids.

The hard-sphere model is, however, not a good reference system for the description of molecular fluids consisting of nonspherical particles. This is particularly clear if one considers the possible phases of such molecular systems. Whereas hard spheres can only form an (isotropic) liquid phase and a crystalline solid phase, there are many nonspherical molecules in nature that can form so-called liquid crystalline phases, that is, phases that have a degree of order that is intermediate between the isotropic fluid and the crystalline solid (see, e.g., Ref. [2]). In fact, several dozens of distinct liquid-crystalline phases have been observed expereimentally. However, in the present review, we focus on the main classes of liquid crystals, i.e. the *nematic* phase, the *smectic-A* phase, the *columnar* phase and the *cholesteric* phase. Schematic drawings of the orientational and translational order characteristic of these phases are shown in Fig 1.1.

In order to extend the theory of simple atomic liquids to the more complex molecular liquids and, a fortiori, to liquid crystals, there is much need for 'exact' results on simple molecular reference fluids. Surprisingly, although there exists no exactly solvable model for atomic liquids, there does, in fact, exist an exactly solvable model for a (nematic) liquid crystal, viz. the Onsager model. [3] In the Onsager model, the liquid-crystal forming molecules are assumed to be infinitely thin, hard spherocylinders. The problem with the Onsager model is that, whereas many simple liquids are, to a first approximation, well described as an assembly of hard spheres, there are only a few liquid-crystal forming molecules with very long, thin hard rods. Once we try to extend our theoretical description of liquid-crystal forming fluids to molecules with less extreme shapes or, for that matter, to liquid-crystalline phases other than the nematic phase, we are again in need of numerical data on molecular "reference fluids" as a substitute for exact results.

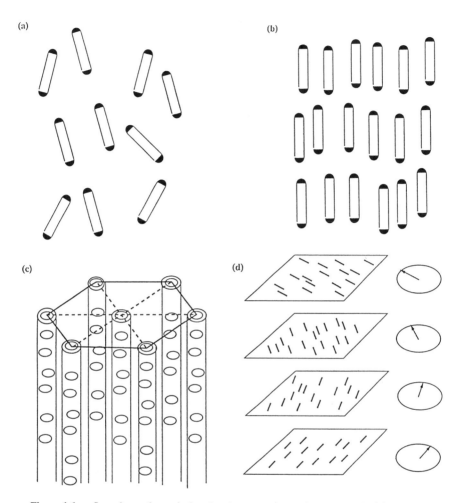

Figure 1.1. Snapshots of a typical molecular arrangement in the nematic (a), smectic-A (b), columnar (c) and cholesteric (d) phase.

Unfortunately, the choice of the appropriate molecular reference fluid is not unique. Even if we restrict our attention to rigid, nonspherical molecules, there are clearly many model systems that could be selected. However, our reference models should satisfy two important criteria: first, they should be sufficiently nontrivial to reproduce the most important classes of liquid crystals known to date, yet they should be sufficiently simple to ensure that they can be easily studied, both theoretically and numerically.

In this review, we discuss in considerable detail a particular class of molecular "reference" systems that meet these criteria, namely *convex, hard-body fluids*. A body is convex if any line-segment connecting two points on the surface of that body is completely contained within that body. During the past decade there has been much progress both in the theoretical description and the numerical simulation of convex hard-body fluids. It is not our aim to give an exhaustive review of progress in this field of research. Rather, we wish to present a coherent overview of the theoretical and numerical techniques that are most widely used in this area of research. The choice of specific examples is, to a large extent, dictated by our own bias. Yet, we have tried as much as possible to refer to related work by other authors. We realize, however, that our review of the relevant literature will contain serious gaps and omissions. We apologize to all those whose contributions we have discussed either inadequately, or not at all.

The material presented in this review is organized in a way that emphasizes the complementary character of theory and simulation. This implies that on every topic we first have a section that discusses the relevant theory and then a section that deals with the appropriate simulation techniques. The latter section will typically contain a few numerical results that are of special interest in the context of the preceding discussion. In this way, we discuss first the static and dynamic properties of hard-core fluids in the isotropic phase, next phase transitions to (liquid)-crystalline phases, and finally some static and transport properties of the nematic liquid-crystalline phase. We do not discuss the dynamical properties of liquid-crystalline phases other than nematic, in view of the paucity of numerical data on such systems.

In this introduction, we give a brief preview of things to come, and explain the philosophy underlying much of the work presented in subsequent sections. For the benefit of the reader who is unfamiliar with computer simulations, we present a brief description of the role of numerical simulation in the study of liquids and liquid crystals. Next, we briefly preview the kind of hard-particle models that we are going to consider, the numerical techniques needed to study such systems, and the theories that are used to describe them.

A. Simulations

Computer simulations sit midway between experimental measurements and theories of condensed matter. Typically, the aim of a theory is to predict the properties of a system in terms of the interactions between molecules. However, these molecular interactions are themselves known only imperfectly, and must be modelled in some way. To test the accuracy of the molecular models separately from that of the theories, it is necessary to obtain reliable "experimental" information about the models. This is accomplished using computer simulation. In recent years, progress has been made using both accurate, "realistic" models of specific molecular systems, and idealized models of wide applicability. In this review, we discuss recent work using models that fall into the latter category.

1. Simulation Strategies

The aim of a computer simulation of a (classical) many-body system is to compute "exactly" the static and dynamic equilibrium properties of the model system. In this context, "exactly" means "to any desired accuracy". There are two distinct factors that limit the accuracy of a simulation. First, simulations of models for bulk liquids or solids are usually performed on rather small systems ($\mathcal{O}(10^2-10^3)$ particles). Even though periodic boundary conditions are usually employed to minimize finite size effects, the properties of such relatively small systems do differ systematically from those of a truly macroscopic sample. In addition to this systematic error introduced in the numerical simulation, we are also faced with statistical fluctuations in the results of our numerical "measurements". In principle, the errors due to finite size effects can be suppressed by going to very large systems, while the statistical fluctuations can be suppressed by performing very long simulations. Clearly, if one is interested in the properties of a "family" of model systems over a wide range of densities, one should not invest all available computing time in one long simulation of one large system. It is even less advisable to plan short simulations of a very large system, in which case the statistical errors would be very much larger than the systematic errors. In general, one should select the model, the system size and the length of the simulations such that a reasonably complete set of simulations of acceptable accuracy can be performed within the available computing budget. In its generality, the preceding statement is vague to the extent of being almost meaningless. However, in the context of simulations of nonspherical, liquid-crystal forming molecules, we can be more precise. Typically, in order to map a phase diagram, one should study the equation of state of several model systems that belong to the same family. In addition, one or more free energy calculation may be required for every model system. As a result, one should expect to perform some 50 runs per model system. For a determination of a complete phase diagram, several hundred simulations will be required. Clearly, with this number of simulations, it is imperative that the individual simulations are "cheap". Again, "cheap" can only be used in the relative sense that any individual run should not consume more than, say, 1% of the available computing budget. As the power of computers continues to grow, much of what is not feasible today will most likely be feasible in a few years time. Yet, it is fair to say that during the period covered by this review, systematic studies of the phase diagram were necessarily limited to rather simple model systems. We should stress, however, that most of the numerical techniques discussed below can be applied to more complex model systems.

B. Hard Particles

Since the 1950s, when computer simulation showed that the hard-sphere model provided a firm base for the study of the statistical mechanics of simple atomic liquids, [4,1], interest has steadily grown in the application of similar techniques to molecular fluids. The wealth of fluid phases found in nature has stimulated attempts to relate phase stability and properties to simple aspects of molecules and their interactions, beginning with elementary considerations of molecular size and shape. It has been known since the time of Onsager [3] that hard nonspherical particles can form an oriented fluid phase if the isotropic liquid is compressed, and that for sufficient elongation, this transition will occur before the system freezes. It is therefore surprising that significant progress in computer simulations of these phases has only been made in the last decade (although serious attempts have been made since the early 1970s [5,6]). The spherocylinder (see Fig. 1.2.a) was probably the first nonspherical hard shape to attract significant interest. This is a cylinder of length L, diameter D, with hemispherical caps of diameter D at each end. As we shall see, this model exhibits an isotropic fluid phase and a solid; for suitable values of L/D nematic and smectic-A liquid crystals can be seen.

A second shape of interest is the spheroid, with perpendicular semi-axes a, b and c. Most intensively studied have been hard ellipsoids of revolution, for which two of the axes are equal, but different in general from the third: $a \neq b = c$. This system forms isotropic fluid and fully ordered solid phases, and, for suitable values of the elongation $e = a/b$, also shows a nematic liquid crystal. Both rod-like (prolate, $e > 1$ see Fig. 1.2.b) and disk-like (oblate, $e < 1$ see Fig. 1.2.c) shapes are possible, and the nematic phases correspond to alignment of the long or short symmetry axes, respectively. An additional, biaxial phase may be exhibited by molecules having unequal axes $a \neq b \neq c$, thus being intermediate in shape between rods and disks.

Different oblate, or tablet-like, shapes can be obtained by slicing the top and bottom off a sphere using two parallel cuts. These "truncated sphere" (see Fig. 1.2.d) shapes may form nematic phases, and additionally can line up to give columnar phases. A further phase having cubic orientational symmetry has also been observed for these shapes. The relation between this "cubatic" phase and the cubatic phases that have been observed experimentally [7] is at this stage uncertain.

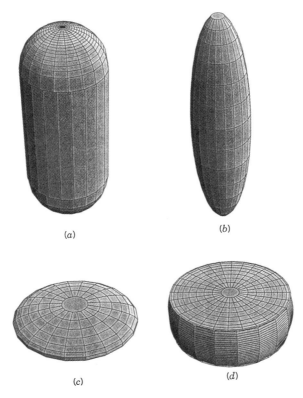

(a)

(b)

(c)

(d)

Figure 1.2. A few well-known hard convex bodies: (a) spherocylinder; (b) prolate ellipsoid of revolution; (c) oblate ellipsoid of revolution; (d) truncated sphere

C. Simulation Methods

Two classes of simulation are in common use: Monte Carlo and molecular dynamics.

In hard-particle Monte Carlo, trial moves are selected using a random number generator, accepted if they do not lead to particle overlap, and rejected if they do. This requires efficient evaluation of a pair overlap function: most usefully a function $F(\mathbf{r}_{ij}, \Omega_i, \Omega_j)$ of the orientations Ω_i and Ω_j and the relative position vector $\mathbf{r}_{ij}$ of a pair of molecules, which takes values $F < 1$ if they overlap, $F > 1$ if they do not, and $F = 1$ at contact. Efficient prescriptions exist [5,6,8,9] to determine F for both spherocylinders and spheroids, and we return to this in Section II.D.1.

The simple prescription of moving particles generates states sampled from the constant NVT ensemble where N is the number of particles, V the sample volume and T the temperature. For particles interacting via infinite repulsive potentials, the configurational integral and all static configurational properties are independent of the temperature T. Extensions of the sampling prescription can be used to generate states from other ensembles, for example constant NPT, where P, the pressure, is prescribed. Again, in the special case of hard particles, configurational properties depend on the ratio P/T rather than on P and T separately.

In hard-particle molecular dynamics, Newton's equations are solved; the aim is to locate the next time of collision between a pair of particles. Between collisions, the configuration is advanced using free flight dynamics; each molecule moves with constant linear and angular momenta. At the point of collision, impulsive forces determined by the conservation laws and the contact condition (i.e., whether the colliding surfaces are rough or smooth) determine the postcollisional momenta. Both free flight and collision dynamics also depend on the choice of molecular masses and moments of inertia. The technique requires efficient evaluation of the pair overlap function F and its time derivative $\dot{F}$, so as to locate the exact time of collision by standard root-finding methods. Typically the constant NVE ensemble is probed, where E is the energy. For hard particles, kinetic energy and total energy are the same, and *static* configurational properties are independent of E. They also do not depend on the chosen mass and moment-of-inertia distribution. The *dynamical* properties, however, do depend on these values; the masses, moments of inertia and total energy determine translational and rotational time scales. It is possible to adapt the molecular dynamics algorithm to probe other ensembles. Intermittent velocity randomization can be included to give constant NVT sampling. A constant pressure form of dynamics has been described for hard particles [10] but a simpler procedure for sampling the constant NPT ensemble is to carry out intermittent volume changes according to the standard Monte Carlo prescription, in between periods of normal dynamics.

Further details of simulation techniques may be found elsewhere. [4,11] In all simulations of bulk phases, periodic boundary conditions are used to eliminate the effects of surfaces. There have been few systematic studies of the effects of periodic box size and shape on the stability and properties of phases, and this is of some concern when simulating systems that exhibit long-range correlations. The general rule seems to be to choose a box as large as possible, given the constraints of limited computer time, and for translationally disordered fluid phases to adopt one of the more nearly spherical geometries: truncated octahedral or rhombic dodecahedral shapes. For solids, or smectic liquid crystals, this may not be appropriate, and cuboidal boxes may be more suitable. In the simulations reported here, both cuboidal and truncated octahedral periodic boundary conditions have been employed. Typical system sizes are in the range

$N = 100$–1000. In molecular dynamics simulations, typical production run lengths are $(0.5$–$1.6) \times 10^6$ collisions in total, depending on density; this corresponds to run times $t_{run} \sim 2000$–$15,000 t_c$, where t_c is the mean time between collisions per molecule. In Monte Carlo work, run lengths are of the order of 10^4–10^6 moves per particle. These parameters are modest by today's standards; a typical run at one state point might take a few hours on a fast desktop workstation, or a few minutes on a supercomputer.

D. Theories

With such simple models, we are clearly going to be interested in comparing with theoretical predictions rather than experiment. Two major classes of theory are especially powerful when applied to hard-particle systems. Kinetic theories make specific predictions for transport coefficients and other dynamical properties in terms of collisional averages. Density functional theories, of which the approaches of van der Waals and Onsager may be considered special cases, are used to predict phase stability and properties, given an approximation scheme for the direct correlation function. Both methods have been extensively tested on the hard-sphere fluid, and their advantages and limitations in this area are well known. Their extension to nonspherical systems, however, has been very limited. In the following sections we attempt to give a perspective view of these theories, and the role of simulation in testing them out.

PART ONE: THE ISOTROPIC PHASE

II. STATIC PROPERTIES

A. Static Properties in the Isotropic Phase

Convexity is the central characteristic that makes hard convex body (HCB) fluids amenable to analysis. It is this property that allows a unique determination of the distance between two such particles. This in turn enables a relative simple description of the hard-core interactions in such systems. The simplest example of a HCB is of course the hard sphere and it should come as no surprise that most calculations of equilibrium thermodynamic properties of HCB systems such as free energies and pressures rest on the foundation provided by earlier analyses of hard spheres. [12] There are, however, features in the HCB systems that have no analog in hard-sphere (HS) systems and these pertain to the orientational degrees of freedom and their canonical momenta. Having said this, it should come as no surprise that most of the successes in the analysis of fluids of HCBs have been obtained in studies of those properties that have a hard-sphere counter-part, such as pressure, chemical potential or the isotropic part of the pair correlation function at contact. Furthermore, most of the stumbling blocks have arisen in the calculation of properties that depend explicitly on the mutual orientation of HCBs, such as pair and direct correlation functions and structure factors. In this section, we summarize some of the progress made in the analysis of scalar properties (pressures, chemical potentials) and vector properties (orientational pair and direct correlation functions) in the theory of isotropic fluids of HCBs.

B. Theory

1. The System

We consider a system consisting of N HCBs in a container of volume V (at a number density $\rho = N/V$) and at a temperature T. The ith particle in the system has a mass m, a center of mass position vector, $\mathbf{r}_i$, a moment of inertia I, an orientation vector $\hat{\mathbf{u}}_i$ (for unaxial rotors) and Euler angles Ω_i (for biaxial or asymmetric tops), a center-of-mass velocity $\mathbf{v}_i$, linear momentum $\mathbf{p}_i$, angular velocity ω_i and angular momentum $\mathbf{L}_i$. Using conventional notation, one obtains a Helmholtz free energy

$$F_N(V,T) = -k_B T \ln Q_N(V,T) \tag{2.1}$$

with $Q_N(V,T)$ the canonical partition function

$$Q_N(V,T) = (N!)^{-1} \int d1\,d2\ldots dN \, \exp(-\beta \mathcal{H}) \tag{2.2}$$

and $\mathcal{H}$ the system Hamiltonian

$$\mathcal{H} = \sum_i \frac{1}{2}[m\mathbf{v}_i^2 + I\omega_i^2] + \sum_{i>j} U(\mathbf{r}_{ij}, \Omega_i, \Omega_j) \tag{2.3}$$

Here $\mathbf{r}_{ij}$ denotes the vector emanating from the mass center of particle i and extending to the mass center of particle j

$$\mathbf{r}_{ij} = \mathbf{r}_j - \mathbf{r}_i \tag{2.4}$$

The phase space volume di is taken to be

$$di = d\mathbf{r}_i\, d\mathbf{p}_i\, d\Omega_i\, d\mathbf{L}_i \tag{2.5}$$

and spans the translational and rotational degrees of freedom.

The hard-body potential energy $U(\mathbf{r}_{ij}, \Omega_i, \Omega_j)$ is given by

$$U(\mathbf{r_{ij}}, \Omega_i, \Omega_j) = \begin{cases} \infty, & \text{if } \mathbf{r}_{ij} \in V_{ex}(\Omega_i, \Omega_j) \\ 0, & \text{otherwise} \end{cases} \qquad (2.6)$$

where $V_{ex}(\Omega_i, \Omega_j)$ is the volume excluded to particle i because of particle j (and vice versa).

Classical dynamics described by Eqs. (2.3) and (2.6) allow the particles to move freely subject to the constraint that particle overlap is forbidden. Since the Hamiltonian is separable into position and momentum components and the momentum is described by a homogeneous quadratic form, then $Q_N(T, V)$ reduces to

$$Q_N(V, T) = Z_N(T, V)[\mathcal{V}_T]^{-N}/N! \qquad (2.7)$$

where $\mathcal{V}_T$ is the de Broglie volume as discussed in Appendix A.A, $Z_N(T, V)$ is the configurational integral

$$Z_N(T, V) = \int \prod_i d\mathbf{r}_i \, d\Omega_i \, \exp(-\beta U^N) \qquad (2.8)$$

and U^N is the N particle potential energy. The microscopic structure of a fluid and the thermodynamic properties can also be expressed in terms of the pair correlation function (pcf), $g(1, 2)$,

$$g(1, 2) = V^2 \int d\mathbf{r}_3 \, d\Omega_3 \ldots d\mathbf{r}_N \, d\Omega_N \, \exp(-\beta U^N)/Z_N(T, V) \qquad (2.9)$$

where 1 and 2 as arguments of $g(1, 2)$ now pertain to the positional coordinates.

C. Thermodynamic Properties

The thermodynamic properties which characterize the HCB system are the internal energy ($E = (\frac{1}{2})Nk_BT(3 + r)$, for a system with r rotational degrees of freedom), the pressure P and the chemical potential μ. The derivation of the pressure and the chemical potential from the canonical ensemble is standard [13] and we merely present the results with a few words of clarification. The pressure can be given in a virial form

$$\beta P = \rho + (3V)^{-1} \sum_{i>j} \langle \mathbf{r}_{ij} \cdot \mathbf{F}_{ij} \rangle \qquad (2.10)$$

where $\mathbf{F}_{ij}$ is the force on particle i due to particle j. For a system comprised of hard smooth particles, the force can be replaced by [14]

$$\beta \mathbf{F}_{12} = \hat{\mathbf{s}}\delta(s - 0^+) \tag{2.11}$$

where $\hat{\mathbf{s}}$ is the outward surface normal from particle 1 to particle 2 and s is the surface-to-surface separation measured along $\hat{\mathbf{s}}$. Accordingly, Eq (2.10) becomes

$$\beta P = \rho + (1/6)\rho^2 \int d\mathbf{r}_{12}\, d\Omega_1\, d\Omega_2 \mathbf{r}_{12} \cdot \hat{\mathbf{s}}\delta(s - 0^+)g(1,2) \tag{2.12}$$

All of the transport and most of the equilibrium properties to be discussed will involve the pcf at a distance infinitesimally removed (0^+) from the contact surface. We can take advantage of the ubiquitous presence of the contact surface by adopting a convex body coordinate system with coordinates s and $\hat{\mathbf{s}}$ in place of $\mathbf{r}_{12}$ and to this we now turn.

To change coordinates from the vector $\mathbf{r}_{12}$ to the unit vector $\hat{\mathbf{s}}$ and the surface-to-surface separation s, we begin by representing the center-to-center vector by

$$\mathbf{r}_{12}(\hat{\mathbf{s}}, s) = \xi_1(\hat{\mathbf{s}}) - \xi_2(-\hat{\mathbf{s}}) + s\hat{\mathbf{s}} \tag{2.13}$$

where $\xi_1(\hat{\mathbf{s}})$ and $\xi_2(-\hat{\mathbf{s}})$ emanate from the mass center of particles 1 and 2, respectively. The radius vector ξ_j for a general convex body can be written in terms of the support function [15]

$$h_j = \hat{\mathbf{s}} \cdot \xi_j(\hat{\mathbf{s}}) \tag{2.14}$$

and by means of the support function, all the geometric properties of the convex body can be derived (see Appendices A.B, A.C and A.D for details). The geometry of the situation is illustrated in Fig. 2.1

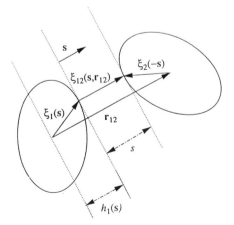

Figure 2.1.   The geometry of two convex bodies. The symbols are explained in the text.

The Jacobian for the transformation from $\mathbf{r}_{12}$ to $\hat{\mathbf{s}}$ and s is

$$
\begin{aligned}
d\mathbf{r}_{12} &= |\, (\partial\mathbf{r}_{12}/\partial s) \cdot (\partial\mathbf{r}_{12}/\partial\theta) \times (\partial r_{12}/\partial\phi)\,|\ ds\,d\theta\,d\phi \\
&= |\,\hat{\mathbf{s}} \cdot (\partial\mathbf{r}_{12}/\partial\theta) \times (\partial r_{12}/\partial\phi)\,|\ ds\,d\theta\,d\phi \\
&= S^{12}(\mathbf{s},\Omega_1,\Omega_2) \sin\theta\ ds\,d\theta\,d\phi = S^{12}(s,\Omega_1,\Omega_2)\,d\mathbf{s} \quad (2.15)
\end{aligned}
$$

where θ and ϕ are the polar and azimuthal angles that $\hat{\mathbf{s}}\,(\mathbf{s}=s\hat{\mathbf{s}})$ makes with respect to a fixed but otherwise arbitrary coordinate system. The abbreviated notation $d\mathbf{s}$ denotes $ds\,d\hat{\mathbf{s}}$. Equation (2.15) is a crucial result as it relates the volume element $d\mathbf{r}_{12}$ on a spherical contour to the contours appropriate for a general convex body. If we now return to Eq. (2.12) we can write directly

$$
\beta P = \rho + (2\pi/3)\rho^2 \int d\Omega_1\,d\Omega_2 h_{12}G(s=0,\Omega_1,\Omega_2) \quad (2.16)
$$

with

$$
h_{12} = h_1 + h_2 \quad (2.17)
$$

and

$$
G(s,\Omega_1,\Omega_2) = S^{12}(s,\Omega_1,\Omega_2)g(s,\Omega_1,\Omega_2) \quad (2.18)
$$

An exact expression for the chemical potential of HCBs can be derived by considering the reversible work required to insert a particle in a hard-body fluid [13,16,17] and the result is

$$
\beta\mu = \ln\rho\mathcal{V}_T + 4\pi\rho \int_{q_{\min}}^{1} dq\,d\Omega_1\,d\Omega_2 h_1 G_q(s=0,\Omega_1,\Omega_2) \quad (2.19)
$$

where $\mathbf{r}_{12}(\hat{\mathbf{s}},s) = q\xi_1(\hat{\mathbf{s}}) - \xi_2(-\hat{\mathbf{s}}) + s\hat{\mathbf{s}}$. Note that G_q depends on the scaling parameter, q, of particle 1 through $g(1,2)$ and S_q^{12}. The lower bound to the scaling variable corresponds to the ratio of axes lengths for the scaled particle: $q_{\min} = $ -long axis/short axis. Both Eqs. (2.16) and (2.19) reduce to the hard- sphere (HS) limit when we take $h_{12} = \sigma$ (the HS diameter) and $S^{12} = \sigma^2$. To proceed further with our analysis of HCBs, we need to be more explicit about the pcf.

1. Relationship of Pair Correlation Functions to Thermodynamic Properties

In 1963, Steele [18] suggested that $g(1,2)$ be expressed as a spherical harmonic expansion

$$
g(1,2) = 4\pi \sum_{j,j',m} g_{jj'm}(r)Y_{j,m}(\Omega_1)Y_{j',-m}(\Omega_2) \quad (2.20)
$$

where $r = |\mathbf{r}_{12}|$ and with Ω_1,Ω_2 being measured with respect to $\mathbf{r}_{12}$. Equation (2.20) can also be expressed as an invariant or scalar product of $\mathbf{r}$ and the orientations $\hat{\mathbf{u}}_1$ and $\hat{\mathbf{u}}_2$, as suggested by Blum. [19–21] For hard-body systems, the expansion coefficients $g_{jj'm}(r)$ of Eq. (2.20) have been determined by integral equation methods [22,23] and by computer simulation [24,25–27]. Although the direct approach afforded by Eq. (2.20) has the advantage of clarity, the convergence characteristic are particularly poor [26,28] as the short range correlations are explicit functions of the surface-to-surface separation and not the center-to-center distance. To improve the convergence properties of the expansion of $g(1,2)$, several modifications have been made.

In an application to the Gaussian overlap approximation for hard ellipsoids, Kabadi and Steele [29,30] represented $g(1,2)$ with a distance variable scaled by the orientation dependent ellipsoid diameter. For fluids of hard spherocylinders Kabadi and Steele [31] devised a center-to-center separation coordinate which reflected the mutual orientation of a pair of molecules. Ghazi and Rigby [32] continued this line of inquiry and found that the convergence of the Kabadi expansion for hard spherocylinders was greatly accelerated from that achieved by Eq. (2.20) when another orientation dependent coordinate was used.

For a system comprised of a hard ellipse in a fluid of hard disk, Kumar et al. [33] expanded the ellipse-disk pcf as

$$g(1,2) = \sum_j g_j(s) P_j(\hat{\mathbf{u}} \cdot \hat{\mathbf{s}}) \tag{2.21}$$

where $P_j(\hat{\mathbf{u}} \cdot \hat{\mathbf{s}})$, the Legendre polynomials, were a function of the orientation of ellipse ($\hat{\mathbf{u}}$) with respect to the surface normal ($\hat{\mathbf{s}}$). However, the Legendre polynomials, $P_j(\hat{\mathbf{u}} \cdot \hat{\mathbf{s}})$ were nonorthogonal as the surface integrations were taken over the convex surfaces described by the addition of a disc to an ellipse and the weight function which accompanies this integration was $S^{12}(s, \hat{\mathbf{u}} \cdot \hat{\mathbf{s}})$. When one orthogonalizes the Legendre polynomials with respect to the weight function, one finds that the first anisotropic contribution to $g(1,2)$ vanishes at zero density and is directly proportional to the particle nonsphericity: the desired properties. But the set of orthogonalized polynomials representing $g(1,2)$ depend explicitly on $\hat{\mathbf{s}}$ (since the integrations are taken over nonspherical surfaces whose nonsphericity changes with $\hat{\mathbf{s}}$). The selection of $\hat{\mathbf{s}}$-dependent orthogonal basis functions was a possibility for simple systems such as an infinitely dilute solution of a single HCB, but it is not practical for the case where several orientation angles are required.

To avoid these complications, Kumar et al. [33] and Talbot et al. [34] followed the suggestion of Kabadi and Steele [31] and expressed $S^{12}(s, \Omega_1, \Omega_2)$ $g(1,2)$ (or G of Eq. (2.18)) in a set of spherical harmonics, $Y_{jm}(\Omega)$,

$$G(s, \Omega_1, \Omega_2) = 4\pi \sum_{j,j',m} G_{jj'm}(s) Y_{j,m}(\Omega_1) Y_{j',-m}(\Omega_2) \tag{2.22}$$

For simplicity we have restricted attention to $C_{\infty v}$ molecules (with two rotational degrees of freedom). The orientation angles in Eq. (2.22) are expressed with respect to the surface normal directed along the minimum surface-to-surface separation. A disadvantage of the method of Eq. (2.22) is that G is anisotropic even in the limit of zero density, but to its credit, the expansion functions are the familiar spherical harmonics and are independent of $\hat{s}$.

The motivation for the orthogonal function expansion of $G(s, \Omega_1, \Omega_2)$ is in part subjective. Both Eqs. (2.16) and (2.19) suggest that to determine structural properties for HCB systems, surface integrals of the pcf are required. But in the case of the pressure, the integral of $\hat{g}$ is weighted by the excluded volume Jacobian $h_{12}S^{12}$ so that the pressure is related to the excluded volume average of g. This, in turn, suggests that perhaps one should expand $h_{12}G(1,2)$ in spherical harmonics rather than $G(1,2)$ alone. To this we counter that in the theory of transport properties, the prefactor of $h_{12}G(1,2)$ does not generally arise and it is $G(1,2)$ that is common to most transport coefficient integrands. [35] As our concerns are with both thermodynamic and dynamic properties, the expression of the pcf will reflect our subjective choice toward surface area averaging as opposed to volume averaging.

By means of Eq. (2.22) and the remarks of the previous paragraph, one defines the isotropic surface-averaged pcf [33]

$$g_{\mathrm{iso}}(s) = \frac{\int d\Omega_1 \, d\Omega_2 G(s, \Omega_1, \Omega_2)}{\int d\Omega_1 \, d\Omega_2 S^{12}(s, \Omega_1, \Omega_2)} = \frac{G_{000}(s)}{S^{12}(s)} \tag{2.23}$$

and as a result, one can express any surface average as

$$< B\delta(s-s^*) > = 4\pi \int d\Omega_1 \, d\Omega_2 G(s, \Omega_1, \Omega_2) B = 4\pi g_{\mathrm{iso}}(s^*) S^{12}(s^*) < B >_{s^*} \tag{2.24}$$

i.e., the product of the isotropically averaged pcf $g_{\mathrm{iso}}(s^*)$, the isotropically averaged surface area $S^{12}(s^*)$ and a conditional average $< B >_{s^*}$ on the convex surface $s = s^*$, thus,

$$< B >_s = \frac{\int d\Omega_1 \, d\Omega_2 G(s, \Omega_1, \Omega_2) B(s, \Omega_1, \Omega_2)}{\int d\Omega_1 \, d\Omega_2 G(s, \Omega_1, \Omega_2)} \tag{2.25}$$

The pressure, Eq. (2.16), can be expressed in these terms

$$\beta P/\rho = 1 + (2\pi/3)\rho < h_{12} >_c g_c S_c \qquad (2.26)$$

where the subscript "c" on $< h_{12} >_c$,g_c $(= g_{iso}(s = 0))$ and S_c denotes the average on the contact $s = 0$ surface.

In contrast, certain equilibrium properties such as the isothermal compressibility, κ_T, and the static orientational correlation factor g_2 (i.e., the parameter that determines the integrated intensity of depolarized light scattering due to collective orientational fluctutaions), are expressed in terms of the *full* pcf rather than its *contact value*. Starting from the familiar relationships [13,36]

$$\rho k_B T \kappa_T = 1 + \rho \int d\mathbf{r}\, d\Omega_1\, d\Omega_2\, [g(\mathbf{r}, \Omega_1, \Omega_2) - 1] \qquad (2.27)$$

$$g_2 = 1 + \rho \int d\mathbf{r}\, d\Omega_1\, d\Omega_2\, P_2(\hat{\mathbf{u}}_1 \cdot \hat{\mathbf{u}}_2)[g(\mathbf{r}, \Omega_1, \Omega_2) - 1] \qquad (2.28)$$

one can apply the convex body coordinates to convert these two equations to

$$\rho k_B T \kappa_T = 1 + 4\pi\rho \int d\Omega_1\, d\Omega_2 \int_0^\infty ds S^{12}[g(1,2) - 1]$$
$$- (4\pi/3)\rho \int d\Omega_1\, d\Omega_2 S^{12}(s = 0, \Omega_1, \Omega_2) h_{12} \qquad (2.29)$$

$$g_2 = 1 + 4\pi\rho \int d\Omega_1\, d\Omega_2 \int_0^\infty ds P_2(\hat{\mathbf{u}}_1 \cdot \hat{\mathbf{u}}_2) S^{12}[g(1,2) - 1]$$
$$- (4\pi/3)\rho \int d\Omega_1\, d\Omega_2 P_2(\hat{\mathbf{u}}_1 \cdot \hat{\mathbf{u}}_2) S^{12}(s = 0, \Omega_1, \Omega_2) h_{12} \qquad (2.30)$$

Both Eqs. (2.29) and (2.30) consist of two parts: a long range term reflecting the decay of the s-dependent pcf and a short range contact term arising from excluded volume considerations. The isothermal compressibility simplifies further to represent this explicit separation

$$\rho k_B T \kappa_T = 1 - \rho V_{12} + 4\pi\rho \int_0^\infty ds S^{12}(s)[g_{iso}(s) - 1] \qquad (2.31)$$

where V_{12} is the excluded volume for a pair of HCBs.

The general remarks made so far do not bring us appreciably closer to the calculation of equilibrium properties of the systems of HCBs. Analytical techniques have had little success in providing the full angle dependence of the pcf for a fluid of HCBs as only the contact orientational pcf for a single HCB in a HS fluid has been determined to date. [17] Most of the work to date on HCB systems in the isotropic phase has focussed on the scalar properties (such as virial coefficients) and the contact properties (such as the pressure and the isotropic pcf) [12] rather than the anisotropic pcf.

2. Scalar Properties

a. *Virial Coefficients.* The virial coefficients of HCB systems, even of single component fluids, increase rapidly in complexity as the number of particles in each cluster increases. Whereas B_2 is particularly straightforward,

$$B_2 = (2\pi/3) \int d\Omega_1 d\Omega_2 h_{12} S^{12}(s = 0, \Omega_1, \Omega_2)$$
$$= (1/2) V_{12} \qquad (2.32)$$

B_3 is more complicated and numerical procedures are needed for its evaluation. [37–41] For example,

$$B_3 = \int \lambda^2 d\lambda \eta^2 d\eta d\Omega_1 d\Omega_2 d\Omega_3$$
$$\times h_{12} S^{12}(s_{12} = 0, \{\Omega\}) h_{13} S^{13}(s_{13} = 0, \{\Omega\}) \chi(2, 3) \qquad (2.33)$$

where $\mathbf{r}_{23} = \lambda \mathbf{r}_{12}(\hat{\mathbf{s}}_{12}) - \eta \mathbf{r}_{13}(\hat{\mathbf{s}}_{13})$ is the center-to-center separation for the λ-scaled 1,2 surface and the η-scaled 1,3 surface. $\chi(2, 3)$ is unity if bodies 2 and 3 overlap and zero otherwise. Evaluation of B_4 and B_5 requires numerical methods. [37,38,40] For elongated prolate ellipsoids, B_4 is negative. In contrast, the first seven virial coefficients of hard spheres are known to be positive. Values of B_2 for various ellipsoids have been collected by Boublik. [12] Certainly the most important property of the virial coefficients of prolate HCBs is

$$\lim_{\text{shape anisotropy} \to \infty} B_n/(B_2^{n-1}) \to 0 \qquad (2.34)$$

This point is discussed in more detail in Section II.D.4.

b. *Equations of State.* Approximate equations of state have been derived on the basis of Scaled Particle Theory (SPT) [12,14,42–45] and re-summed virial expansions [46–48]. Both of these approaches begin with an assumption as to how the solution should behave. In SPT one guesses how the pcf depends on the scaling length whereas in the virial expansion resummations, one guesses at the density dependence of the pressure or the contact pcf. Exact constraints, such as virial coefficients or other limiting behaviors, fix the constants in the assumed equation of state (or isotropic contact **g**). The philosophy behind the scaled-particle theory is explained in more detail in the context of phase transitions in hard-core fluids (Section IV.A.2). A reasonably up to date account of scaled-particle theory in the context of isotropic hard-core fluids can be found in Ref. [12]. Boublik and Nezbeda have played a significant role in the development of the equations of state of HCBs and their work has been complemented by Wojcik and Gubbins, [46] Naumann et al. [47] and Song and Mason. [48] From the standpoint of accommodating data on many systems, the approach of Song and Mason appears to be most successful. However, none of the "phenomenological" approaches provides insight into the nature of a first order isotropic to nematic transition and, in that sense, no "first principles" equation of state is available.

The studies of pressures (by analytical approximate means) have also led to theories of the contact pcf and in particular, its volume average

$$g_{\text{vol}} = (4\pi/3) \int d\Omega_1 \, d\Omega_2 G(s=0,\Omega_1,\Omega_2) h_{12}/V_{12} \qquad (2.35)$$

rather than the surface averaged pcf of Eq. (2.23). For the purpose of forming a perspective on the basic algebraic forms of the isotropic pcfs, we sketch the SPT approach used by Boublik, [14] who found the pressure and g_{vol} to be

$$\beta P/\rho = [1 + 3\alpha y + 3\alpha^2 y^2] \frac{1}{1-\rho^*} \qquad (2.36)$$

$$g_{\text{vol}} = \left[1 + \frac{3\alpha y}{1+3\alpha} + \frac{3\alpha^2 y^2}{1+3\alpha}\right] \frac{1}{1-\rho^*} \qquad (2.37)$$

Here y is the Barboy–Gelbart [49–51] density variable

$$y = \rho^*/(1-\rho^*) \qquad (2.38)$$

which is discussed in more detail in Section IV.A.2, and α the nonsphericity parameter

$$\alpha = 4\pi R_1 S_1/(3V_1) \qquad (2.39)$$

where R_1 and S_1 are, respectively, $1/4\pi$ times the mean curvature and the surface area of a single body and V_1 is the volume. (For spheres $\alpha = 1$.)

The goal of Boublik's SPT was to determine the contact pcf, which was related to the reversible work (i.e., the chemical potential) for the insertion of a q-scaled particle in a fluid. For a point particles, the PV work is related to the probability of finding a point cavity and this, in turn, is related to the free volume ($= V - NV_1 = V(1 - \rho^*)$). By means of the exact relationships between the contact pcf and the work to insert a point particle in a fluid, exact conditions are placed on the volume average of both $g(q = 0)$ and $dg(q)/dq|_{q=0}$. A third exact condition follows from the work necessary to increase the volume of a macroscopic HCB from V_q to $V_q + dV_q$. This third condition links the pressure to $g(q = \infty)$. When these three conditions are incorporated into an assumed functional of the scaling parameter, Boublik obtained Eqs. (2.36) and (2.37). Equation (2.36) reduces to the Percus–Yevick [52] (PY) result (i.e., obtained via the compressibility relation) [13] when $\alpha = 1$, viz.,

$$(\beta P(\mathrm{HS})/\rho)_{\mathrm{PY}-c} = \frac{(1 + \rho^* + \rho^{*2})}{(1 - \rho^*)^3} \tag{2.40}$$

For small anisotropies ($\alpha \simeq 1$) and low densities, Eqs. (2.36) and (2.37) are useful. Heuristic modifications of Boublik's results for highly nonspherical HCBs (5:1 particles), based on the Carnahan–Starling hard sphere limiting result [53], improved the accuracy but the HCB equations of state and contact pcfs prove to be less accurate than their HS analogues. [38–40] At present, the most accurate equation of state and contact pcf are due to Song and Mason [48] who found that

$$g_{\mathrm{vol}} = \frac{1 - \gamma_1 \rho^* + \gamma_2 \rho^{*2}}{(1 - \rho^*)^3} \tag{2.41}$$

where

$$\gamma_1 = 3 - \frac{1 + 6\alpha + 3\alpha^2}{1 + 3\alpha} \tag{2.42}$$

and

$$\gamma_2 = 3 - \frac{1 + 2.6352\alpha + 7\alpha^2}{1 + 3\alpha} \tag{2.43}$$

Although the isotropic contact pcfs of fluids of HCBs can be estimated with some accuracy using the Song and Mason result, this area of research is by no means closed, as systematic and accurate first principle results are not at hand.

3. Vector Properties

a. Virial Coefficients. Ordinarily, the virial coefficients are not considered to be vector properties. However, in the context of the liquid crystal work to be presented in the following sections, we can anticipate some reinterpretations. B_2, B_3 as well as all the higher virial coefficients depend on the mutual orientation of the bodies. This dependence could have been anticipated from Eq. (2.32) since we expressed B_2 as an integral over an orientation dependent integrand,

$$B_2 = \int d\Omega_1 \, d\Omega_2 B_2(\Omega_1, \Omega_2) \qquad (2.44)$$

with

$$B_2(\Omega_1, \Omega_2) = (2\pi/3)h_{12}S^{12}(s = 0, \Omega_1, \Omega_2) \qquad (2.45)$$

The angle dependence of B_2 was first derived by Onsager [3] for spherocylinders and by Isihara [54] for ellipsoids. Mulder [55] determined $B_2(\Omega_1, \Omega_2)$ analytically for spheroplatelets (a biaxial spherocylinder). Tjipto-Margo and Evans used orthogonal function expansions to express $B_2(\Omega_1, \Omega_2)$ and $B_3(\Omega_1, \Omega_2, \Omega_3)$ for uniaxial ellipsoids [41] and $B_2(\Omega_1, \Omega_2)$ for biaxial ellipsoids. [56]

b. *Contact Orientational Correlations.* She et al. [17,57] employed SPT to find the dependence of the contact pcf on the orientation of the solvent (taken to be a HS of radius a) with respect to a solute (a single but arbitrary HCB). The contact pcf for a fully scaled (q = 1) HCB was found to be function of $\hat{s}$ (the HCB-atom contact surface normal), the solvent packing fraction and various measures of the geometry of the HCB. Specifically,

$$g(s = 0, x) = 1 + 4\rho^* g_{HS} + \sum_{j=1}^{3} a_j(x)/(1 + h(x)/a)^j \qquad (2.46)$$

where

$$x = \hat{u} \cdot \hat{s} \qquad (2.47)$$

and the $a_j(x)$ coefficients are given elsewhere. [57] Equation (2.46) was derived using a SPT with four exact conditions: $g(s = 0, \hat{s}, q = 0)$, $(dg(s = 0, \hat{s}, q)/dq)|_{q=0}$, $(d^2g(s = 0, \hat{s}, q)/dq^2)|_{q=0}$ and $g(s = 0, \hat{s}, q = \infty)$. Only the second derivative term introduces orientation dependence into $g(s = 0, \hat{s}, q)$. All theories based on constraints for $g(s, q = 0)$ and $dg(s, q = 0)/dq$ will predict the pcf to be isotropic on the contact surface.

When the scaled particle is allowed to become spherical, the contact pcf and the pressure can be derived for the HS fluid within the context of the four-condition SPT described above, thus [33,57]

$$g_{HS} = \frac{1 - (1/4)\rho^* + (1/2)\rho^{*2} - (1/8)\rho^{*3}}{(1 - \rho^*)^2(1 - (3/4)\rho^* + (1/2)\rho^{*2})} \tag{2.48}$$

and

$$(\beta P_{HS}/\rho)_{SPT} = 1 + 4\rho^* g_{HS} \tag{2.49}$$

This four-condition SPT provides the exact isotropic and anisotropic second viral coefficient and at low density g_{HS} has the expansion

$$g_{HS} \simeq 1 + 2.5\rho^* + 4.3755\rho^{*2} \tag{2.50}$$

Equation (2.50) is in disagreement with the findings of Tully-Smith and Reiss, [58] who find that a similar four-condition SPT predicts incorrect second and third virial coefficients. The resulting contact properties of the four-condition SPT are close to but not identical to that derived from the PY theory using the pressure equation [13]

$$(\beta P_{HS}/\rho)_{SPT} \simeq (\beta P_{HS}/\rho)_{PY-p} = \frac{1 + 2\rho^* + 3\rho^{*2}}{(1 - \rho^*)^2} \tag{2.51}$$

Basically, an n-condition SPT and the PY theory have no formal equivalence; although the calculated equations of state are similar, this is more fortuitous than substantive. SPT in itself does not suggest any particular method of closure of the hierarchy (say by means of the choice of an appropriate length scaling functional) and so to this exent there is no unique SPT. Clearly the exact conditions are unique, however. At present, SPT has yet to provide a theory for the direct correlation function or for that matter, for a simple theory of the pcf. Reiss and Casberg [59] have calculated the HS pcf using the ideas of SPT but this version of SPT bears little resemblance to the original SPT. Certainly much remains to be done in the utilization of SPT to understand the properties of fluids of HCBs.

4. Integral Equation Methods

The pcf $g(1,2)$ and its companion, the total correlation function $h(1,2)$

$$h(1,2) = g(1,2) - 1 \tag{2.52}$$

(not be confused with the support function, h_{12}) can be approximated as a solution to an integral equation. The integral equations and the approximations for $g(1,2)$ are succinctly stated in terms of the direct correlation function $c(1,2)$, defined by the Ornstein–Zernike equation

$$h(1,2) = c(1,2) + \rho \int d3 h(1,3) c(2,3) \qquad (2.53)$$

For fluids of HCBs, the PY approximation is

$$c(1,2) = y(1,2)\chi(1,2) \qquad (2.54)$$

and this is to be compared with the HNC approximation

$$c(1,2) = y(1,2)f(1,2) + y(1,2) - 1 - \ln y(1,2) \qquad (2.55)$$

where we have introduced $y(1,2)$, the background correlation function

$$y(1,2) = \exp(+\beta U(1,2))g(1,2) \qquad (2.56)$$

and $f(1,2)$ the Mayer f-bond

$$f(1,2) = \exp(-\beta U(1,2)) - 1 \qquad (2.57)$$

In the HS system, $c(1,2)$ has been found to have a small positive tail outside the hard core for the HNC theory whereas $c(1,2)$ vanishes outside the hard core for the PY theory.

Methods of solution of the integral equations for the anisotropic pcf of nonspherical molecules are given by Gray and Gubbins. [60] Stemming from the work of Chen and Steele, [22] a considerable literature is now developing regarding the spherical harmonic expansion of $g(1,2)$. No reported calculations of $g(1,2)$ take advantage of the explicitly convex nature of the particles and hence the techniques presented in the preceding pages. That which is known about the expansion properties of $g(1,2)$ and $c(1,2)$ prior to 1988 is summarized by Nezbeda et al. [61] More recently Labik et al. [62] compared the $g_{jj'm}(r)$ expansion functions for hard dumbbells obtained from integral equation theories (PY, HNC and Bridge function methods). Talbot et al. [24] also compared the results of HNC and PY closures on the $g_{jj'm}(r)$ for hard ellipsoids. Generally all the integral equation predictions of $g_{jj'm}(r)$ are in basic agreement with each other and with MD simulations. Although the PY theory was less accurate than the HNC and Bridge function theories, one might argue that the selection amongst theories could be based on practicality, which would, in turn, would always favor the use of the PY theory.

A more sensitive measure of orientational correlations than the $g_{jj'm}(r)$ is required to discriminate between the growing assortment of integral equations. One such measure involves the expansion coefficients $c_{jj'm}(r)$ of $c(1,2)$

$$c(1,2) = \sum_{j,j',m} c_{jj'm}(r) C_{j,m}(\Omega_1) C_{j',-m}(\Omega_2) \qquad (2.58)$$

Perera et al. [25] calculated the volume integral of the Legendre polynomial parts of $c(1,2)$

$$c_{jj0} = 4\pi \int_0^\infty dr r^2 \, d\hat{\mathbf{u}}_1 \, d\hat{\mathbf{u}}_2 P_j(\hat{\mathbf{u}}_1 \cdot \hat{\mathbf{u}}_2) c(1,2)$$

$$= 4\pi \int_0^\infty dr r^2 c_{jj0}(r) \qquad (2.59)$$

In the PY approximation for 3:1 hard ellipsoids, Perera [25] found that $c_{000}(r)$ behaved like the corresponding HS function; for $j \geq 2$, c_{jj0} was nearly density independent and in marked contrast with the strong density dependence from the HNC findings. Furthermore, the PY predictions for c_{jj0} ($j \geq 2$) displayed less density dependence than that derived from the two term virial expansion for $c(1,2)$. [63] This decided difference between the anisotropies in $c(1,2)$ has a profound influence on the issue of liquid crystal formation in the context of PY and HNC theories. The anisotropy in $c(1,2)$, derived by the PY theory, is so weak that the PY theory fails to predict a liquid crystal transition at any realizable density. Thus, the PY theory, broadly accepted as a good indicator of radial correlations in the HS fluids [12,60] shows a serious breakdown in the analysis of orientational correlations in fluids of HCBs.

D. Simulations

In this section, we discuss numerical simulations of hard-body fluids in the isotropic phase. The material in this section is organized as follows. First, we discuss those aspects of simulation techniques that are peculiar to hard-body systems, or otherwise not completely standard. We devote considerable attention to a systematic description of the various tests that can be used to detect a hard-core overlap of two (convex) bodies, mainly because, in the existing literature, the discussion of this subject is quite fragmented. Next, we briefly summarize the essential features of the computation of the first few virial coefficients and the equation of state of hard-body fluids. Following this technical introduction, we review the results that have been obtained using these techniques. In view of the large amount of numerical data on hard-body fluids that have been reported in the literature, we focus the discussion of the simulation results on those features that are, in some sense, peculiar to hard-body fluids.

1. Overlap Criteria

The first step in a computer study of a model system for a liquid crystal forming substance, is the selection of the actual model. As was explained in the Introduction, we prefer to study models that are computationally "cheap". At the same time, the model should be sufficiently rich to give rise to a nontrivial phase diagram. And finally, it is obviously attractive if the model belongs to a "family" of models that includes, as special or limiting cases, systems about which much is already known. For example, hard spheres can be considered as a special case of both hard ellipsoids and of hard spherocylinders. Similarly, the exactly solvable Onsager model of thin, hard rods, is again a limiting case of both models.

In this review, we consider several families of model systems, viz. hard ellipsoids (both uni-axial and bi-axial), hard spherocylinders, hard platelets (truncated spheres) and hard fused spheres. As we shall see below, all these model systems exhibit interesting static or dynamic behavior. In this section, we show that these models are all convenient from a computational point of view. As we are discussing hard-core models, the computation of the potential energy of the system can be reduced to a series of tests for overlap between pairs of molecules i and j with orientations Ω_i and Ω_j, at a relative distance $\mathbf{r}_{ij}$.

Usually, the test for overlap between two particles can be reduced to a test of the sign of one, or several, functions of the relative coordinates of a pair of particles. The choice of these functions is, in general, not unique but is dictated by computational convenience. For instance, we shall find that for ellipsoids there are (at least) two, quite different tests for overlap that are best used in combination. Below we discuss the overlap criteria that have been used for the model systems decribed in this review. In addition, we briefly refer to some other model systems.

We should, however, first explain that we are really only discussing a *subclass* of all possible hard-core interactions. In the most general case, one can construct a hard-core model by simply defining a pair potential-energy function $u(\mathbf{r}_{ij}, \Omega_i, \Omega_j)$ to be infinite for some finite, connected domain of coordinate values and zero elsewhere. Once this function is specified, we can construct the excluded volume of a pair of particles. In general, this excluded volume will not be convex. More importantly, in general it will not be meaningful to speak about the *shape* of the individual hard-core particles. In other words, although the *excluded volume* of a pair of particles can be visualized as an object in space, the individual particles cannot. It should be stressed that the idea that individual particles have a shape of their own, is a classical one that has little meaning at the molecular level. Hence, there is nothing wrong with a hard-core model that cannot be interpreted in terms of the overlap of two well-defined geometrical objects. In fact, a popular example of such a nondecomposable model is the Gaussian hard-core model [39,64]. However, in this review, we limit ourselves to hard-core models where the individual particles have a well defined shape. The reason for restricting ourselves to such models is two-fold: first, it is easier to develop a mental picture of the factors that determine the static and dynamic behavior of hard-body systems, if we can visualize that system. Second, in nature, hard-core systems are most closely approximated by colloidal particles. For these particles, that can often be *seen* by electron microscopy, it is not unreasonable to attribute a shape to individual particles. Finally, for most of the models that we discuss, the shape of the individual particles is convex. This choice is motivated only by the fact that the theoretical description of both static and dynamic properties is often much simpler for convex than for nonconvex hard particles.

a. Spheres and Composite Particles. It is convenient to start our description of overlap criteria with the simplest case, namely hard spheres. Two hard spheres of radius R_1 and R_2, respectively overlap if the distance r_{12} between the centers of these spheres is less than $\sigma_{12} \equiv R_1 + R_2$. In a simulation, we usually do not compare r_{12} with σ_{12}, but r_{12}^2 with σ_{12}^2, because the latter test is computationally cheaper. For future reference, it is important to note that the hard-sphere overlap test can be considered as a sequence of tests for *nonoverlap*. The test could be broken down into three steps namely $\Delta_1^2 \equiv \sigma_{12}^2 - x_{12}^2 < 0$, $\Delta_2^2 \equiv \Delta_1^2 - y_{12}^2 < 0$, and $\Delta_3^2 \equiv \Delta_2^2 - z_{12}^2 < 0$. Only if all three tests are *not satisfied* do we have overlap between the two spheres. In fact, as any pair of particles that fails the final test must also fail the previous two, it may be computationally cheaper to carry out only the final test. However, that is not the issue here. What we wish to show is that it is possible to break up our test into subtests that allow us to decide, at an early stage, whether a given pair of particles does *not* overlap. Later, on when we consider more complex overlap tests, we will see that it is advisable to have a cheap test for nonoverlap as the first "filter" in the test sequence.

As a specific example of such a screening, consider a composite particle consisting of several hard spheres. Such fused hard sphere models have been used to model rod-like mesogens. [65] Let us assume that we wish to know if two molecules, both consisting of n identical hard spheres, are overlapping. Clearly, the test for overlap between these two composite molecules can be broken down into n^2 hard-sphere tests (for convenience, we assume that we do not have to worry about intra-molecular overlaps). Now, we see that the nature of the subtests is different from the hard-sphere case. As soon as we find overlap between *any* pair of hard spheres, we know that the two molecules overlap and we can terminate the test sequence. However, in order to make sure that the two molecules do not overlap, we have to run through the complete sequence of n^2 tests.

However, if the two molecules are sufficiently far apart, it is possible to tell in advance that there can be no overlap. For instance, one could construct for every molecule a circumscribed sphere that contains all spheres in that molecule. Clearly, no overlap between the two molecules is possible if the circumscribed spheres of these molecules do not overlap. Thus, we can obviate n^2 overlap tests by one "nonoverlap" test. This example illustrates the respective role of overlap and nonoverlap tests. *Nonoverlap* tests are used to ensure that we do not perform expensive overlap tests on molecules that are far apart. But once we know that we really must carry out the complete test, then it is better to have a series of *overlap* tests, because this sequence can be terminated as soon as any such test is satisfied.

In many cases, this combination of nonoverlap and overlap tests is not applied to a single pair of molecules, but to all neighbors of a given molecule that has undergone a trial displacement. First, we "short-list" the possible overlap partners by using a nonoverlap test. Next, we apply the overlap test on the short-listed neighbors. As soon as an overlap is detected, we know that we can reject the trial move. In this sense, the construction of the well-known Verlet neighborlist [66] is simply an example of a short-list produced by a nonoverlap filter. For anisometric hard-core molecules, the corresponding nonoverlap tests are used to construct a "nonspherical" Verlet list.

b. Spherocylinders. Let us now consider a slightly more complex hard-core model, namely the spherocylinder. Just as a sphere can be defined as the set of points that are within a distance R from a given origin (namely the center of the sphere), so the spherocylinder can be thought of as the set of points that are within a distance R from a line segment of length L. Clearly, we can draw around every point on this line segment a sphere of radius R that contains all points that are within a distance R from that point. Hence, a spherocylinder can be considered as the union of all spheres around points on a line segment L. We can thus consider a spherocylinder as the volume that is swept out by a sphere of radius R that is moved along a line segment of length L. The test for overlap between two spherocylinders can be constructed by computing the shortest distance between the two line segments that form the "core" of the spherocylinders. If this distance of closest approach is less than $D_{12} \equiv R_1 + R_2$, the two spherocylinders overlap. This distance of closest approach is therefore the central quantity to be computed in an overlap test for spherocylinders. In fact, some of the steps needed to compute the distance of closest approach between two spherocylinders, are also needed in the construction of other overlap criteria to be discussed below. We therefore break down the construction of the overlap test for spherocylinders into a number of elementary steps, namely:

1. the construction of the point of closest approach between two lines;
2. the construction of the perpendicular distance-vector between two lines;
3. the determination of the distance of closest approach between two line segments in a plane.

Although the overlap test for spherocylinders usually skips the second step, it is useful to include it, both for future reference and because it makes the whole procedure more transparent. Step 1 is sufficient to determine the overlap of line segments in two dimensions. Combined with step 3, it allows us to test for overlap between two-dimensional spherocylinders. Steps 1 and 2 will turn out to be useful in the test for overlap between hard platelets in three dimensions.

POINT OF CLOSEST APPROACH BETWEEN TWO LINES. Our aim is to determine the minimum distance between two finite line-segments i and j, with orientations $\hat{\mathbf{u}}_i$ and $\hat{\mathbf{u}}_j$ and centers $\mathbf{r}_i$ and $\mathbf{r}_j$. Let us first consider the minimum distance between these two segments, in the limit that their length is infinite. In that case, we can describe any point on line i parametrically as

$$\mathbf{r}_i(\lambda) = \mathbf{r}_i + \lambda \hat{\mathbf{u}}_i$$

while line j is given by

$$\mathbf{r}_j(\mu) = \mathbf{r}_j + \mu \hat{\mathbf{u}}_j$$

The vector distance between these two points is given by

$$\mathbf{r}_{ij}(\lambda, \mu) = (\mathbf{r}_i - \mathbf{r}_j) + \lambda \hat{\mathbf{u}}_i - \mu \hat{\mathbf{u}}_j \qquad (2.60)$$

Next, we wish to determine those values of λ and μ for which the distance r_{ij} is minimal. A simple method to find these values of λ and μ is the following. Construct the dot product of $\mathbf{r}_{ij}(\lambda, \mu)$ with both $\hat{\mathbf{u}}_i$ and $\hat{\mathbf{u}}_j$. The shortest distance vector must be perpendicular to both $\hat{\mathbf{u}}_i$ and $\hat{\mathbf{u}}_j$. Hence, we should solve the following simultaneous equations:

$$(\mathbf{r}_i - \mathbf{r}_j) \cdot \hat{\mathbf{u}}_i = -\lambda \hat{\mathbf{u}}_i \cdot \hat{\mathbf{u}}_i + \mu \hat{\mathbf{u}}_j \cdot \hat{\mathbf{u}}_i$$
$$(\mathbf{r}_i - \mathbf{r}_j) \cdot \hat{\mathbf{u}}_j = -\lambda \hat{\mathbf{u}}_i \cdot \hat{\mathbf{u}}_j + \mu \hat{\mathbf{u}}_j \cdot \hat{\mathbf{u}}_j \qquad (2.61)$$

Solving these equations for λ and μ, we obtain

$$\begin{pmatrix} \lambda_0 \\ \mu_0 \end{pmatrix} = \frac{1}{1 - (\hat{\mathbf{u}}_i \cdot \hat{\mathbf{u}}_j)^2} \begin{pmatrix} -\hat{\mathbf{u}}_i \cdot \mathbf{r}_{ij} + (\hat{\mathbf{u}}_i \cdot \hat{\mathbf{u}}_j)(\hat{\mathbf{u}}_j \cdot \mathbf{r}_{ij}) \\ +\hat{\mathbf{u}}_j \cdot \mathbf{r}_{ij} - (\hat{\mathbf{u}}_i \cdot \hat{\mathbf{u}}_j)(\hat{\mathbf{u}}_i \cdot \mathbf{r}_{ij}) \end{pmatrix} \qquad (2.62)$$

where we have used the shorthand notation $\mathbf{r}_{ij} \equiv \mathbf{r}_{ij}(\lambda = 0, \mu = 0)$. We have assumed that the line segments i and j are not parallel. In fact, the parallel case is simpler, and is discussed separately. The above expression for λ_0 and μ_0 allows us to carry out a testing for overlap between two line segments of length L_i and L_j in two dimensions. In that case, we simply have to verify that $|\lambda_0| \leq L_i/2$ and $|\mu_0| \leq L_j/2$.

PERPENDICULAR DISTANCE VECTOR BETWEEN TWO LINES. We wish to know the shortest distance between two line-segments in three dimensions. We can decompose this vector in a component perpendicular to $\mathbf{u}_i$ and $\hat{\mathbf{u}}_j$ and a component in the plane of $\mathbf{u}_i$ and $\hat{\mathbf{u}}_j$. By varying λ and μ, we can only change the latter distance. Hence, our problem reduces to the computation of the fixed perpendicular distance between the lines i and j and of the minimal in-plane distance between the two line segments. For future reference, it is convenient to construct the perpendicular distance vector as follows. From the unit vectors $\hat{\mathbf{u}}_i$ and $\hat{\mathbf{u}}_j$, we construct three orthogonal unit vectors $\hat{\mathbf{u}}_+$, $\hat{\mathbf{u}}_-$ and $\hat{\mathbf{u}}_\perp$, defined as

$$\hat{\mathbf{u}}_+ \equiv \frac{1}{\sqrt{2}} \frac{\hat{\mathbf{u}}_i + \hat{\mathbf{u}}_j}{(1 + (\hat{\mathbf{u}}_i \cdot \hat{\mathbf{u}}_j))^{1/2}} \qquad (2.63)$$

$$\hat{\mathbf{u}}_- \equiv \frac{1}{\sqrt{2}} \frac{\hat{\mathbf{u}}_i - \hat{\mathbf{u}}_j}{(1 - (\hat{\mathbf{u}}_i \cdot \hat{\mathbf{u}}_j))^{1/2}} \qquad (2.64)$$

$$\hat{\mathbf{u}}_\perp \equiv \hat{\mathbf{u}}_+ \times \hat{\mathbf{u}}_- \qquad (2.65)$$

The perpendicular vector distance between lines i and j is then given by

$$\mathbf{r}_{ij}^\perp \equiv (\mathbf{r}_{ij} \cdot \hat{\mathbf{u}}_\perp)\hat{\mathbf{u}}_\perp \qquad (2.66)$$

DISTANCE OF CLOSEST APPROACH BETWEEN TWO LINE SEGMENTS IN A PLANE. Finally, we must compute the distance of closest approach between two line segments in a plane. Of course, if both $|\lambda_0| \leq L_i/2$ and $|\mu_0| \leq L_j/2$, the two line segments intersect, and the in-plane distance is zero. In that case, the total distance is given by $|\mathbf{r}_{ij}^\perp|$. However, we have to consider the more general case that the in-plane distance between the line segments is nonzero. For this purpose, it is convenient to take as the origin of our (two-dimensional) coordinate frame, the intersection point between the lines i and j projected in a plane spanned by $\hat{\mathbf{u}}_i$ and $\hat{\mathbf{u}}_j$. In this frame, the center of segment i is located at $-\lambda_0\hat{\mathbf{u}}_i$ and the center of segmenty j is at $-\mu_0\hat{\mathbf{u}}_j$. The squared distance between two arbitrary points on the lines given by $\gamma\hat{\mathbf{u}}_i$ and $\delta\hat{\mathbf{u}}_j$, is

$$(r_{ij}^{\parallel}(\gamma, \delta))^2 = \gamma^2 + \delta^2 + 2\gamma\delta(\hat{\mathbf{u}}_i \cdot \hat{\mathbf{u}}_j) \qquad (2.67)$$

The curves of the constant in-plane distance are ellipses with major axes along the lines $\gamma = \delta$ and $\gamma = -\delta$. Figure 2.2 shows a contour plot of the (squared) in-plane distance in the (γ, δ)-plane. When computing the shortest distance between two line segments, we should search for the minimum of the parabolic function shown in the figure, subject to the conditions $|\gamma + \lambda_0| \leq L_i/2$ and $|\delta + \mu_0| \leq L_j/2$. This constraint defines a rectangle in the (γ, δ)-plane. The procedure to find the distance of closest approach is now as follows.

1. If the origin is contained in the rectangle, the line segments intersect and the in-plane distance is zero.

2. Otherwise, find the allowed values of γ and δ that are closest to the origin: γ_{min} and δ_{min}. As, in the present case, the origin does not correspond to an allowed (γ, δ)-combination, either $|\gamma_{min}|$ or $|\delta_{min}|$, or both, are not equal to zero. Without loss of generality, we assume that $|\gamma_{min}| \geq |\delta_{min}|$ (otherwise, we simply relabel γ and δ). We now fix γ at the value γ_{min}. Next, we minimize $(r_{ij}^{\parallel}(\gamma_{min}, \delta))^2$ with respect to δ. From Eq. (2.67) it follows that the minimum distance is reached for $\delta_{min}' = -2\gamma_{min}(\hat{\mathbf{u}}_i \cdot \hat{\mathbf{u}}_j)$.

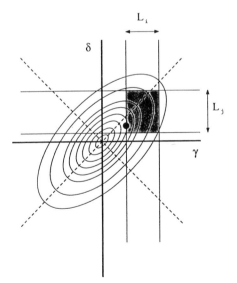

Figure 2.2. Contour lines of equal distance between two line segments of length L_i and L_j in a plane. The geometrical interpretation of this figure is discussed in the text

3. We now test if $|\delta'_{min} + \mu_0| \le L_j/2$. If so, we have found the value of δ that minimizes the in-plane distance. Otherwise, we choose the allowed value of δ that is closest to δ'_{min}.

Using Eq. (2.67), we can now simply evaluate the squared, in-plane distance. As we have already computed the (squared) perpendicular distance, we know the distance of closest approach of the two line segments i and j.

c. *Platelets.* We can use many of the results obtained in the previous section to construct a test for overlap between two infinitely thin platelets with radii R_i and R_j. In what follows, we denote the coordinate of the center of either disk by $\mathbf{r}_i$ (or $\mathbf{r}_j$), while the unit vectors $\hat{\mathbf{u}}_i$ and $\hat{\mathbf{u}}_j$ give the orientation of the symmetry axes of the disks.

As in the previous section, we define $\hat{\mathbf{u}}_\perp$, the unit vector perpendicular to both $\hat{\mathbf{u}}_i$ and $\hat{\mathbf{u}}_j$. If we project both disks on a plane normal to $\hat{\mathbf{u}}_\perp$, all points in the plane of disk i project onto a line with direction

$$\mathbf{v}_i \equiv \hat{\mathbf{u}}_\perp \times \hat{\mathbf{u}}_i$$

Similarly, the plane of disk j projects onto a line with direction

$$\mathbf{v}_j \equiv \hat{\mathbf{u}}_\perp \times \hat{\mathbf{u}}_j$$

The projections of disks i and j are line segments with a length $2R_i$ and $2R_j$ respectively. Clearly, a necessary (but not sufficient) condition for overlap of the two disks is that these line segments cross. But the problem of the intersection of two line segments in a plane was already discussed in Section II.D.1 above. In the present case, it is more convenient to express λ (μ), the distance between the center of disk i (j) and the intersection-line of the planes of disk i and j, in terms of the unit vectors $\hat{\mathbf{u}}_i$ and $\hat{\mathbf{u}}_j$, rather than the auxiliary vectors $\mathbf{v}_i$ and $\mathbf{v}_j$, defined above. It is easy to show that

$$\begin{pmatrix} \lambda^2 \\ \mu^2 \end{pmatrix} = \frac{1}{1 - (\hat{\mathbf{u}}_i \cdot \hat{\mathbf{u}}_j)^2} \begin{pmatrix} (\hat{\mathbf{u}}_j \cdot \mathbf{r}_{ij})^2 \\ (\hat{\mathbf{u}}_i \cdot \mathbf{r}_{ij})^2 \end{pmatrix} \tag{2.68}$$

If either $\lambda^2 > R_i^2$ or $\mu^2 > R_j^2$, no overlap is possible. Otherwise, the plane of disk j intersects disk i over a line segment of length

$$\Delta_i = \sqrt{R_i^2 - \lambda^2}$$

and

$$\Delta_j = \sqrt{R_j^2 - \mu^2}$$

Clearly, in order to have overlap between disks i and j, we must satisfy the following condition:

$$\Delta_i + \Delta_j < |\mathbf{r}_{ij} \cdot \hat{\mathbf{u}}_\perp| \tag{2.69}$$

This concludes our test for overlap between two infinitely thin hard platelets. In practice, this overlap criterion is rewritten in such a way that the "expensive" computation of square roots is avoided. For details, see Ref. [67].

d. Truncated Spheres. A convenient convex hard-core model for plate-like molecules with a finite thickness is obtained by taking a sphere of diameter D and slicing off the top and the bottom at a distance $\pm L/2$ from the equatorial plane. In the limit $L/D \to 0$, the truncated sphere reduces to an infinitely thin, hard platelet whereas, in the limit $L \to D$, we obtain a hard sphere. At first sight, the truncated sphere would appear to be a needlessly complex model for a disk-like molecule. It would seem more natural to model such particles by short cylinders. However, from a computational point of view, truncated spheres are more convenient because the test for overlap between two such particles can be decomposed into a finite sequence of simple subtests. For cylinders the corresponding tests appear to be less simple, except when the cylinders are parallel. This may explain why only approximate simulations of freely rotating, hard cylinders have been reported in the literature. [68]

Although the test for overlap between truncated spheres consists of simple steps, the test as a whole is rather elaborate. The reason is that different tests are needed to detect overlap of the two spherical rims, of the two flat circular faces, and of a flat face and a circular rim. Below, we discuss all these tests in succession. In addition, the procedure also includes a few tests that allow us to ascertain, at an early stage, either that i and j *cannot* overlap, or that they *must* overlap.

As before, we consider two particles i and j, at positions $\mathbf{r}_i$ and $\mathbf{r}_j$ and with orientations $\hat{\mathbf{u}}_i$ and $\hat{\mathbf{u}}_j$. For convenience, we assume that the truncated spheres have the same diameter D and the same thickness L. The generalization of the overlap tests to dissimilar truncated spheres is straightforward. Below, we go through the sequence of overlap tests, step by step.

1. *Sphere test.* We first test if particles i and j are close enough to overlap. A necessary conditions is

$$r_{ij}^2 < D^2$$

Only if this test is satisfied do we proceed with further overlap tests.

2. *Rim-rim test.* Next, we consider the cosine of the angle that $\mathbf{r}_{ij}$ makes with $\hat{\mathbf{u}}_i$. If the absolute value of this cosine is less than L/D, the $\mathbf{r}_{ij}$ *must* intersect the spherical rim of particle i. Similarly, if the absolute value of the cosine of the angle that $\mathbf{r}_{ij}$ makes with $\hat{\mathbf{u}}_j$ is less than L/D, then $\mathbf{r}_{ij}$ *must* intersect the spherical rim of particle j. If both conditions are satisfied, the test for overlap between the truncated spheres i and j reduces to a test for overlap between two spheres of diameter d. However, this test was already performed in the previous step. Hence, if

$$\text{Max}[(\mathbf{r}_{ij}.\hat{\mathbf{u}}_i)^2, (\mathbf{r}_{ij}.\hat{\mathbf{u}}_j)^2] < r_{ij}^2(L/D)^2$$

then i and j must overlap

3. *Too close or too far?* Next, we consider $\cos \theta_{ij} \equiv \hat{\mathbf{u}}_i.\mathbf{u}_j$. We distinguish between the case $|\cos \theta_{ij}| < (L/D)$ and $|\cos \theta_{ij}| > (L/D)$. We can easily compute $r_{\min}(\theta_{ij})$, the shortest possible distance that i and j can approach one another without overlapping. If $|\cos \theta_{ij}| < (L/D)$, then $r_{\min}(\theta_{ij}) = L/2 + D/2$. Hence if $|\mathbf{r}_{ij}|$ is less than $r_{\min}(\theta_{ij})$, then the particles *must* overlap. Conversely, if the distance between the center of one particle and the equatorial plane of the other is larger than $r_{\min}(\theta_{ij})$, then the particles *cannot* overlap. Next, consider the case where $|\cos \theta_{ij}| > (L/D)$. In that case, the smallest possible distance between two nonoverlapping particles is

$$r_{\min}(\theta_{ij}) = L/2(1 + |\cos \theta_{ij}|) + \sqrt{D^2 - L^2}/2|\sin \theta_{ij}|$$

Again, the particles *must* overlap if

$$|\mathbf{r}_{ij}| < r_{\min}(\theta_{ij})$$

and they *cannot* overlap if

$$\text{Max}[(|\mathbf{r}_{ij}.\hat{\mathbf{u}}_i)|, |(\mathbf{r}_{ij}.\hat{\mathbf{u}}_j)|]] > r_{\min}(\theta_{ij})$$

4. *Circular faces overlap?* If we have passed all the tests thus far, two tests remain. The first is a test for overlap between one of the (circular) faces of particle i with one of the faces of particle j. As we have already discussed the test for overlap between two infinitely thin disks, we do not repeat the criteria that have been derived in Section II.D.1.

5. *Face-rim overlap?* The only kind of overlap that we have not yet excluded is between the circular face of one platelet and the spherical rim of the other. To derive the criterion for such an overlap, let us first consider the intersection of a *sphere* of diameter D around $\mathbf{r}_i$ with a circular disk of radius $R_f \equiv \sqrt{D^2 - L^2}/2$ that constitutes the nearest face of particle j. The distance between $\mathbf{r}_i$ and the plane of this face is $d_\perp = |\mathbf{r}_{ij} \cdot \mathbf{u}_j| -$

$L/2$. The sphere around j intersects the plane in a circle of radius R_c $= \sqrt{(D/2)^2 - d_\perp^2}$. The distance between the center of this circle and the axis of particle j is $\sqrt{r_{ij}^2 - (\mathbf{r}_{ij} \cdot \mathbf{u}_j)^2}$. Overlap between the sphere and the circular disk is only possible if the latter distance is less than $R_f + R_c$. Now we take into account that particle i is a truncated sphere rather than a full sphere. This imposes an additional constraint on the overlap criterion, namely that at least some points of the intersection between the sphere around $\mathbf{r}_i$ and the nearest circular face of j must be within a distance $L/2$ from the equatorial plane of i. In fact, this test can be simplified because we have already eliminated overlap between the circular faces of i and j. This means that if *any* point in the intersection of the sphere and the platelet is contained between the top and bottom faces of i, then *all* points in the intersection must satisfy the same criterion, because otherwise the circular faces of i and j must intersect. The choice of the point on which to apply the test is then merely a matter of convenience. In practice, we take the projection of $\mathbf{r}_i$ on the plane of j, unless this point is outside the radius of the circular face of j. In the latter case, we take the point where the perimeter of the circular face of particle j intersects the projection of $\mathbf{r}_{ij}$ in the plane of this face. This test is performed to test if the rim of i intersects the face of j or vice versa.

This completes our description of the test for overlap between two truncated spheres. This test demonstrates how tests for overlap and for non-overlap can be combined in a systematic manner.

e. Ellipsoids. We discuss two different approaches towards the determination of overlap between ellipsoidal hard particles. The first is due to Vieillard-Baron. [5,69] The second is due to Perram and Wertheim. [9] The reason for including both techniques is the fact that it is computationally attractive to use them in combination as explained below.

THE VIEILLARD-BARON CRITERION. The starting point of the Vieillard-Baron criterion (VB) is the equation describing the locus of points on the surface of an ellipsoid

$$F_A(\mathbf{r}) = (\mathbf{r} - \mathbf{r}_A) \cdot \mathbf{A}^{-1} \cdot (\mathbf{r} - \mathbf{r}_A) - 1 = 0 \qquad (2.70)$$

where $\mathbf{r}_A$ is the location of the center of the ellipsoid A, and $\mathbf{A}$ is the matrix $\mathbf{A} = \sum_{i=1}^{3} R_i^2 \hat{\mathbf{a}}_i \otimes \hat{\mathbf{a}}_i$, where R_i is half the length of the ith axis and $\{\hat{\mathbf{a}}_i\}_{i=1,2,3}$ is a set of unit vectors along the axes. Introducing so-called homogeneous projective coordinates x_μ $\mu = 0, 1, 2, 3$, through $r_\nu = x_\nu/x_0$, $\nu = 1, 2, 3$, the ellipsoid equation can be written as

$$\sum_{\mu=0}^{3}\sum_{\nu=0}^{3}\mathcal{A}_{\mu\nu}x_{\mu}x_{\nu} = 0 \tag{2.71}$$

where the 4×4 matrix $\mathcal{A}$ is defined by

$$\begin{aligned}
\mathcal{A}_{00} &= -1 - \mathbf{r}_A \cdot \mathbf{A}^{-1} \cdot \mathbf{r}_A & \\
\mathcal{A}_{0j} &= (\mathbf{r}_A \cdot \mathbf{A}^{-1})_j, & j &= 1,2,3 \\
\mathcal{A}_{i0} &= (\mathbf{A}^{-1} \cdot \mathbf{r}_A)_i, & i &= 1,2,3 \\
\mathcal{A}_{ij} &= \mathbf{A}_{ij}^{-1}, & i,j &= 1,2,3
\end{aligned} \tag{2.72}$$

Given two ellipsoids A and B in the above-mentioned representation, we can construct the so-called pencil of conics passing through them, defined by the equation

$$\sum_{\mu=0}^{3}\sum_{\nu=0}^{3}(\mathcal{A}_{\mu\nu} + \lambda\mathcal{B}_{\mu\nu})x_{\mu}x_{\nu} = 0, \quad \lambda \in \mathbb{C} \tag{2.73}$$

This pencil contains four (possibly multiply) degenerate conics, for those λ such that

$$P(\lambda) \equiv \det(\mathcal{A} + \lambda\mathcal{B}) = 0 \tag{2.74}$$

The properties of these degenerate conics determine whether A and B have any real points in common. These properties in turn follow from the roots of $P(\lambda)$. We can now state the relevant rule: A and B do *not* have any real points in common (i.e., do *not* intersect) if and only if all roots of $P(\lambda)$ are real and *not* all are negative.

We thus need a root determination scheme. There are several choices available but we prefer the one given in the classical algebra text by Weber. [70] Given a normalized fourth degree polynomial,

$$N(\lambda) = \lambda^4 + n_3\lambda^3 + n_2\lambda^2 + n_1\lambda + n_0 \tag{2.75}$$

we can convert into the canonical form by the substitution $\lambda = \tau - \frac{1}{4}n_3$, yielding

$$C(\tau) = \tau^4 + c_2\tau^2 + c_1\tau + c_0 \tag{2.76}$$

with

$$\begin{aligned}
c_2 &= -\frac{3}{8}n_3^2 + n_2 \\
c_1 &= \frac{1}{8}n_3^3 - \frac{1}{2}n_3n_2 + n_1 \\
c_0 &= -\frac{3}{256}n_3^4 + \frac{1}{16}n_3^2 n_2 - \frac{1}{4}n_3 n_1 + n_0
\end{aligned} \tag{2.77}$$

The discriminant D of the canonical polynomial is given by

$$27D = 4\left(c_2^2 + 12c_0\right)^3 - \left(2c_2^3 - 72c_2c_0 + 27c_1^2\right)^2 \tag{2.78}$$

The necessary conditions for the reality of the roots are

(i) $D > 0$

(ii) $c_2 < 0$

(iii) $c_2^2 - 4c_0 > 0$

while the condition that not all roots are negative is met when

(iv) *at least* one of the coefficients $\{n_3, n_2, n_1\} < 0$

The only input needed for the application of the overlap criterion are thus the coefficients $\{n_k\}$ of the normalized charcateristic polynomial $P(\lambda)$ for the specific ellipsoidal particles in question. Here we will give the explicit results for two cases:(i) nonisomorphic ellipsoids of revolution; (ii) isomorphic general ellipsoids.

(i) *Nonisomorphic ellipsoids of revolution.* For ellipsoids of revolution, it is useful to define the matrices

$$\mathbf{A}^{-1} = \alpha\mathbf{1} + \gamma\hat{\mathbf{a}} \otimes \hat{\mathbf{a}}, \qquad \mathbf{B}^{-1} = \beta\mathbf{1} + \delta\hat{\mathbf{b}} \otimes \hat{\mathbf{b}} \tag{2.79}$$

where

$$\begin{aligned}
\alpha &= R_{A,1}^{-2}, & \gamma &= R_{A,3}^{-2} - R_{A,1}^{-2} \\
\beta &= R_{B,1}^{-2}, & \delta &= R_{B,3}^{-2} - R_{B,1}^{-2}
\end{aligned} \tag{2.80}$$

and $\hat{\mathbf{a}}$ and $\hat{\mathbf{b}}$ are unit vectors along the major symmetry axis of the two particles.

Defining $\mathbf{r} = \mathbf{r}_B - \mathbf{r}_A$, $\Delta_A = \det \mathbf{A}^{-1}$ and $\Delta_B = \det \mathbf{B}^{-1}$ and introducing the auxiliary terms

$$\begin{aligned}
P_A &= -2\alpha\beta\delta - \beta\gamma\delta + \beta\gamma\delta(\hat{\mathbf{a}} \cdot \hat{\mathbf{b}})^2 - 3\alpha\beta^2 \\
&\quad -\beta^2\gamma + \alpha\Delta_B r^2 - \Delta_B + \gamma\Delta_B(\mathbf{r} \cdot \hat{\mathbf{a}})^2 \\
P_B &= -2\alpha\beta\gamma - \alpha\gamma\delta + \alpha\gamma\delta(\hat{\mathbf{a}} \cdot \hat{\mathbf{b}})^2 - 3\alpha^2\beta \\
&\quad -\alpha^2\delta + \beta\Delta_A r^2 - \Delta_A + \delta\Delta_A(\mathbf{r} \cdot \hat{\mathbf{b}})^2 \\
P_{AB} &= \alpha\beta\gamma\delta\left\{r^2 - r^2(\hat{\mathbf{a}} \cdot \hat{\mathbf{b}})^2 + 2(\mathbf{r} \cdot \hat{\mathbf{a}})(\mathbf{r} \cdot \hat{\mathbf{b}})(\hat{\mathbf{a}} \cdot \hat{\mathbf{b}})\right\} \\
&\quad +\alpha\beta^2\gamma r^2 + \alpha\beta^2\gamma(\mathbf{r} \cdot \hat{\mathbf{a}})^2 - \alpha\Delta_B r^2 \\
&\quad +\alpha^2\beta\delta r^2 + \alpha^2\beta\delta(\mathbf{r} \cdot \hat{\mathbf{b}})^2 - \beta\Delta_A r^2 \\
&\quad +\Delta_A + \Delta_B + 2\alpha^2\beta^2 r^2 - \gamma\Delta_B(\mathbf{r} \cdot \hat{\mathbf{a}})^2 - \Delta_A(\mathbf{r} \cdot \hat{\mathbf{b}})^2 \tag{2.81}
\end{aligned}$$

The coefficients of the normalized characteristic polynomial are then given by

$$n_3 = -P_A/\Delta_B$$
$$n_2 = -(P_A + P_B + P_{AB})/\Delta_B \qquad (2.82)$$
$$n_1 = -P_B/\Delta_B$$
$$n_0 = \Delta_A/\Delta_B$$

(ii) *Isomorphic general ellipsoids.* For the isomorphic, but general (i.e., $R_1 \neq R_2 \neq R_3$) ellipsoids, the results can be given in a more compact form. Defining $l_i = R_i^{-2}$, ϵ_{ijk} the totally antisymmetric three-tensor and assuming summation over repeated Cartesian indices, we get

$$n_0' = -l_1 l_2 l_3$$

$$n_1' = l_1 l_2 l_3 \, l_i (\mathbf{r} \cdot \hat{\mathbf{b}}_i)^2 - l_1 l_2 l_3 - \frac{1}{2} l_i l_j l_k (\hat{\mathbf{b}}_i \cdot \hat{\mathbf{a}}_j \wedge \hat{\mathbf{a}}_k)^2$$

$$n_2' = l_1 l_2 l_3 \, l_i \mid \epsilon_{ijk} \mid (\mathbf{r} \cdot \hat{\mathbf{a}}_j \wedge \hat{\mathbf{b}}_k)^2 + \frac{1}{2} l_i^2 l_j^2 \mid \epsilon_{ijk} \mid (\mathbf{r} \cdot \hat{\mathbf{a}}_k \wedge \hat{\mathbf{b}}_k)^2$$

$$- \frac{1}{2} l_i l_j l_k (\hat{\mathbf{a}}_i \cdot \hat{\mathbf{b}}_j \wedge \hat{\mathbf{b}}_k)^2 - \frac{1}{2} l_i l_j l_k (\hat{\mathbf{b}}_i \cdot \hat{\mathbf{a}}_j \wedge \hat{\mathbf{a}}_k)^2$$

$$n_3' = l_1 l_2 l_3 \, l_i (\mathbf{r} \cdot \hat{\mathbf{a}}_i)^2 - l_1 l_2 l_3 - \frac{1}{2} l_i l_j l_k (\hat{\mathbf{a}}_i \cdot \hat{\mathbf{b}}_j \wedge \hat{\mathbf{b}}_k)^2 \qquad (2.83)$$

where the primed coefficients are related to the normalized ones through the relation

$$n_k = n_k'/n_0', \quad k = 1, 2, 3 \qquad (2.84)$$

THE PERRAM–WERTHEIM CRITERION. Given two ellipsoids A and B, we form the family of interpolating functions

$$F(\mathbf{r}, \lambda) = \lambda F_A(\mathbf{r}) + (1 - \lambda) F_B(\mathbf{r}) + 1 \qquad (2.85)$$

using the definition (2.70) for the equations defining the surface of an ellipsoid. Introduce

$$F(A, B) = \max_{\lambda \in [0,1]} \min_{\mathbf{r}} F(\mathbf{r}, \lambda) \qquad (2.86)$$

The following property of the quantity $F(A, B)$ is now proposed to hold

$$F(A, B) = \begin{cases} > 1, & A \text{ and } B \text{ are } nonoverlapping \\ 1, & A \text{ and } B \text{ are exteriorly tangent} \\ < 1, & A \text{ and } B \text{ overlap} \end{cases} \qquad (2.87)$$

A sketch of the proof of this, at first sight remarkable property runs as follows. The minimum over all space of $F(\mathbf{r}, \lambda)$ is found by solving

$$2\nabla F(\mathbf{r}, \lambda) = \lambda \mathbf{A}^{-1} \cdot (\mathbf{r} - \mathbf{r}_A) + (1 - \lambda) \mathbf{B}^{-1} \cdot (\mathbf{r} - \mathbf{r}_B) = 0 \qquad (2.88)$$

We can easily solve for this minimum, which we denote by $\mathbf{r}(\lambda)$

$$\mathbf{r}(\lambda) - \mathbf{r}_A = (1 - \lambda)\mathbf{A} \cdot \mathbf{C} \cdot \mathbf{r}_{AB} \quad \text{or} \quad \mathbf{r}(\lambda) - \mathbf{r}_B = -\lambda\mathbf{B} \cdot \mathbf{C} \cdot \mathbf{r}_{AB} \quad (2.89)$$

where $\mathbf{r}_{AB} = \mathbf{r}_B - \mathbf{r}_A$ and

$$\mathbf{C} = (\lambda\mathbf{B} + (1 - \lambda)\mathbf{A})^{-1} \quad (2.90)$$

We can now eliminate $\mathbf{r}(\lambda)$ to find

$$\min_{\mathbf{r}} F(\mathbf{r}, \lambda) = F(\mathbf{r}(\lambda), \lambda) = \lambda(1 - \lambda)\mathbf{r}_{AB} \cdot \mathbf{C} \cdot \mathbf{r}_{AB} \quad (2.91)$$

Consider the path $\wp(\lambda) = \{\mathbf{r}(\lambda) \mid \lambda \in [0, 1]\}$. This path runs from $\mathbf{r}_0 = \mathbf{r}_B$ to $\mathbf{r}_1 = \mathbf{r}_A$ (cf. 2.89), the interpolating function vanishing at these endpoints, that is, $F(\mathbf{r}_B, 0) = F(\mathbf{r}_A, 1) = 0$. The following cases can now be distinguished:

- If A and B do *not* overlap, the path $\wp$ passes through the region *outside* both A and B, where F_A and F_B are both > 0. Since F is a convex combination of F_A and F_B it follows that $F(\mathbf{r}(\lambda), \lambda) > 1$ in this region and hence $F(A, B) > 1$, as proposed.
- If A and B overlap then, inside the region of overlap $A \cap B$, $F(\mathbf{r}, \lambda) < 1$ independent of λ, since both F_A and F_B are less than zero there. This implies that $F(\mathbf{r}(\lambda), \lambda) < 1$ for all λ and hence $F(A, B) < 1$.
- Finally, if A and B touch exteriorly at $\mathbf{r}^*$, we have that $F(\mathbf{r}^*, \lambda) = 1$ for all λ as $F_A(\mathbf{r}^*) = F_B(\mathbf{r}^*) = 0$, so $F(\mathbf{r}(\lambda), \lambda) \leq 1$. This precludes the path from going outside both A and B and hence there must be a λ^* such that $\mathbf{r}(\lambda^*) = \mathbf{r}^*$ so that $F(A, B) = F(\mathbf{r}^*, \lambda^*) = 1$.

The only implicit assumption we still have to prove is that $F(A, B)$ is in fact unique. This is easily accomplished by twice differentiating

$$\frac{d^2}{d\lambda^2} F(\mathbf{r}(\lambda), \lambda) = -2\mathbf{r}_{AB} \cdot \mathbf{C} \cdot \left(\lambda\mathbf{A}^{-1} + (1 - \lambda)\mathbf{B}^{-1}\right) \cdot \mathbf{C} \cdot \mathbf{r}_{AB} < 0 \quad (2.92)$$

The inequality follows because all the matrices in the expression are positive definite. We have now shown that $F(\mathbf{r}(\lambda)\lambda)$ is indeed concave on the interval $[0, 1]$ hence the maximum $F(A, B)$ is unique.

We now turn to the explicit calculation of $F(\mathbf{r}(\lambda)\lambda)$ which involves the computation of the matrix $\mathbf{C}$. Define

$$\mathbf{\Gamma} = \mathbf{C}^{-1} = \lambda\mathbf{A} + \mu\mathbf{B} \quad (2.93)$$

where we have introduced the shorthand $\mu = 1 - \lambda$. We make use of the fact that $\mathbf{\Gamma}$ is a root of its own characteristic equation

$$Q(\xi) = \det \mathbf{\Gamma} - \xi\mathbf{1} = -\xi^3 + w_2\xi^2 - w_1\xi + w_0 \quad (2.94)$$

allowing us to find

$$\mathbf{C} = \mathbf{\Gamma}^{-1} = \frac{1}{w_0} \left\{ \mathbf{\Gamma}^2 - w_2 \mathbf{\Gamma} + w_1 \mathbf{1} \right\} \qquad (2.95)$$

The algebra involved in computing the coefficients w_j is much simpler than in the Vieillard-Baron case and we can give the result for two arbitray ellipsoids. Writing $A_j = R_{A,j}^2$ and $B_j = R_{B,j}^2$, we have

$$
\begin{aligned}
w_0 &= \lambda^3 B_1 B_2 B_3 + \lambda^2 \mu \left(\frac{1}{2} B_i B_j A_k (\hat{\mathbf{b}}_i \wedge \hat{\mathbf{b}}_j \cdot \hat{\mathbf{a}}_k)^2 \right) \\
&\quad + \lambda \mu^2 \left(\frac{1}{2} A_i A_j B_k (\hat{\mathbf{a}}_i \wedge \hat{\mathbf{a}}_j \cdot \hat{\mathbf{b}}_k)^2 \right) + \mu^3 A_1 A_2 A_3 \\
w_1 &= \lambda^2 (B_1 B_2 + B_1 B_3 + B_2 B_3) \qquad\qquad (2.96) \\
&\quad + \lambda \mu \left\{ (A_1 + A_2 + A_3)(B_1 + B_2 + B_3) - A_i B_j (\hat{\mathbf{a}}_i \cdot \hat{\mathbf{b}}_j)^2 \right\} \\
&\quad + \mu^2 (A_1 A_2 + A_1 A_3 + A_2 A_3) \\
w_2 &= \lambda (B_1 + B_2 + B_3) + \mu (A_1 + A_2 + A_3)
\end{aligned}
$$

Using Eq. (2.91), $F(\mathbf{r}(\lambda), \lambda)$ is now easily calculated. Its maximization still has to be carried out numerically, but given the concavity of this function, an ultrafast routine like the Brent method [71] can be used.

OPTIMIZING THE OVERLAP TEST.   It turns out to be computationally efficient to mix the two overlap criteria in such a manner that nonoverlaps are detected as economically as possibly. The following three-stage process is designed to do just this.

1. Test for overlap of the circumscribed spheres. If these do not overlap accept the move. Else
2. Evaluate the function $F(\mathbf{r}(\frac{1}{2}), \frac{1}{2})$ from the Perram–Wertheim criterion. If this is larger than unity we know that the ellipsoids do not overlap and can accept the move. Else
3. Perform the Vieillard-Baron test, which with its four subcriteria is the most involved.

The first two tests are "nonoverlap" tests, while the third is an "overlap" test.

f. Parallel Hard Particles.   In a number of cases, it is of interest to study the properties of model systems with restricted orientations. Most common among these restricted orientation models are systems of parallel (hard) particles. Such model systems can be thought of as a limiting case of a system of freely rotating particles in a strong aligning field. The main reason to study systems of parallel particles is that both the theoretical analysis and the numerical simulation of such systems is usually simpler than that of their freely rotating counterparts. This is particularly clear in the case of aligned ellipsoids. Any configuration of a system of aligned ellipsoids can be transformed into a configuration of hard spheres via a simple affine transformation. For instance, spheroids with an axial ratio α that are aligned along the z-axis, can be transformed to the correspond-ing hard-sphere system via the affine transformation $z \to z' = z/\alpha$. All systems of aligned ellipsoids (both uni-axial and bi-axial) are therefore equivalent to the hard-sphere system. Hence, the behavior of any system of aligned ellipsoids can be deduced from the behavior of the hard-sphere system at the same packing fraction.

However, for hard particles with other shapes, the aligned system is not trivially related to some known reference system. Here, we briefly review the overlap criteria for two model systems consisting of aligned particles.

ALIGNED SPHEROCYLINDERS.   Consider a system of sphero-cylinders aligned along the z-axis. For convenience, we assume that all particles have the same diameter D, while the cylindrical part has a length L. In order to test whether two particles i and j overlap, we first compute z_{ij}, their distance in the z-direction, and $r_{ij}^{\perp}$, their distance in the xy-plane. Overlap between i and j is only possible if the following conditions are both satisfied:

$$z_{ij} < L + D \quad \text{and} \quad r_{ij}^{\perp} < D$$

Let us assume that these conditions are indeed satisfied. Then, if

$$z_{ij} < L$$

the cylindrical parts must overlap. In fact, if we study hard, parallel right cylinders (i.e., cylinders without a hemi-spherical cap), this is our final test. For spherocylinders, however, if $z_{ij} > L$, then we must test if the hemi-spherical caps overlap. The corresponding test is

$$(r_{ij}^{\perp})^2 + (z_{ij} - L)^2 < D^2$$

Clearly, the test for overlap between aligned spherocylinders is consider-ably simpler than the corresponding test for freely oriented spherocylin-ders.

ALIGNED TRUNCATED SPHERES.   Next, consider truncated spheres with di-ameter D and thickness L, aligned along the z-axis. The test for overlap is extremely simple. If

$$z_{ij} > L$$

no overlap is possible. Otherwise, we simply apply the test for overlap between two spheres, that is,

$$r_{ij}^2 < D^2$$

g. *Other Shapes.* Finally, we briefly review a few other convex hard-core models that have been discussed in the literature. Although some of these bodies appear to have a simpler shape than the models discussed above, the test for overlap may be more involved.

RIGHT CYLINDERS. A case in point is a model consisting of right cylinders of length L and diameter D. The test for overlap between two such particles can be decomposed into several subtests. The first test is similar to the computation of the distance of closest approach between two spherocylinders. However, we can only use this test if the shortest vector between the axes of the two cylinders intersects both cylinders. Otherwise, we must test if the flat faces of the cylinders overlap (see Section. II.D.1). In the final test, we check if the flat face of one cylinder intersects the cylindrical part of the other. This test can be reduced to an overlap between coplanar ellipses (see Section. II.D.1). Clearly, this sequence of tests is feasible. However, it is more complex than the test for overlap between two spherocylinders.

OBLATE SPHEROCYLINDERS. An oblate model that could serve as an alternative to the truncated sphere is the so-called oblate spherocylinder. [46] This shape is obtained by moving the center of a sphere with diameter L over a circular disk with diameter D. In this case, the time-consuming step in the overlap test is the computation of the distance of closest approach between the toroidal rims of two oblate spherocylinders. Wojcik and Gubbins solve this problem by using an iterative minimization scheme. [46] Although such a scheme may be quite efficient in practice, we have limited ourselves to test schemes that are guaranteed to terminate in a finite number of steps.

"UFOs". Siders and co-workers [72,73] have considered an alternative model for an oblate molecule, namely the intersection volume of two equal hard spheres at a distance $r < D$ (where D is the diameter of the spheres). This object, called UFO by Siders et al., might, in some cases, serve as an alternative for oblate spheroids. The tests for overlap between UFOs are fairly straightforward.

GAUSSIAN CORE. Thus far, we have only considered models in which the individual particles have a well-defined geometry. However, as mentioned earlier, this is a subclass of all hard-core interactions. The use of hard-core interaction that cannot be interpreted in terms of the "shape" of the individual particles has advantages and disadvantages. The advantage is that it is often possible to write very simple expressions for the potential energy function. The disadvantage is that much of our intuitive understanding of local packing effects is lost if we use a model in which individual particles do not have a "shape". Best known among these nongeometrical hard-core models is the so-called Gaussian-core model. [39,64] In this model, the form of the potential energy function is similar to that of hard spheres, that is,

$$u(r_{12}) = \begin{cases} 0, & r_{12} > \sigma_0 \\ \infty & r_{12} < \sigma_0 \end{cases}$$

But, unlike the hard-sphere case, σ_0 now depends on the orientations ($\hat{\mathbf{u}}_1$ and $\hat{\mathbf{u}}_2$) of the molecules, and the orientation of the vector $\mathbf{r}_{12}$ joining the centers of mass of the molecules:

$$\sigma(\mathbf{r}_{12}, \hat{\mathbf{u}}_1, \hat{\mathbf{u}}_2) = \sigma_0 \left(1 - \frac{\chi}{2} \left[\frac{(\hat{\mathbf{r}}_{12} \cdot \hat{\mathbf{u}}_1 + \hat{\mathbf{r}}_{12} \cdot \hat{\mathbf{u}}_2)^2}{1 + \chi \hat{\mathbf{u}}_1 \cdot \hat{\mathbf{u}}_2} + \frac{(\hat{\mathbf{r}}_{12} \cdot \hat{\mathbf{u}}_1 - \hat{\mathbf{r}}_{12} \cdot \hat{\mathbf{u}}_2)^2}{1 - \chi \hat{\mathbf{u}}_1 \cdot \hat{\mathbf{u}}_2} \right] \right)^{-1/2}$$

$$(2.97)$$

where χ is a measure of the nonsphericity of the molecule. For two parallel molecules lying side by side, $\sigma_\perp = \sigma_0$. If the same molecules are positioned end to end, $\sigma_\parallel = \sigma_0[(1 + \chi)/(1 - \chi)]^{1/2}$. Conversely, if we *define* the aspect ratio of the molecule as $\kappa \equiv \sigma_\parallel / \sigma_\perp$, then

$$\chi = \frac{\kappa^2 - 1}{\kappa^2 + 1}$$

For two parallel Gaussian hard-core molecules, the excluded volume is identical to that of parallel ellipsoids with the same aspect ratio. However, the Gaussian hard-core interaction and the hard ellipsoid interaction are not identical for nonparallel molecules. In fact, Perram et al. [74] have shown that the Gaussian hard-core model has a simple interpretation in terms of the function $F(\mathbf{r}, \lambda)$ defined in Eq. (2.85). Namely, that the overlap criterion in the Gaussian hard-core model corresponds to the criterion $F(\mathbf{r}, 1/2) = 1$. We recall that the latter criterion is the "quick" Perram–Wertheim test described in Section II.D.1. From the analysis below Eq. (2.85), it then follows immediately that the Gaussian overlap criterion is always an *upper bound* to the ellipsoid overlap criterion. Moreover, Perram et al. show that the difference between the two models becomes more pronounced with increasing nonsphericity of the molecules or, in the case of unlike molecules, when the molecules become more dissimilar. In fact, the available simulation data on hard-ellispoids and hard Gaussian-overlap models [39,64,75] appear to pass this test.

2. Virial Coefficients

Once the criterion for hard-core overlap between two convex hard bodies has been specified, we are in a position to study the properties of the model under consideration by computer simulation. Such simulations (either MC or MD) are usually performed on a model consisting of several hundreds to several thousands of particles. First, however, we briefly discuss a few-body calculation, namely the numerical evaluation of the virial coefficients of the system. The virial coefficients B_n with $n = 1, 2, 3, \ldots$, are the expansion coefficients of the compressibility factor $Z = PV/Nk_BT$ in powers of the number density $\rho \equiv N/V$:

$$Z = 1 + B_2\rho + B_3\rho^2 + \cdots \tag{2.98}$$

where we have used the fact that $B_1 = 1$. Clearly, knowledge of the first few virial coefficients would allow us to predict the equation of state of the model system at moderately low densities. This is important, because it provides an independent test of our MC or MD calculations, in which the equation of state is computed in a completely different fashion. The virial coefficients can be expressed in terms of sums of (multidimensional) integrals of the Mayer f-functions (see, e.g. Ref. [1]). Ree and Hoover [76] have shown that, for hard-core particles, the number of "diagrams" can be greatly reduced and that the resulting expressions lend themselves to numerical (Monte Carlo) evaluation. Although we do not discuss the Ree–Hoover scheme in any detail, we wish to point out that different implementations of the MC scheme are possible. In order to clarify this point, we need to know only one thing about the Ree–Hoover scheme to compute B_n, namely that the quantity that must be sampled is a function of the coordinates of n molecules and that a necessary (although not sufficient) condition for this function to be nonzero is that particle i overlaps with j, j with k, $\ldots$ and l with i, where $i, j, k, \ldots, l$ is any permutation of $\{1, 2, 3, \ldots, n\}$. The schemes to compute the B_n are based on an algorithm to generate, in an unbiased way, configurations where 1 overlaps with 2,2 with 3, $\ldots$, $n-1$ with n. The contribution of this configuration to the nth virial coefficient can then be evaluated by a series of tests for overlap (e.g., between n and 1) and nonoverlap, that are described in Ref. [76]. We only wish to point out that there are two distinct ways of generating the "open chain" configuration $1 - 2 - \cdots - n$. The simplest (and most common) is a dynamic scheme in which normal "Metropolis" sampling is performed on the coordinates of all n particles. Whenever a trial move "breaks" the chain from 1 to n, it is rejected. Although this scheme is simple, it has the disadvantage that there is appreciable correlation between successive configurations and, more importantly, the sampling becomes less efficient as the molecules become more anisometric. The second scheme, which does not suffer form these drawbacks, is the "static" scheme. Here, a new, random chain conformation is generated from scratch at every step. In order to achieve this, one must be able to generate, with the correct probability, all configurations of molecule k that overlap with $k - 1$. For extremely elongated molecules, the static scheme is, to our knowledge, the only feasible method to compute the higher virial coefficients. In order to explain the difference between the static and the dynamic schemes, consider the computation of the third virial coefficient of line segments in two dimensions. The third virial coefficient can be computed from the probability that line segment 3 overlaps with 1, given that 1 overlaps with 2 and 2 with 3. In the dynamic scheme, we would prepare the three lines such that the pairs 1–2 and 2–3 overlap. Next, we would perform a random trial displacement or rotation of one of the particles. If that trial moves maintains the 1–2 and the 2–3 overlap, it is accepted. Otherwise it is rejected. Of course, subsequent configurations are rather strongly correlated. Hence, there is not much point in testing for overlap between 3 and 1 after *every* trial move. In the static scheme, we fix particle 1. Next, we generate a trial orientation for particle 2 with a probability that is proportional to the pair-excluded volume of 1 and 2 for that particular orientation. Finally, we place the center of mass of 2 anywhere in this excluded volume. This procedure guarantees that 1 and 2 overlap and also that all overlapping configurations are generated with the correct statistical weight. We repeat the same procedure to insert particle 3 in such a way that it overlaps with 2. Finally, we test for overlap between 1 and 3. In this static scheme, there are no correlations between subsequent configurations. When comparing the static and dynamic schemes, we should bear in mind that the dynamic scheme is easier to implement than the static scheme. In particular, care should be taken in the static scheme that all trial configurations are generated with the correct weight. In contrast, in the dynamic scheme, a simple test for overlap suffices to accept or reject a trial configuration.

3. *Equation of State*

There exist several techniques to compute the pressure in a hard-core system. Although these techniques are, in principle, equivalent, they appear rather different. In molecular dynamics simulations, the most convenient starting point for computation of the pressure is based on the virial. This expression is based on the observation that, in a bounded N-particle system,

$$\sum_{i=1}^{N} \mathbf{p}_i \cdot \mathbf{r}_i$$

is bounded, where $\mathbf{p}_i$ is the momentum of particle i and $\mathbf{r}_i$ denotes the center of mass position of that particle. As the virial itself is bounded, its average time derivative is zero. This condition yields an expression that relates the average collisional momentum transfer along the line joining the center of mass of two collision partners to the pressure in the system:

$$P/(\rho k_B T) = 1 + \frac{1}{d} < \Delta \mathbf{P}_{ij} \cdot \mathbf{r}_{ij} > \qquad (2.99)$$

where $\Delta \mathbf{P}_{ij}$ denotes the collisional momentum transfer between particles i and j, while $\mathbf{r}_{ij}$ denotes the vector joing the centers of mass of these particles and d is the dimensionality of the system. The angular brackets denote time averaging. In a hard-core molecular dynamics program, as the collisional momentum transfer has to be computed anyway, the computation of the pressure requires very little overhead. For more details, see Ref. [77].

Clearly, the above method to measure the pressure cannot be used in Monte Carlo simulations. A convenient technique to measure the pressure in such simulations is based on the fact that pressure is equal to (minus) the volume derivative of the Helmholtz free energy F

$$P = - (\partial F / \partial V)_{NT} \qquad (2.100)$$

We can approximate the pressure by a ratio of finite differences:

$$P \approx - (\Delta F / \Delta V)_{NT} \qquad (2.101)$$

Equation. (2.101) is a convenient starting point for a numerical scheme to measure P. To this end, we must compute the free energy difference between a system at volume V and the same system at a smaller volume $V' = V + \Delta V$. The free energy of a system of N molecules at volume V is given by $F = -k_B T \ln Q(N, V, T)$, with

$$Q(N, V, T) = \frac{q^N(T) V^N}{N!} \int_0^L \cdots \int_0^L d\mathbf{r}^N \, d\mathbf{\Omega}^N \, \exp(-\beta U(\mathbf{r}^N, \mathbf{\Omega}^N)) \qquad (2.102)$$

where $\mathcal{V}_T$ is the part of the partition funtion that results from integration over the momenta (see Appendix A.A). It is convenient to rewrite Eq. (2.102) in a slightly different way. Let us assume that the system is contained in a cubic box with diameter $L = V^{1/3}$. We now define scaled coordinates $\mathbf{s}^N$, by

$$\mathbf{r}_i = L\mathbf{s}_i$$

for $i = 1, 2, \ldots, N$. If we now insert these scaled coordinates in Eq. (2.102) we obtain

$$Q(N, V, T) = \frac{V^N}{\mathcal{V}_T^N N!} \int_0^1 \cdots \int_0^1 d\mathbf{s}^N \, d\mathbf{\Omega}^N \, \exp(-\beta U(\mathbf{s}^N, \mathbf{\Omega}^N; L)) \quad (2.103)$$

In Eq (2.103), we have written $U(\mathbf{s}^N, \mathbf{\Omega}^N; L)$ to indicate that U depends on the *real* rather than the *scaled* distances between the particles. The expression for the Helmholtz free energy of the system is

$$\begin{aligned} F(N, V, T) &= -k_B T \ln Q \\ &= -k_B T \ln(\frac{V^N}{\mathcal{V}_T^N N!}) \\ &\quad -k_B T \ln \left(\int d\mathbf{s}^N \, d\mathbf{\Omega}^N \, \exp(-\beta U(\mathbf{s}^N, \mathbf{\Omega}^N; L)) \right) \end{aligned} \quad (2.104)$$

It is now straightforward to write down the expression for the pressure given in Eq. (2.101):

$$\begin{aligned} -\frac{\Delta F}{\Delta V} &= \frac{k_B T}{\Delta V} \ln(Q(V')/Q(V)) \\ &= \frac{k_B T}{\Delta V} \ln \left(\frac{V'^N \int d\mathbf{s}^N \exp(-\beta U(\mathbf{s}^N, \mathbf{\Omega}^N; V'))}{V^N \int d\mathbf{s}^N \exp(-\beta U(\mathbf{s}^N, \mathbf{\Omega}^N; V))} \right) \end{aligned} \quad (2.105)$$

or

$$P = P_{\mathrm{id}} - \frac{k_B T}{\Delta V} \ln < \exp(-\beta \Delta U(\mathbf{s}^N)) >, \quad (2.106)$$

where $\Delta U \equiv U(\mathbf{s}^N, \mathbf{\Omega}^N; V) - U(\mathbf{s}^N, \mathbf{\Omega}^N; V')$. Equation. (2.106) may be interpreted as the acceptance probability of a *virtual* Monte Carlo move in which the volume is decreased from V to V'. Equation. (2.106) is valid if *all* nonoverlapping configurations of a system with volume V' correspond to nonoverlapping configurations of the system with the larger volume V. For convex hard bodies, this is indeed the case. For nonconvex hard bodies, one should measure the acceptance ratio of virtual volume changes from V to V', and vice versa. For more details, see Ref. [67].

An alternative scheme to determine the equation of state of a hard-body fluid in a Monte Carlo simulation, is to perform such simulations at constant pressure, [78,79] rather than at constant volume. If the pressure of a system of N particles is fixed at P, then the probability density to find that system in a particular configuration of the N molecules (as specified by $\mathbf{s}^N, \mathbf{\Omega}^N$) *and* a given volume V is given by

$$\mathcal{P}(V; \mathbf{s}^N, \mathbf{\Omega}^N)$$

$$= \frac{V^N \exp(-\beta PV) \exp(-\beta U(\mathbf{s}^N, \mathbf{\Omega}^N; L))}{\int_0^\infty dV' \, V'^N \exp(-\beta PV') \int d\mathbf{s}^N \, d\mathbf{\Omega}^N \exp(-\beta U(\mathbf{s}^N, \mathbf{\Omega}^N; L'))}$$

We can carry out Metropolis sampling on the reduced coordinates $\mathbf{s}^N, \mathbf{\Omega}^N$ *and* on the volume V, with a weight function $\rho(\mathbf{s}^N, V)$ proportional to

$$\exp(-\beta\{U(\mathbf{s}^N, \mathbf{\Omega}^N; V) + PV - N\beta^{-1} \ln V\})$$

In the constant N, P, T MC method, V is simply treated as an additional coordinate, and trial moves in V must satisfy the same rules as trial moves in $\mathbf{r}$ (in particular, we should maintain the symmetry of the underlying Markov chain). Let us assume that our trial moves consist of an attempted change of the volume from V to $V' = V + \Delta V$, where ΔV is a random number uniformly distributed between over the interval $[-\Delta V_{max}, +\Delta V_{max}]$. In the Metropolis scheme, such a random, volume changing move will be accepted if

$$\exp(-\beta[U(\mathbf{s}^N, \mathbf{\Omega}^N; V') - U(\mathbf{s}^N, \mathbf{\Omega}^N; V) + P(V' - V) - N\beta^{-1} \ln(V'/V)]) > \mathcal{R}$$
$$(2.107)$$

where $\mathcal{R}$ is a random number, uniformly distributed over the interval $[0, 1]$.

4. Some Results

It is not our aim to review the large amount of numerical data that have been gathered on the static properties of isotropic hard-body fluids. In fact, several papers exist that discuss the comparison of the numerical results for virial coefficients and for equations of state of such fluids with various approximate theories. In particular, the excellent review by Boublik and Nezbeda [12] contains a compilation of the first five virial coefficients of prolate and oblate spherocylinders and ellipsoids, and of the first seven virial coefficients of hard spheres. Subsequently, several papers have appeared that discuss the equation of state or virial coefficients (of other hard-body fluids [38,39,64,80,81]. There are fewer systematic studies of the structure of hard-body fluids. However, a useful compilation of some of the numerical data can be found in a paper by Nezbeda et al.[61]

a. Virial Coefficients of Spherocylinders. Onsager's model plays a unique role in the theory of liquid crystals, because it is the only exactly solvable model with full translational and orientational degrees of freedom which exhibits a transition to the nematic phase. However, as explained in Section IV.A.2, the Onsager theory is only valid in the limit $L/D \to \infty$, while most thermotropic liquid crystals have effective L/D ratios of 3 to 5. The reason why the Onsager theory cannot be used to describe molecular systems with such "small" L/D values is the following: an essential assumption in its derivation is that, in the expansion of the free energy in powers of the density, all virial coefficients B_n with $n > 2$ may be neglected. This condition is satisfied if $B_n/B_2^{n-1} \ll 1$. Onsager gave a qualitative argument to show that for large L/D, $B_3/B_2^2 \sim (D/L) \ln(L/D)$. Hence, in the limit $L/D \to \infty$, the reduced third virial coeffient does indeed vanish. Onsager made the plausible assumption that the higher virial coefficients can also be neglected in the same limit. Of course, the question arises if this assumption about the higher virial coefficients is indeed correct, and if so, how large L/D must be to observe this asymptotic behavior. Unfortunately, the virial coefficients compiled in Ref. [12] are limited to particles with a length to width ratio between 10 and 0.1. We therefore computed the third through fifth virial coefficient of hard spherocylinders as a function of L/D, for L/D between 1 and 10^5, [82,83] using the techniques described in Section II.D.2.

Figure 2.3 shows the L/D dependence of B_3/B_2^2, B_4/B_2^3 and B_5/B_2^4 for hard spherocylinders with L/D between 1 and 10^5. As can be seen from Fig. 2.3, the computed virial coefficients do indeed become small for large L/D. However, for B_4 and B_5 this decrease only sets in at rather large L/D values. This effect is seen more clearly by dividing the reduced virial coefficients by a factor proportional to the value that one should expect if the asymptotic L/D dependence was valid for all L/D; that is, $(D/L) \ln(L/D)$ for B_3 and D/L for B_4 and B_5. The resulting "scaled" virial coefficients are shown in Fig. 2.4. Apparently, for B_3 the asymptotic behavior already sets in for small L/D. Not so for B_4 and

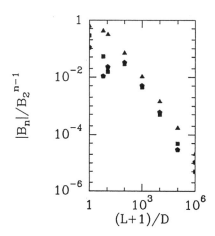

Figure 2.3. Virial coefficients spherocylinders. Filled triangles, B_3/B_2^2, filled squares, $|B_4|/B_2^3$ and filled pentagons, B_5/B_2^4. For $L/D \to \infty$, B_3 and B_5 are positive while B_4 is negative. For $L/D \to 0$, all three virial coefficients are positive.

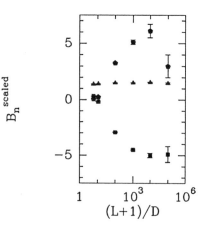

Figure 2.4. Scaled virial coefficients spherocylinders. To show the approach of the virial coefficients of hard spherocylinders to their asymptotic behavior, we plot $B_3/B_2^2 \times (L/D)/\ln(L/D)$ (▲), $B_4/B_2^3 \times (L/D)$ (■) and $B_5/B_2^4 \times (L/D)$ (●). As $L/D \to \infty$, these "scaled" virial coefficients are all expected to approach a constant value.

B_5. As can be seen from Fig. 2.4, the approach to the asymptotic behavior is only observed beyond $L/D = \mathcal{O}(10^2)$. The reason is that, whereas only one (positive) cluster integral contributes to B_3, B_4 and B_5 contain several contributions from diagrams of different sign and different asymptotic L/D dependence. The partial cancellation of these contributions is responsible for the behavior of B_4 and B_5 for small L/D. In the limit $L/D \to \infty$, only the leading diagram survives. This diagram is positive for odd n and negative for even n. Almost identical behavior is observed for the virial coefficients of hard ellipsoids. [84,85].

E. Structure

The pair structure of the hard ellipsoid fluid has been examined by Talbot and co-workers. [24,34] Here the interest lies in spherical harmonic expansions of the orientation dependence of $g(r, \mathbf{e}_1, \mathbf{e}_2)$, the success or otherwise of theories in fitting to the simulation results, and the use of alternative expansions, for example in the surface–surface distance. The surface–surface distance is calculated by a procedure due to Wertheim and Talbot. [86] For completeness we sketch it here.

1. Surface–Surface Distance for Ellipsoids

Consider two ellipsoids of revolution, with centre of mass positions $\mathbf{r}_i$ and $\mathbf{r}_j$ and with unit vectors along the symmetry axes $\mathbf{e}_i$ and $\mathbf{e}_j$. Both ellipsoids have semi-axes a and b. The surface of i is defined by

$$(\mathbf{r} - \mathbf{r}_i) \cdot \mathbf{A}_i^{-1} \cdot (\mathbf{r} - \mathbf{r}_i) = 1$$

where

$$\mathbf{A}_i = b^2 \mathbf{1} + (a^2 - b^2)\mathbf{e}_i\mathbf{e}_i \qquad (2.108)$$

$$\mathbf{A}_i^{-1} = b^{-2}\mathbf{1} + (a^{-2} - b^{-2})\mathbf{e}_i\mathbf{e}_i \qquad (2.109)$$

At any point $\mathbf{r}$ on the surface, the outward unit normal $\mathbf{n}_i$ lies in the direction $\mathbf{A}_i^{-1} \cdot (\mathbf{r} - \mathbf{r}_i)$. The point on the surface having a prescribed normal $\mathbf{n}_i$ is given by

$$\xi_i = \mathbf{r} - \mathbf{r}_i = \mathbf{A}_i \cdot \mathbf{n}_i / \sqrt{\mathbf{n}_i \cdot \mathbf{A}_i \cdot \mathbf{n}_i}$$

Similar expressions apply for j. The problem of finding the shortest distance between two ellipsoids reduces to finding three vectors, $\xi_{ij} = \mathbf{r}_{ij} + \xi_i - \xi_j, \mathbf{n}_j$ and $-\mathbf{n}_i$ all parallel. These vectors are shown in Fig. 2.1.

This suggests several iterative schemes, all based on an initial guess ξ_{ij} at the surface–surface vector. One simple approach is to calculate new values of ξ_i and ξ_j from

$$\xi_i' = -\frac{\mathbf{A}_i \cdot \xi_{ij}}{\sqrt{\xi_{ij} \cdot \mathbf{A}_i \cdot \xi_{ij}}} \quad \text{and} \quad \xi_j' = \frac{\mathbf{A}_j \cdot \xi_{ij}}{\sqrt{\xi_{ij} \cdot \mathbf{A}_j \cdot \xi_{ij}}} \qquad (2.110)$$

that is, the surface points on each ellipsoid that would have ξ_{ij} as a normal. Then a new estimate $\xi_{ij}' = \mathbf{r}_{ij} + \xi_i' - \xi_j'$ would be mixed with the old ξ_{ij} in some proportion for the next iteration. In other words, we are iteratively solving the equation

$$\mathbf{C} \cdot \xi_{ij} = \mathbf{r}_{ij} \quad \text{with} \quad \mathbf{C} = \mathbf{1} + \frac{\mathbf{A}_i}{\sqrt{\xi_{ij} \cdot \mathbf{A}_i \cdot \xi_{ij}}} + \frac{\mathbf{A}_j}{\sqrt{\xi_{ij} \cdot \mathbf{A}_j \cdot \xi_{ij}}} \qquad (2.111)$$

for ξ_{ij} given that $\mathbf{r}_{ij}$, $\mathbf{A}_i$, and $\mathbf{A}_j$ are fixed. An alternative approach is to write the equation as $\xi_{ij} = \mathbf{C}^{-1}\mathbf{r}_{ij}$ and solve this by iteration, setting at each stage $\xi_{ij}' = \mathbf{C}^{-1}(\xi_{ij}) \cdot \mathbf{r}_{ij}$. This seems to be slow to converge.

Starting from an initial guess at ξ_{ij} and the outward normals $\mathbf{n}_i = -\mathbf{n}_j = -\xi_{ij}/|\xi_{ij}|$, calculate revised values ξ_i' and ξ_j' from Eq. (2.110). From this, compute $\xi_{ij}' = \mathbf{r}_{ij} + \xi_i' - \xi_j'$ and new values $\mathbf{n}_i' = -\mathbf{n}_j' = -\xi_{ij}'/|\xi_{ij}'|$. The mixing factor for the next iteration is determined as follows. Construct the matrix

$$\mathbf{R} = \frac{\mathbf{A}_i - \xi_i\xi_i}{\sqrt{\mathbf{n}_i \cdot \mathbf{A}_i \cdot \mathbf{n}_i}} + \frac{\mathbf{A}_j - \xi_j\xi_j}{\sqrt{\mathbf{n}_j \cdot \mathbf{A}_j \cdot \mathbf{n}_j}}$$

and from it the matrix

$$\mathbf{K} = \frac{\text{Tr}(\mathbf{R})(\mathbf{1} - \mathbf{n}_i\mathbf{n}_i) - \mathbf{R}}{\text{Tr}(\text{Cof}(\mathbf{R}))}$$

where Tr stands for trace and $\text{Cof}(\mathbf{R})$ is the matrix of cofactors of $\mathbf{R}$. Then the old and new normals are mixed together in proportion $\mathbf{n}_i + \chi\mathbf{n}_i'$ where

$$\chi = \frac{|\xi_{ij}'|\xi_{ij}' \cdot \mathbf{K} \cdot \xi_{ij}'}{\xi_{ij}'^2 - (\xi_{ij}' \cdot \mathbf{n}_i)^2}$$

for the next iteration.

Talbot et al. [24] investigated the expansion of $g(r, \mathbf{e}_1, \mathbf{e}_2)$ in rotational invariants, discussed in Section II.C.1, for a range of ellipsoid shapes and densities. They compared their results with the standard theories of liquid structure, hypernetted chain (HNC) and Percus–Yevick (PY). A general result was that both these theories seem to give reasonable descriptions of the first few expansion functions for prolate ellipsoids with $e = 2, 3, 5$, with HNC being slightly superior. Moreover, the accuracy of the theories at a given packing fraction did not depend strongly on the elongation of the molecules, an encouraging and perhaps surprising result. The most stringent test is at high packing fraction, and for $e = 2$ the fluid phase extends at least up to $\rho/\rho_{cp} = 0.8$, well above the freezing point for the hard-sphere fluid. Indeed, at this density, solutions to the HNC equations could not be obtained. Nonetheless, agreement with PY is still fair for most of the expansion functions. The exception is the function $g_{222}(r)$, which seems to be very sensitive to inaccuracies in the theory, especially at short range.

For oblate ellipsoids with $e = \frac{1}{3}$, at the single density studied, $\rho/\rho_{cp} = 0.5$, the HNC theory is once more in reasonable agreement with simulation results. PY, however, is rather poor for this case for several of the expansion functions.

A disadvantage of expanding $g(r, \mathbf{e}_1, \mathbf{e}_2)$ is that the higher coefficients do not decrease quickly in magnitude, that is, the expansion does not converge very well. Talbot et al. [34] investigated the consequences of expanding the function $G(s, \mathbf{e}_1, \mathbf{e}_2)$ discussed in Section II.C.1, where s is the surface–surface distance. This seems to bring the expected benefits of more rapid convergence. Moreover, once the (exactly calculable) effects of the surface Jacobian $S(s)$ are taken into account, the expansion functions seem to be quite short-ranged in s. This breakdown succeeds in highlighting certain physical features of the packing in dense liquids, such as the preference for prolate ellipsoids to pack side by side. Various properties characterizing pairs in contact, such as the pressure and the collision rate, are easily related to the appropriate functions in the limit $s \to 0$. This approach also leads to a description of the structure in terms of the isotropically averaged pair distribution function $g_{iso}(s)$, which, at least at short distances, is very similar to the hard-sphere pair distribution function for the same packing fraction and molecular volume. Finally, in this study, the use of surface–surface functions as a route to the orientational correlation parameter g_2 was examined. It was shown that the function $g_2(s)$ can be separated into a short-range excluded volume component and a longer range term. There is a substantial contact contribution to g_2. However, it was also evident that, in both centre–centre and surface–surface decompositions, the longer range component makes a significant contribution to the integrated value g_2 measured, for example, in depolarized light scattering experiments. For details, we refer the reader to Ref. [34].

III. DYNAMICAL PROPERTIES

A. Transport Phenomena in Isotropic Phases

The time evolution of a dynamical quantity in a system comprised of HCBs arises from the action of free streaming (uninterrupted translation and rotation) and collisions. Free streaming is implemented into the dynamics by operators akin to the infinitesimal generators of free translation and rotation. In hard-particle systems, collisions are strictly two-body events and for convex bodies, the colliding bodies touch at only one point at any instant. The effect of a collision on a property is represented by binary collision operators and owing to the nature of impulsive force laws, the physical content of a binary collision operator is exceedingly simple. The binary collision operator determines if two particles are approaching and are at a contact separation. If these conditions are satisfied, then the incoming particle momenta are replaced by the restituting momenta, derived on the basis of the conservation laws for linear and angular momentum and kinetic energy. Since the colliding bodies are smooth and convex, the collisional impulse is directed along the surface normal at the point of impact.

The development of a kinetic theory of rigid convex bodies was due to Curtiss and his co-workers [87–96]. Equations of motion, BBGKY hierarchies and collision operators derived by the Curtiss school were natural extensions of the Chapman–Enskog hard-sphere kinetic theory. [97,98] Much of the progress in the calculation of dynamical properties of HCB systems has resulted from exploiting the analogies with hard-sphere systems. Even the accommodation of the angular momentum and the orientational angles appropriate to HCB collisions has been influenced by the techniques developed for hard spheres. As a result of these efforts, transport coefficients and spectra of fluids of HCBs can be obtained nearly to the same level of rigor as that for hard-sphere systems. The development of HCB transport theory has not been greatly influenced by the adoption of special coordinate systems which restrict analysis to a certain shape (e.g., an ellipsoid or a spherocylinder). Rather the techniques developed by Curtiss, Dahler and Hoffman allow transport properties, for hard spheres and for HCBs of arbitrary shape, to be treated on the same footing.

The HCB kinetic theory has several basic ingredients:

1. the definition of the Liouville operator (hereafter called the pseudo-Liouville operator since it will prove to be non self-adjoint); [99]
2. a systematic method for the derivation of formal [100] and at the same time operational [101] expressions that permit the computation of transport coefficients and time correlation functions;
3. an exact way to perform the collision integrals over the momentum and orientational angles decribing the phase space for a two body encounter. [95]

B. Time Evolution Operators

The time evolution of HCB systems was cast in a binary collision operator formalism by Cole et al. [99] and here we begin with a summary of their findings. Consider the time evolution of an arbitrary dynamical variable, $A(\Gamma)$ where Γ denotes the phase space coordinates of the N particle system. By Liouville's theorem, $A(\Gamma)$ evolves in time according to

$$\begin{aligned} A(\Gamma;t) &= \exp(iLt)A(\Gamma;0) \\ &= A(\exp(iLt)\Gamma) = A(\Gamma(t)) \end{aligned} \tag{3.1}$$

When the potentials are soft and differentiable, the operator iL is a linear differential operator and Eq. (3.1) generates a series expansion in time. However, when the forces are impulsive, there is no meaning to an expansion of Eq. (3.1) in terms of the spatial derivatives of delta function forces. Instead one must reinterpret the formal time evolution operator and separate iL into a free streaming, iL_0, and an interaction $(iL - iL_0)$ part. Likewise, the dynamics are to be treated using this same separation,

$$\exp(iLt) = \exp(iL_0 t) + \int_0^t d\tau \, \exp(iL\tau)(iL - iL_0) \, \exp(iL_0(t - \tau)) \tag{3.2}$$

where the streaming operator (for rotor and spherical tops) is given by

$$iL_0 = \sum_j (\mathbf{v} \cdot \nabla_{\mathbf{r}} + (\omega \times \hat{\mathbf{u}}) \cdot \nabla_{\hat{\mathbf{u}}})_j \tag{3.3}$$

Next, we "coarse-grain" time, that is, we divide time into bins of width ϵ, and stipulate that it is during the interval ϵ that a collision occurs. In this view, one replaces the collection of terms in the $(iL - iL_0)$ part of Eq. (3.2) with three terms: first, a free propogation operator, $\exp(iL_0(t - \tau))$, which places the system at the beginning of the desired time bin; second, an operator which constructs the difference between the post- and precollision values of the system during the time interval ϵ; and third, an operator $\exp(iL\tau)$ which propagates the system under the influence of the full Liouvillian. The explicit time evolution operator satisfying these conditions is

$$\exp(iL(\epsilon)t) = \exp(iL_0 t) + \int_0^t d\tau \, \exp(iL\tau)T(\epsilon) \, \exp(iL(t-\tau)) \qquad (3.4)$$

where

$$T(\epsilon) = \delta\left(s - \frac{1}{2}|ds/dt|\epsilon\right)|ds/dt|H(-ds/dt)(\mathbf{b}-1) \qquad (3.5)$$

s is the surface-to-surface separation of two bodies measured along $\hat{\mathbf{s}}$, the surface normal. ds/dt is the velocity along $\hat{\mathbf{s}}$ (to be given explicitly later) and $\mathbf{b}$ is an operator which converts precollision momenta to their post-collision values and vice versa. The binary collision operator, $T(\epsilon)$, registers a collision provided the particles are approaching (negative ds/dt) and reside infinitestimally outside the contact surface. [102] The curious $\frac{1}{2}\epsilon$ in Eq. (3.5) arises as a consequence of the mean value theorem which specifies that one locate the collision to be half-way through the coarse-grained time period ϵ. Finally, the Liouville operator is given by the $\epsilon \to 0$ limit. We suppress ϵ in all the equations to follow but note that the ϵ limit must be taken at the end of the calculation rather than at the outset.

Equation (3.4) defines an expansion of the system evolution operator which allows one to evaluate time correlation functions of the form

$$< B(0)|A(t) >= \int d\Gamma f^N(\Gamma)B(\Gamma) \, \exp(iLt)A(\Gamma) \qquad (3.6)$$

where $f^N(\Gamma)$ is the N particle equilibrium distribution function. For soft potentials, a parts integration of Eq. (3.6) yields

$$< B(0)|A(t) >= \int d\Gamma A(\Gamma) \, \exp(-iLt)f^N(\Gamma)B(\Gamma) \qquad (3.7)$$

and both Eqs. (3.6) and (3.7) are equivalent. To make hard particles abide by Eqs. (3.6) and (3.7), some effort is needed and this issue has been discussed on several occasions by Ernst et al. [102,103] for spherical systems. The complication for hard-body systems arises because of the combined effect of the streaming operators and the discontinuous nature of f^N. When two particles overlap, $f^N = 0$, whereas at a separation infinitesimally exterior to the excluded volume surface, f^N is nonzero.

The action of iL_0 on f^N is to introduce delta functions that have the appearance of terms in the collision operator. In other words, when we derive a result equivalent to Eq. (3.7) for HCB systems, the binary collision operator will have to be altered. Provided that the f^N function is to the left of iL, no infinities are produced by the action of iL_0. In order to retain the property of Eq. (3.7), we must replace $\exp(-iLt)f^N(\Gamma)$ by its Hermitian adjoint, (denoted by $\dagger$), defined through the relation

$$\int d\Gamma B(\Gamma)X(\Gamma)A(\Gamma) = \int d\Gamma (X(\Gamma)B(\Gamma))^\dagger A(\Gamma) \tag{3.8}$$

where $X(\Gamma)$ is an arbitrary operator and $A(\Gamma)$ and $B(\Gamma)$ two arbitrary functions of the phase space coordinates.

To construct the adjoint of the hard body Liouville operator, consider the $< B|iL|A >$ matrix element. First we integrate the streaming operator,

$$\int d\Gamma f^N(\Gamma)B(\Gamma)iLA(\Gamma)$$

$$= \int d\Gamma(-A(\Gamma)iL_0(B(\Gamma)f^N(\Gamma)) + f^N(\Gamma)B(\Gamma)TA(\Gamma)) \tag{3.9}$$

and we note that [99]

$$\int d\Gamma f^N(\Gamma)B(\Gamma)TA(\Gamma) = \int d\Gamma A(\Gamma)T^\dagger B(\Gamma)f^N(\Gamma) \tag{3.10}$$

where

$$T^\dagger = (ds/dt)\delta(s)(H(ds/dt)\mathbf{b} + H(-ds/dt)) \tag{3.11}$$

Combining Eqs. (3.9), (3.10) we find

$$\int d\Gamma f^N(\Gamma)B(\Gamma)iLA(\Gamma) = \int d\Gamma A(\Gamma)(iL(f^N(\Gamma)B(\Gamma)))^\dagger \tag{3.12}$$

so that the adjoint of iL is given by

$$(iL)^\dagger = -iL_0 + T^\dagger \tag{3.13}$$

and accordingly

$$< B(0)|A(t) > = \int d\Gamma f^N(\Gamma)B(\Gamma)\exp(iLt)A(\Gamma)$$

$$= \int d\Gamma A(\Gamma)\exp((-iL_0 + T^\dagger)t)[f^N(\Gamma)B(\Gamma)]^\dagger \tag{3.14}$$

A consequence of Eq. (3.14) is that the time evolution of the distribution function is described by a Liouville operator different from that used to describe the time evolution of a dynamical variable and this difference is reflected directly in the equations of motion for a dynamical variable

$$dA(\Gamma;t)/dt = (iL_0 + T)A(\Gamma;t) \qquad (3.15)$$

and for the full N particle distribution function $f^N(\Gamma;t)$

$$(\partial/\partial t)f^N(\Gamma;t) = (-iL_0 + T^\dagger)f^N(\Gamma;t) \qquad (3.16)$$

It then follows that the first member of the BBGKY hierarchy, $f(1;t)$, the singlet distribution function obeys

$$(\partial/\partial t)f(1;t) = -iL_0 f(1;t) + \int_0^t d2 T_{12}^\dagger f(1,2;t) \qquad (3.17)$$

where $f(1,2;t)$ is the pair distribution function. Here the reduced distribution functions are defined by

$$f(1,2,...n;t) = (N!/(N-n)!) \int d(n+1)...dN f^N(\Gamma;t) \qquad (3.18)$$

This completes our analysis of the time evolution of an arbitrary dynamical variable and the phase space distribution function.

C. Time Correlation Functions

There are several ways to calculate a general transport coefficient in the Enskog (uncorrelated collision) approximation. Broadly, these procedures are based on a linear response theory in which the "disturbance" is a shear gradient, a temperature gradient or a concentration gradient and the response is a momentum, energy or a diffusion current, respectively. [97,98] For the specific case of a linear response theory for the shear viscosity, one would calculate the stress tensor as a consequence of the imposed shear gradient and the coefficient relating stress to strain is the shear viscosity. The "Enskog" method for the calculation of a transport coefficient or more generally those appproaches based on the use of the distribution function were shown by Ernst et al. [102,104,105] to be equivalent to the flux–flux time correlation function approaches. All of the usual facets of Enskog kinetic theory (basis functions expansions, the factorization ansatz of the precollisional $f(1,2;t)$) [106] also arise in the analysis of flux–flux time correlation functions.

Some of the vestiges of distribution function Enskog kinetic theory can be avoided and the resulting calculations simplified by the use of a Mori projection operator. In this approach, practiced by Forster, [107] Mazenko, [108] Mazenko and Yip, [109] Evans [101] and summarized succinctly by Hansen and McDonald, [1] one applies the Mori method (with one or two primary variables) to derive a generalized Langevin equation for the flux–flux time correlation function. Apart from some simple thermodynamic functions, the decay rates for the flux–flux time correlation function are the transport coefficients. We will illustrate this method by the calculation of a few single particle properties, hydrodynamic collective properties and purely position dependent orientational properties.

From the above paragraphs, it would appear that there are two distinct schools of thought: those involved with the construction of BBKGY hierarchies for the singlet and pair distribution function (motivated by the desire to derive improved Boltzmann–Enskog equations of motion) and those concerned with the derivation of generalized Langevin equations for the flux time correlation functions. Sung and Dahler [110–112] have taken a middle road and have derived generalized Langevin equations for the tagged particle distribution function and the untagged members of the surrounding fluid. The Mori techniques used by Sung and Dahler are precisely the same techniques that we use here. It seems to us that the transport theory is both simpler to use and to understand if one takes a limited description and focusses attention on the flux time correlation functions rather than on the phase space distribution function. We illustrate this method by the calculation of a few single particle properties, hydrodynamic collective properties and purely position dependent orientational properties.

1. Single Particle Properties

The simplest transport properties to understand for a fluid of nonspherical molecules are the self diffusion coefficient D_s

$$D_s = (1/3) \int_0^\infty dt < \mathbf{v}(0) \cdot \mathbf{v(t)} > = \int_0^\infty dt\, C_v(t) = k_B T/(mf_s) \quad (3.19)$$

and the rotational diffusion coefficient D_ω

$$D_\omega = (1/2) \int_0^\infty dt < \omega(0) \cdot \omega(\mathbf{t}) > = \int_0^\infty dt\, C_\omega(t) = k_B T/(If_\omega) \quad (3.20)$$

where $C_v(t)$ and $C_\omega(t)$, are the linear and angular momentum time correlation functions, respectively. Accompanying the diffusion coefficients are the friction coefficients f_s and f_ω which are also used in the preceding two equations.

We illustrate the Mori method by a calculation of f_ω. The choice of f_ω rather than f_s was made in order to demonstrate the role of nonsphericity which clearly governs angular momentum relaxation but is only of secondary importance for linear momentum relaxation. The Mori method allows one to derive an exact generalized Langevin equation for $C_\omega(t)$ [113,114]

$$(\partial/\partial t)C_\omega(t) = -f_\omega^E C_\omega(t) - \int_0^t d\tau \nu_\omega(\tau)C_\omega(t - \tau) \qquad (3.21)$$

In the parlance of Hansen and MacDonald, [1] f_ω^E is the direct term,

$$f_\omega^E = - <\omega \cdot |T|\omega> / <\omega^2> \qquad (3.22)$$

also the single variable Enskog friction, and $\nu_\omega(\tau)$ is the indirect term

$$\nu_\omega(\tau) = - <\omega \cdot |T \exp(iQ_\omega LQ_\omega \tau)Q_\omega T|\omega> / <\omega^2> \qquad (3.23)$$

with

$$Q_\omega = 1 - |\omega> [I/(2\beta)] <\omega| \qquad (3.24)$$

and is a measure of the correlations usually ignored in Enskog theory. By means of Eqs. (3.21)–(3.24), one can associate the total friction with the sum of the direct and the indirect terms

$$f_\omega = f_\omega^E + \int_0^\infty dt \nu_\omega(t) \qquad (3.25)$$

The Enskog (or direct) term f_ω^E involves a single binary collision operator, whose matrix elements can be determined easily. This term measures the uncorrelated impulses on the surface of the tagged particle as weighted by the anisotropic pair correlation function of the system.

The indirect term, $\nu_\omega(t)$, is more complex as it contains both the dilute gas and the dense gas/liquid classes of correlations. The gas phase correlations omitted from a single variable Enskog theory and contained in $\nu_\omega(t)$ reflect two effects: (1) that ω is not an exact eigenfunction of T and T couples ω (fortunately only weakly) to higher Sonine polynomials of ω and v (see Refs. [115,116]) and (2) that a colliding pair may suffer a chattering collision. [99,117] A chattering collision arises when a particular trajectory has more than one impulsive encounter. In the context of a binary collision expansion, the chattering sequences arise from products of T operators such as $T_{12} \exp(iL_0t)T_{12}$, that is, an initial 1,2 impact with the hard core, free propagation for a time t, followed by another 1,2 impact. It is precisely this class of collisions that are eliminated from all hard-sphere binary collision expansions [118,119] although these same collision sequences are particularly important for rotational energy relaxation cross sections. [117] Neither the Sonine nor the chattering corrections to f_ω alters the density dependence from its Enskog value.

The liquid phase correlations present in $\nu_\omega(t)$ reflect the role of solvent induced recollisions of a specific pair of molecules [113,114,120] and this process has a direct influence on the density dependence of f_ω. To incorporate some of the solvent induced correlated collisions in $\nu_\omega(t)$, one must represent the resolvent $\exp(iQ_\omega L Q_\omega t)$ (alias the recollision operator) with basis functions having the same symmtery as $T\omega$. One such function is ϕ_ω,

$$\phi_\omega = \mathbf{r}_{12} \times \hat{\mathbf{s}}\Delta(s) \qquad (3.26)$$

where the "delta" function has the property that

$$\Delta(s) = \begin{cases} 1, & \text{if } 0 < s < d \\ 0, & \text{otherwise} \end{cases} \qquad (3.27)$$

and d is a small (but noninfinitesimal) distance. ϕ_ω is square normalizable function that behaves like a torque exerted by the solvent on the contact surface of the tagged particle. Using the time reversal properties of T, one can show that

$$\nu_\omega(\tau) = [< \omega \cdot |iL|\phi_\omega > / < (\phi_\omega)^2 >]^2 < \phi_\omega(0) \cdot \phi_\omega(t) > / < \omega^2 > \qquad (3.28)$$

where all of the static matrix elements can be determined exactly and $< \phi_\omega(0) \cdot \phi_\omega(t) >$ can be approximated using diffusional (Smoluchowski) modelling. [113,114] Thus, the rotational friction follows from Eq. (3.25) where f_ω^E is given by Eq. (3.22) and $\nu_\omega(t)$ from Eq. (3.28) The predictions of Enskog theory and the utility of the recollision corrections to the rotational friction will be assessed in subsequent sections.

2. Collective Properties

There are many transport coefficients for polyatomic fluids [121] and in principle all of these should be determined. For the present purposes, we will demonstrate how the Mori, generalized Langevin method described in the previous section provides a mechanism for the calculation of the shear viscosity η and the thermal conductitvity λ of a pure fluid of HCBs. [101]

Several methods have been applied to the calculation of the transport properties of fluids of HCBs. Theodosopulu and Dahler, [122–124] Jagannathan, et al. [125] and Cole et al., [126] Cole and Evans, [35] and Evans [101] have provided distinct means for the determination of transport coefficients. In the work of the Dahler group, kinetic equations were derived for the phase space density and these equations were solved by moment methods developed by Grad. [127] In the Cole, Evans and Hoffman works, [35] the transport properties were expressed as a time correlation function and the method of Ernst (developed for hard spheres) [104,105] was applied to transform the time correlation function expressions into distribution function schemes so that the usual Enskog factorizations and expansions could be made. For both the Dahler and the Ernst methods, the operating expressions for the evaluation of the transport coefficients are complicated and in the end, the relationship of the computed transport coefficients to the hard-sphere limit is not always clear. A direct time correlation function approach, [1,101] based on the Mori method is, in our opinion, simpler and leads to results of some generality. Cole et al. [126] Cole and Evans, [35] and Theodosopulu and Dahler [122–124] found that "new" collisional integrals arose in the transport coefficient calculations which incorporated the nonlocal nature of the collision event. In the present approach, these non-local corrections follow as matrix elements of the binary collision operator.

We begin with a calculation of the shear viscosity, η. Accordingly we introduce the collective variable whose decay defines η : the $\mathbf{k}$-space current density [1]

$$\mathbf{v}(\mathbf{k}, t) = \sum_i \mathbf{v}_i(t) \, \exp(i\mathbf{k} \cdot \mathbf{r}_i(t)) \tag{3.29}$$

along with its transverse part,

$$\mathbf{J}(\mathbf{k}, t) = i\mathbf{k} \times \mathbf{v}(\mathbf{k}, t) \tag{3.30}$$

From linear response theory, the time correlation function of the fluctuating current density obeys a phenomonological decay law

$$m\rho(\partial/\partial t) < \mathbf{J}(-\mathbf{k}, 0) \cdot \mathbf{J}(\mathbf{k}, t) >= -\eta k^2 < \mathbf{J}(-\mathbf{k}, 0) \cdot \mathbf{J}(\mathbf{k}, t) > \tag{3.31}$$

with a decay rate of $k^2 \eta / (m\rho)$. So, to calculate η one must derive an analogous but microscopic equation for the time correlation function of the fluctuating current density and associate the decay rate with $k^2 \eta / (m\rho)$.

In the Mori, generalized Langevin method, one writes the kinetic equation for the transverse momentum density time correlation function as [128]

$$(\partial/\partial t)C(\mathbf{k}, t) = -\nu_1 C(\mathbf{k}, t) - \int_0^t d\tau \nu_2(\tau) C(\mathbf{k}, t - \tau) \tag{3.32}$$

with

$$\nu_1(\mathbf{k}) = - < \mathbf{J}(-\mathbf{k})|iL|\mathbf{J}(\mathbf{k}) > /C(\mathbf{k}) \tag{3.33}$$

$$\nu_2(\mathbf{k}, \tau) = - < \mathbf{J}(-\mathbf{k})|iL \exp(iQLQ\tau)iQL|\mathbf{J}(\mathbf{k}) > /C(\mathbf{k}) \quad (3.34)$$

$$C(\mathbf{k}) = C(\mathbf{k}, t = 0) \quad (3.35)$$

$$Q_J = 1 - |\mathbf{J}(\mathbf{k}) > [C(\mathbf{k})]^{-1} < \mathbf{J}(-\mathbf{k})| \quad (3.36)$$

and

$$C(\mathbf{k}, t) = < \mathbf{J}(-\mathbf{k}, 0) \cdot \mathbf{J}(\mathbf{k}, t) > \quad (3.37)$$

By comparison of the phenomonological decay law, Eq. (3.31), with the generalized Langevin result, Eq. (3.32), one obtains

$$\eta = \lim_{k,z \to 0} \eta(k, z)$$

$$\eta(k, z) = m\rho(\nu_1(\mathbf{k}) + \int_0^\infty dt \, \exp(-zt)\nu_2(\mathbf{k}, t))/k^2 \quad (3.38)$$

where η is the zero frequency $(z = 0)$, zero wave vector $(k = 0)$ limit of the k, z dependent shear viscosity. For soft potentials, $\nu_1(\mathbf{k})$ vanishes and $\nu_2(\mathbf{k}, t)$ is the wave vector dependent time correlation function of the fluctuating part of the stress tensor. For hard bodies, $\nu_1(\mathbf{k})$ is a nonvanishing, positive definite collisional term (the "direct" part of η) which can be determined exactly. $\nu_2(\mathbf{k}, t)$, the "indirect" part of $\eta(k, z)$, can be determined by approximating the resolvent operator and we do this in the spirit of traditional kinetic theory methods (such as in Grad's method of moments [127] or early Enskog theory [97,98]).

The simplest approach to unravel the memory function is to represent the projected time evolution operator, $\exp(iQ_JLQ_Jt)$, by a single basis function, $\phi(\mathbf{k})$,

$$\phi(\mathbf{k}) = iL_0\mathbf{J}(\mathbf{k}) \quad (3.39)$$

which is the convective derivative of the flux in question. In this single Hermite polynomial, Enskog approximation, appropriate at dilute gas densities, $\nu_2(\mathbf{k}, t)$ becomes

$$\nu_2(\mathbf{k}, t) = < \mathbf{J}(-\mathbf{k})|iL|\phi(\mathbf{k}) >< \phi(-\mathbf{k})| \exp(iQLQt)|\phi(\mathbf{k}) >$$
$$\times < \phi(-\mathbf{k})|iQL|\mathbf{J}(\mathbf{k}) > /(\Omega^4(\mathbf{k})C(\mathbf{k})) \quad (3.40)$$

where

$$\Omega^2(\mathbf{k}) = < \phi(-\mathbf{k})| \exp(iQLQt)|\phi(\mathbf{k}) > \quad (3.41)$$

Since $\mathbf{J}(\mathbf{k})$ and $\phi(\mathbf{k})$ have different time reversal symmetries, then

$$< \mathbf{J}(-\mathbf{k})|iL|\phi(\mathbf{k}) >= - < \phi(\mathbf{k})|iQL|\mathbf{J}(\mathbf{k}) >^* \tag{3.42}$$

and if the projected time dependence of $\phi(\mathbf{k})$ is dominated by collisional relaxation rather than by convection, then

$$< \phi(-\mathbf{k})| \exp(iQLQt)|\phi(\mathbf{k}) >\simeq< \phi(-\mathbf{k})| \exp(Tt)|\phi(\mathbf{k}) > \tag{3.43}$$

On this basis, $\nu_2(\mathbf{k}, t)$ becomes

$$\begin{aligned}
\nu_2(\mathbf{k}, t) &= |< \phi(-\mathbf{k})|iL|\mathbf{J}(\mathbf{k}) > |^2 \\
&\quad \times < \phi(-\mathbf{k})| \exp(Tt)|\phi(\mathbf{k}) > /[\Omega^4(\mathbf{k})C(\mathbf{k})] \\
&= [1 + q_\eta(\mathbf{k})]^2 < \phi(-\mathbf{k})| \exp(Tt)|\phi(\mathbf{k}) > /C(\mathbf{k}) \tag{3.44}
\end{aligned}$$

with

$$q_\eta(\mathbf{k}) =< \phi(-\mathbf{k})|T|\phi(\mathbf{k}) > /\Omega^2(\mathbf{k}) \tag{3.45}$$

To complete the reduction of $\nu_2(\mathbf{k}, t)$, we represent the inverse of T by a single function, $\phi(\mathbf{k})$ itself, so that

$$\int_0^\infty dt \, \exp(-zt) < \phi(-\mathbf{k})| \exp(Tt)|\phi(\mathbf{k}) >= \Omega^2(\mathbf{k})/(z + \nu_3(\mathbf{k})) \tag{3.46}$$

with

$$\nu_3(\mathbf{k}) = |< \phi(-\mathbf{k})|T|\phi(\mathbf{k}) > |/\Omega^2(\mathbf{k}) \tag{3.47}$$

and finally [101]

$$\int_0^\infty dt \, \exp(-zt)\nu_2(\mathbf{k}, t) = [(1 + q_\eta(\mathbf{k}))\Omega(\mathbf{k})]^2/[C(\mathbf{k})(z + \nu_3(\mathbf{k}))] \tag{3.48}$$

Thus, the shear viscosity follows from the Eq. (3.38) as the sum of the direct part, given by Eq. (3.33) and the indirect part, given by Eq. (3.48).

The thermal conductivity, λ, is determined by precisely the same approach as used for the shear viscosity. [101] Now we associate $\mathbf{J}(\mathbf{k}, t)$ with the local kinetic energy density, $\mathbf{J}_E(\mathbf{k}, t)$

$$\mathbf{J}_E(\mathbf{k}, t) = \sum_i \left(\frac{1}{2}\right) [mv_i^2 + I\omega_i^2 - 5k_BT] \exp(i\mathbf{k} \cdot \mathbf{r}_i(t)) \tag{3.49}$$

which obeys a linear transport equation

$$m\rho C_v(\partial/\partial t) < \mathbf{J}_E(-\mathbf{k}, 0)\cdot\mathbf{J}_E(\mathbf{k}, t) >= -k^2\lambda < \mathbf{J}_E(-\mathbf{k}, 0)\cdot\mathbf{J}_E(\mathbf{k}, t) > \tag{3.50}$$

As a result of an analysis completely analogous to that applied to the shear viscosity, we find that λ is the zero frequency, zero wave vector limit of the full $\lambda(k, z)$, thus

$$\lambda = \lim_{k,z \to 0} \lambda(k, z)$$

$$\lambda(k, z) = m\rho C_v(\nu_1(\mathbf{k}) + \int_0^\infty dt \, \exp(-zt)\nu_2(\mathbf{k}, t))/k^2 \qquad (3.51)$$

where $\nu_1(\mathbf{k})$ is given by Eq. (3.33) and $\nu_2(\mathbf{k}, t)$ by Eq. (3.48) with $\mathbf{J}_E$ replacing $\mathbf{J}$. The shear viscosity and the thermal conductivity obtained from this procedure are summarized in a subsequent section.

3. Positional and Orientational Properties

Both the single particle and collective transport properties discussed in the previous sections were derived from time correlation functions of momentum variables which changed as a consequence of collisions. When dealing with time correlation functions of collisionally conserved dynamical variables, such as positions and orientations, another version of the Mori method is needed and this version uses two rather than one primary variable. [129,130]

Consider a time correlation function of the form

$$C_A(t) = \langle A(0)|A(t) \rangle / \langle A|A \rangle \qquad (3.52)$$

where A is a positional collisional invariant. If two primary variables, A and its first derivative, $iL_0A \equiv \dot{A}$, are employed in a Mori analysis, one obtains an identity for a frequency component of $C_A(t)$ in the form of a two term continued fraction

$$C_A(z) = \int_o^\infty dt \, \exp(-zt)C_A(t)$$
$$= 1/[z + \Omega^2/(z + \nu_1 + \nu_2(z))] \qquad (3.53)$$

Here

$$\Omega^2 = \langle \dot{A}^2 \rangle / \langle A^2 \rangle \qquad (3.54)$$

$$\nu_1 = - \langle \dot{A}|T|\dot{A} \rangle / \langle \dot{A}^2 \rangle \qquad (3.55)$$

$$\nu_2(z) = \langle \dot{A}|iL[z - iQ_2LQ_2]^{-1}iQ_2L|\dot{A} \rangle / \langle \dot{A}^2 \rangle \qquad (3.56)$$

and

$$Q_2 = 1 - |A \rangle [\langle A^2 \rangle^{-1}] \langle A| - |\dot{A} \rangle [\langle \dot{A}^2 \rangle^{-1}] \langle \dot{A}| \qquad (3.57)$$

The correlation time τ_A associated with the dynamical variable A is the zero frequency ($z = 0$) part of $C_A(z)$ and this is

$$\tau_A = (\nu_1 + \nu_2(z = 0))/\Omega^2 \tag{3.58}$$

Equation (3.58) has the precisely the same form as the equations for the hydrodynamic transport coefficients, Eqs. (3.38) and (3.51), and the friction coefficient, Eq. (3.25).

By means of the Mori-generalized Langevin formalism presented above, one can calculate the orientational correlation times for a fluid. [131] Specifically, consider the orientational correlation time τ_2 associated with the decay of a second rank Legendre polynomial $P_2(\hat{\mathbf{u}} \cdot \hat{\mathbf{z}})$ (or P_2 for short) specifying the orientation $\hat{\mathbf{u}}$ of a rotor-like molecule with respect to $\hat{\mathbf{z}}$, an arbitrarily chosen laboratory frame. To perform this calculation, we associate A in Eq. (3.53) with P_2 and $\dot{A}$ with its time derivative

$$\dot{A} = iL_0 P_2 = \omega \cdot (\hat{\mathbf{u}} \times \nabla_u) P_2 \tag{3.59}$$

If we wish to represent the effects of free precession in the ν_2 term, we express the resolvent operator in terms of the projected convective derivative of $\dot{A}$,

$$B = Q_2 \ddot{A} \tag{3.60}$$

and so

$$\tau_2 = (I/6k_B T)(\nu_1 + \nu_2(z = 0)) \tag{3.61}$$

where

$$\nu_1 = f_\omega^E \tag{3.62}$$

$$\nu_2(z = 0) = |<\dot{A}|iL|B>|^2/[<B|T|B><\dot{A}^2>] \tag{3.63}$$

Equation (3.61) is applied to the calculation of the single particle and collective orientational correlation times in the following sections.

4. Collision Integrals

In the previous section, we summarized procedures in which transport properties were cast in the form of collision integrals (or matrix elements of binary collision operators). Fortunately, the momentum portions of the binary collision integrals can be performed exactly as shown by Hoffman. [95] Actually the method espoused by Hoffman applies to matrix elements adjoint of the collision operator (and to the so-called Chapman–

Cowling bracket integrals [97] $[A, B]$ which are related to the binary collision operator by the identity [35]

$$[A, B] = - < A(1)|T_{12}^\dagger|(B(1) + B(2)) > = < (B(1) + B(2))|T_{12}|A(1) > \tag{3.64}$$

That a method exists for the exact calculation of the momentum integrals
is crucial to the development of the kinetic theory. For without Hoffman's procedure, our general collision integrals would require numerical
methods for their evaluation. Hoffman's procedure has been applied (and
amplified) elsewhere. [132] Here we merely summarize its essentials.

Consider a collision of two smooth HCBs in which each HCB possesses
three linear and two angular momentum degrees of freedom. Only one velocity component plays a crucial role in the dynamics of the combined ten
dimensional momentum space and that component is ds/dt, the relative
velocity of the contact points

$$ds/dt = \hat{\mathbf{s}} \cdot \mathbf{g} \tag{3.65}$$

with

$$\mathbf{g} = \mathbf{v}_2 + \omega_2 \times \xi_2 - (\mathbf{v}_1 + \omega_1 \times \xi_1) \tag{3.66}$$

ξ_j is a vector that emanates from the mass center of particle j and extends
to the contact point on its surface. After the collision, the component
of $\mathbf{g}$ along $\hat{\mathbf{s}}$, the surface normal at the contact point, changes sign from
its negative precollisional value $(\hat{\mathbf{s}} \cdot \mathbf{g}^*)$ to its positive postcollisional value
$(\hat{\mathbf{s}} \cdot \mathbf{g})$

$$\mathbf{g} = \mathbf{g}^* - 2\hat{\mathbf{s}}\hat{\mathbf{s}} \cdot \mathbf{g}^* \tag{3.67}$$

When the set of ten momenta are expressed in an orthogonal coordinate
system in which $(\hat{\mathbf{s}} \cdot \mathbf{g})$ is a member, the effect of the collision is to reverse
one component (ds/dt) and to leave all the remaining nine momentum
variables unchanged. Recall that when one deals with hard-sphere collisions, the velocity coordinates are normally taken to be the center of
mass $(\frac{1}{2})(\mathbf{v}_1 + \mathbf{v}_2)$ and the relative velocity $\mathbf{v}_r \equiv (\mathbf{v}_2 - \mathbf{v}_1)$ and it is $\mathbf{v}_r$ which
plays the role of $\mathbf{g}$. For nonspherical molecules, Hoffman expressed all
the momentum dependent functions in terms of the nine orthogonal but
otherwise arbitrary coordinates and the one special ds/dt coordinate. Integrations over the nine momentum degrees of freedom could be executed
freely (i.e., independent of collision details) and the one special coordinate remained for integration subject to the particular constraints (e.g., a
restricted integration over the pre- or postcollisional hemisphere).

After having performed the momentum integrals exactly, one is left with
various orientation integrals over the excluded volume surface, $d\hat{\mathbf{u}}_1 d\hat{\mathbf{u}}_2$

$d\mathbf{r}_1 d\mathbf{r}_2$, and these remaining integrations are conducted in the manner discussed in Section II.C. On the basis of these remarks, all the collision integrals can be reduced to surface integrals weighted by the contact pair correlation function. This completes our summary of the kinetic theory of polyatomic fluids.

D. Applications of Enskog Theory

In this section, we summarize a few transport coefficients for a pure fluid of convex bodies and an infinitely dilute solution of a single convex body in a fluid of spheres. The entries were chosen either for their simplicity or for their importance to transport processes in dense fluids.

1. Pure Fluids of HCBs

The simplest transport coefficients are the self-diffusion coefficients for the particle translation and rotation, D_s and D_ω, respectively. These two transport coefficients can be expressed in terms of the friction coefficients which are collisional matrix elements in the Enskog approximation

$$f_s^E = | < \mathbf{v}_1 \cdot |T_{12}|\mathbf{v}_1 > |/ < v_1^2 > \tag{3.68}$$

$$f_\omega^E = | < \omega_2 \cdot |T_{12}|\omega_1 > |/ < \omega_1^2 > \tag{3.69}$$

After some algebra, one obtains

$$f_s^E = \rho g_c v_r \sigma_v \tag{3.70}$$

$$f_\omega^E = \rho g_c v_r \sigma_\omega \tag{3.71}$$

Here $g_c (\equiv g_{iso}(s = 0))$ is the surface averaged contact pair correlation function discussed in Section II, v_r is the relative thermal speed,

$$v_r = [8k_B T/(\pi\mu)]^{1/2} \tag{3.72}$$

μ is the reduced mass for the pair of HCBs ($= m/2$) and σ_v and σ_ω are the linear and angular momentum cross sections, given explicitly by

$$\sigma_v = (2/3)\pi S_c < 1/D >_c \tag{3.73}$$

$$\sigma_\omega = 2\pi(\mu/I)S_c < (\xi_1 \times \hat{s})^2/D >_c$$
$$= 2\pi(\mu/I)S_c < (1 - x^2)(h'(x))^2/D >_c \tag{3.74}$$

where S_c is the average surface area on the contact surface and

$$D^2 = 1 + (\mu/I)[(\xi_1 \times \hat{s})^2 + (\xi_2 \times \hat{s})^2] \tag{3.75}$$

Roughly, $1/D^2$ and $(D^2 - 1)/D^2$ measure the fractional participation of linear and angular momentum in the collision, respectively. Both the linear and the angular momentum friction coefficient and cross sections are related to the collision frequency per particle Z (and the total cross section σ_{tot}) by means of an Enskog sum rule:

$$Z = (3/4)f_v + (1/2)f_\omega = (1/2)\rho g_c v_r \sigma_{\text{tot}} \tag{3.76}$$

$$\sigma_{\text{tot}} = \pi S_c < D >_c \tag{3.77}$$

Z contains a factor of one-half in order to avoid counting the 1–2 and 2–1 collisions as two events.

The frequency dependent shear viscosity and thermal conductivity can be determined by our methods of Section III.C.2 and we obtain in the $k = 0$ limit,

$$\eta(z) = \eta_0(z)(1 + q_\eta)^2 + \eta_1 \tag{3.78}$$

$$\lambda(z) = \lambda_0(z)(1 + q_\lambda)^2 + \lambda_1 \tag{3.79}$$

For the shear viscosity, one has

$$\eta_0 = \lim_{z \to 0} \eta_0(z)$$

$$= (15/16)[mk_BT/\pi]^{\frac{1}{2}}/g_cS_c < 5/D - 2/D^3 >_c \tag{3.80}$$

$$\eta_0(z) = \rho k_B T/(z + v_3) \tag{3.81}$$

$$v_3 = (4\pi/15)\rho v_r g_c S_c < 5/D - 2/D^3 >_c \tag{3.82}$$

$$\eta_1 = (4\pi/15)[mk_BT/\pi]^{\frac{1}{2}}\rho^2 g_c S_c < (2r^2 - (\mathbf{r} \cdot \hat{\mathbf{s}})^2)/D >_c \tag{3.83}$$

$$q_\eta = (4\pi/15)\rho g_c S_c < h_{12}/D^2 >_c \tag{3.84}$$

and for the thermal conductivity,

$$\lambda_0 = \lim_{z \to 0} \lambda_0(z)$$

$$= (49/32)[mk_B^3 T/\pi]^{\frac{1}{2}}/g_cS_c < Q_0/D >_c \tag{3.85}$$

$$\lambda_0(z) = (9k_B/2)\rho k_B T/(z + v_4) \tag{3.86}$$

$$v_4 = (36\pi/49)\rho v_r g_c S_c < Q_0/D >_c \tag{3.87}$$

$$\lambda_1 = (2\pi/3)k_B[mk_BT/\pi]^{\frac{1}{2}}.\rho^2 g_c S_c < Q_1 >_c \tag{3.88}$$

$$q_\lambda = (2\pi/7)\rho g_c S_c < h_{12}(2 + D^{-4} - 2D^{-2}) >_c \tag{3.89}$$

with

$$\mathbf{a}_i = (\mu/I)^{1/2}\hat{\mathbf{s}} \times \xi_i \tag{3.90}$$

$$Q_0 = D^2 + (1/3)(a_1/D)^2 - 2(a_1^4/D^2) + (9/8)(a_1^2 - a_2^2)^2/D^4 \tag{3.91}$$

$$Q_1 = r^2[1 + 2a_2^2(1 + 2a_2^2)/D^2]/D \tag{3.92}$$

The algebraic form of the derived transport coefficients is identical to that found by Jagannathan et al. [125] Each transport coefficient has a frequency independent plateau (arising from the "direct" terms) and a Maxwell model [133,134] relaxing term (arising from the "indirect terms"). In a dilute gas, the transport coefficients are given by $\eta_o(z)$ and $\lambda_o(z)$. Equations (3.78) and (3.79) were written in a form to emphasize the similarities to those for hard-sphere fluids.

Not all dynamical properties of fluids of HCBs necessarily map smoothly onto those of hard spheres. Consider the spherical limit approached by allowing the HCB to become spherical ($D = 1, \mathbf{r} = \hat{\mathbf{s}}\sigma, S_c = \sigma^2, g_c = g_{HS}, \mathbf{a}_i = 0$). In this limit, the shear viscosity, Eq. (3.78) for the fluid of HCBs reduces to that of a hard-sphere fluid. However, the same situation does not prevail for the thermal conductivity. Bodies with a vanishingly small nonsphericity can still transmit angular kinetic energy and thus there is a discontinuous change in λ from that of an atom,

$$\lambda(\text{HS})/\rho = (mk_B^3 T\sigma^2/\pi)^{1/2}[(75/64)(1 + (2\pi/5)X)^2/X + (2\pi/3)X] \quad (3.93)$$

to that of a spherical diatom,

$$\lambda(\text{HCB})/\rho = (mk_B^3 T\sigma^2/\pi)^{1/2}[(49/32)(1 + (2\pi/7)X)^2/X + (2\pi/3)X] \quad (3.94)$$

with the same dimensions of the atom. Here $X(\equiv \rho\sigma^3 g_{HS})$ is a dimensionless contact density. The difference between $\lambda(\text{HS})$ and $\lambda(\text{HCB})$ (in the spherical limit) arises trivially from the differing heat capacities of hard spheres and HCBs and nontrivially from dynamical corrections.

This difference between the thermal conductivities of a spherical HCB and a hard sphere is largest at low density (when $X = 0$ and $\lambda(\text{HCB})/\lambda(\text{HS}) = 1.31$) and decreases monotonically at high densities (when $X \to \infty$ and $\lambda(\text{HCB})/\lambda(\text{HS}) = 0.85$). Evidence for this trend can be found in the MD simulations of Murad et al. [135] on a fused hard sphere model of Cl_2.

The molecular dynamics of rotation in condensed phases is sampled by the single particle and the collective second rank orientational correlation functions, whose integrals give, respectively, the single particle and collective correlation times

$$\tau_2^s = \int_0^\infty dt < P_2(\hat{\mathbf{u}}(0) \cdot \hat{\mathbf{u}}(t)) > \quad (3.95)$$

$$\tau_2^c = \sum_{i,j} \int_0^\infty dt < P_2(\hat{\mathbf{u}}_j \cdot \hat{\mathbf{u}}_i(t)) > / \sum_{i,j} < P_2(\hat{\mathbf{u}}_j \cdot \hat{\mathbf{u}}_j) > \quad (3.96)$$

τ_2^s and τ_2^c are related by the Kivelson–Keyes [131] equation

$$\tau_2^c/\tau_2^s = g_2/j_2 \tag{3.97}$$

where g_2 is the static second rank pair correlation factor (see Eq. (2.28)) and j_2 is the dynamical equivalent of g_2. τ_2^s can be calculated by the procedures of Section III.C.3 and we obtain

$$\tau_2^s = f_\omega^E I/(6k_B T) + (4/3)(1/f_\omega^E)\Lambda \tag{3.98}$$

where

$$\Lambda = 1/(2- < a_1^4/D^3 >_c / < a_1^2/D >_c) = [0.5, 0.9) \tag{3.99}$$

Equation (3.98) depicts a high density branch in which τ_2^s is proportional to the friction ($\simeq \rho g_c$) and a low density branch in which τ_2^s is inversely proportional to friction ($\simeq 1/(\rho g_c)$). The convective upswing in τ_2^s reflects the result that without collisions the second rank projection of the molecular orientation does not relax. Kinetic theory with its hard potentials does not predict a density independent term in τ_2^s which we might interpret as an intercept in a plot of τ_2^s versus $\rho g_c (\simeq \eta/\rho)$ in contrast to the situation involving soft forces. [136] Note that Eq. (3.98) differs from that reported by Cole [35] in two ways: (1) we find no intercept; and (2) the coefficient of the convective term is (4/3) rather than the (5/3) originally quoted. The values for Λ remain the same.

Kinetic theory methods can be directly applied to determine collective properties such as τ_2^c and this calculation can be reduced to a few matrix elements, specifically [132]

$$\tau_2^c = [< \dot{P}_2^c(\hat{\mathbf{u}} \cdot \hat{\mathbf{z}})|T|\dot{P}_2^c(\hat{\mathbf{u}} \cdot \hat{\mathbf{z}}) > / < (P_2^c(\hat{\mathbf{u}} \cdot \hat{\mathbf{z}})^2 >]g_2 I/(6k_B T) \tag{3.100}$$

where

$$P_2^c(\hat{\mathbf{u}} \cdot \hat{\mathbf{z}}) = \sum_i P_2(\hat{\mathbf{u}}_i \cdot \hat{\mathbf{z}}) \tag{3.101}$$

After performing the matrix elements in Eq. (3.100) and comparing the result with the Kivelson–Keyes relation, we can identify

$$j_2 = 1/(1 + dj) \tag{3.102}$$

with

$$dj = 2(\mu/I)\pi < h'(x)h'(x_2)[\hat{\mathbf{u}} \cdot \hat{\mathbf{u}}_2(1 - 2x^2 - 2x_2^2 + 2xx_2\hat{\mathbf{u}} \cdot \hat{\mathbf{u}}_2) + xx_2]/D >_c /\sigma_\omega \tag{3.103}$$

where $x_2 = \hat{\mathbf{s}} \cdot \hat{\mathbf{u}}_2$ and $h'(x) = dh(x)/dx$. This completes our summary of the orientational properties.

2. Infinitely Dilute Solution

The fluid of HCBs with the simplest static and dynamic properties is the infinitely dilute solution comprised of a hard-sphere solvent and a single HCB solute particle. For this case, the macroscopic shear viscosity and thermal conductivity of the fluid is that appropriate to a hard-sphere fluid and the single particle and collective orientational correlation times are identical. The tagged particle friction coefficients (and accompanying cross-sections) for the linear and angular momentum and rotational kinetic energy differ slightly those of single component fluids of HCBs and are [113,114]

$$f_i = \rho v_r g_c \sigma_i \tag{3.104}$$

with $i = v, \omega$, and r, the velocity, angular momentum and rotational energy, respectively. v_r is the relative thermal speed of the hard sphere with respect to the HCB. Cross-sections derived on the basis of Enskog theory are

$$\sigma_v = (4\pi/3)(\mu/m) \int_0^1 dx\, G(s = 0, x)/(D(x)g_c) \tag{3.105}$$

$$\sigma_\omega = 2\pi \int_0^1 dx\, G(s = 0, x)(D^2(x) - 1)/(D(x)g_c) \tag{3.106}$$

$$\sigma_r = 4\pi \int_0^1 dx\, G(s = 0, x)(D^2(x) - 1)/(D^3(x)g_c) \tag{3.107}$$

with $x = \hat{\mathbf{u}} \cdot \hat{\mathbf{s}}$,

$$G(s = 0, x) = g(s = 0, x)S^{12}(s = 0, x) \tag{3.108}$$

$g(s = 0, x)$ the orientation dependent contact pair correlation function, $S^{12}(s = 0, x)$ the surface area element on the HS-HCB excluded volume surface and $D^2(x)$ the HS-HCB translational-to-rotational energy transfer function

$$D^2(x) = 1 + (\mu/I)(\xi_1 \times \hat{\mathbf{s}})^2 = 1 + (\mu h'^2(x)/I)(1 - x^2) \tag{3.109}$$

For a uniaxial convex body with a specified support function, we can calculate Enskog approximations to the three cross sections given above using the SPT orientation dependent contact pair correlation function, Eq. (2.46) and the surface area element S^{12} appropriate to a HCB-sphere surface (see Appendix A.D). Furthermore, by means of the recollision kinetic theory summarized earlier, that is, incorporating the single function given by Eq. (3.26), the rotational friction coefficient can be extended beyond the uncorrelated binary collision approximation. The recollision correlations take the form [113,114]

$$f_\omega \simeq f_\omega^E (1 + \rho^* \exp(5\rho^*)) \qquad (3.110)$$

f_ω includes a positive definite enhancement to the friction based on the influence of "caging" recollisions, which were analyzed using a Smoluchowski equation to describe diffusion in a hard-sphere potential of mean force.

If the convex body is slightly aspherical, then σ_v, σ_ω and σ_r can be determined analytically. For simplicity, consider an ellipsoidal HCB with semi-major axis a and semi-minor axis b in a fluid of hard spheres with diameter σ. We obtain

$$\sigma_v \simeq (4\pi/3)(\mu/m)S_c \qquad (3.111)$$
$$\sigma_\omega \simeq (16\pi/15)S_c\mu(a-b)^2/I \qquad (3.112)$$
$$\sigma_r \simeq 2\sigma_\omega \qquad (3.113)$$

with $S_c = (a + (1/2)\sigma)^2$. For pure fluids of HCBs, these results undergo small changes ($\mu \to m/2, \sigma \to a$) and the distinction between single particle and collective orientational relaxation times becomes important. In the case of slightly aspherical prolate ellipsoids, j_2 becomes

$$1/j_2 \simeq 1 + (2/5) \exp(-[(b/a)^2 - 1]/10) \qquad (3.114)$$

and thus j_2 is less than unity in an Enskog approximation for ellipsoids. Spherocylinders show similar behavior and those results have been given elsewhere. [35,137] This completes our brief summary of transport properties.

E. Simulations

1. Translational and Rotational Diffusion

Translational and rotational motion in the isotropic phase have been investigated for ellipsoids of revolution with $e = 2, 3, \frac{1}{3}$. [34,138] Translational and rotational diffusion coefficients are calculated by integrating the linear and angular velocity autocorrelation functions, respectively:

$$D_T = \frac{1}{3} \int_0^\infty dt \, \langle \mathbf{v}_i(0) \cdot \mathbf{v}_i(t) \rangle$$

$$D_R = \frac{1}{2} \int_0^\infty dt \, \langle \omega_i(0) \cdot \omega_i(t) \rangle$$

In doing this, it is important to check that any slow long-time decay is correctly included. In the case of translational diffusion, it is possible to check the numerical procedure by measuring the limiting gradient of the mean-square displacement with respect to time; this should equal $6D$. Additional checks were carried out in this study, to compare with the well-known diffusion coefficients of hard spheres at similar system sizes. It should be borne in mind that D_T depends significantly on system size, so like must be compared with like.

A major aim of this study was to compare with the predictions of Enskog theory, which is based on the independent binary collision model. As was explained in Section III.D, the Enskog predictions are simply obtained by averaging appropriate geometrical functions for pairs at contact. [35] The results of this study [138] are shown in Figs. 3.1 and 3.2, and several interesting features are present. Firstly, as for hard spheres, D_T/D_T^E first rises, then falls, as density increases. The high density decrease is well understood; caging effects in the dense liquid produce a rapid decay of the velocity autocorrelation function, and indeed a negative rebound region at the highest densities, which cause D_T to be much less than the simple kinetic theory would predict. The same observation applies to ro-

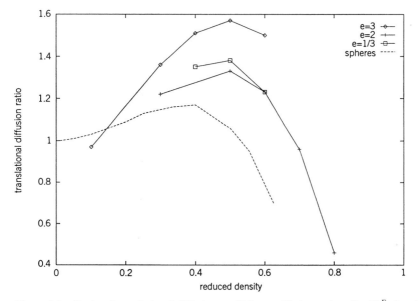

Figure 3.1. Ratio of translational diffusion coefficient to Enskog value, D_T/D_T^E, for ellipsoids of axial ratio $e = 3, 2, 1/3$ and 1 (hard spheres), plotted as a function of reduced density ρ/ρ_{cp} where ρ_{cp} is the close-packed density. The results for hard spheres are taken from Ref. [139].

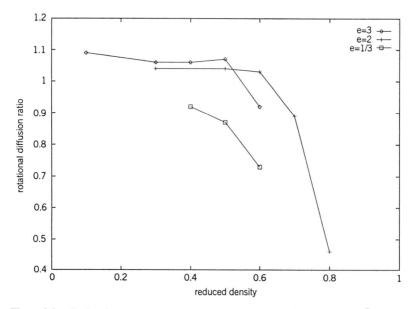

Figure 3.2. Ratio of rotational diffusion coefficient to Enskog value, D_R/D_R^E, for ellipsoids of axial ratio $e = 3$, 2 and 1/3, plotted as a function of reduced density ρ/ρ_{cp} where ρ_{cp} is the close-packed density.

tational diffusion; caging dominates. Further study of caging phenomena in these systems [24] has raised some puzzling features, however. For both $\langle \mathbf{v}_i(0) \cdot \mathbf{v}_i(t) \rangle$ and $\langle \omega_i(0) \cdot \omega_i(t) \rangle$, a minimum in the function is reached after quite a large number of binary collision times, typically 20–$30 \times t_c$. This time is much longer than what is typically observed in a dense fluid of hard spheres and is not really compatible with the idea of a rapid reversal of velocity following a single "rattle" in a cage. Reference [24] discusses the various factors that may contribute to this effect.

At low densities, for hard spheres, D_T approaches its Enskog value, since the independent binary collision model becomes correct in the dilute gas. This is manifestly not true for the ellipsoid systems, although the deviations from unity amount to, at most, 10% for the systems studied in the limit $\rho \to 0$. The translational diffusion coefficient is systematically lower than would be predicted by theory. At the same time, the rotational diffusion coefficient is slightly higher. These discrepancies occur because, even when the density is sufficiently low that molecules meet as isolated pairs, each such event may consist of several correlated collisions. This "chattering" phenomenon was discussed in Section III.C.1. Although chattering collisions are not accounted for by the simplest version of Enskog theory, a more sophisticated version [99] predicts these effects. A qualitative explanation can easily be put forward, referring to Fig. 3.3.

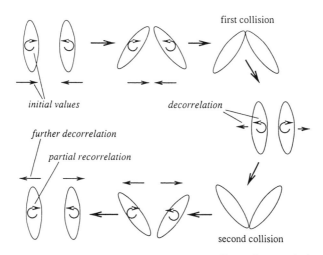

Figure 3.3. Double collision between two hard bodies, illustrating correlating and decorrelating effects on linear and angular velocities (schematic).

Here we focus on the "double-hit" collisions, the ones that provide the largest correction to the simple Enskog prediction. Many of these will involve a side-by-side geometry, as shown in the figure, and we take as an example a near-symmetrical case with both molecules approaching each other at similar speeds. The effect of the first collision is to reverse both the indicated linear and angular velocities. This contributes to the decorrelation effects seen as the initial decay of both $\langle \mathbf{v}_i(0) \cdot \mathbf{v}_i(t) \rangle$ and $\langle \omega_i(0) \cdot \omega_i(t) \rangle$. This part is correctly given by simple Enskog theory. However a second collision modifies both linear and angular velocities further. The effect on the linear velocity is to enhance the results of the first collision; the velocities become *more* decorrelated with their original values, on average. The effect on the angular velocities is to partially negate the results of the first collision, and to *restore* some of the original correlation. Consequently, $\langle \mathbf{v}_i(0) \cdot \mathbf{v}_i(t) \rangle$ decays at long times more quickly than expected, giving $D_T < D_T^E$, while $\langle \omega_i(0) \cdot \omega_i(t) \rangle$ decays at long times more slowly than expected, giving $D_R > D_R^E$. This description is, of course, qualitative. In reality, there will be a complete range of geometries and initial conditions in three dimensions, over which an average must be taken.

The most interesting feature in Fig. 3.1, however, is the rise of D_T/D_T^E above unity at intermediate densities. In the hard-sphere case, this is associated with the celebrated $t^{-3/2}$ algebraic long-time tails in the velocity autocorrelation function [139–141] arising from coupling with hydrodynamic vortex modes. It seems likely that another effect is operating in the case of the ellipsoids, swamping the vortex coupling, since the enhancement in

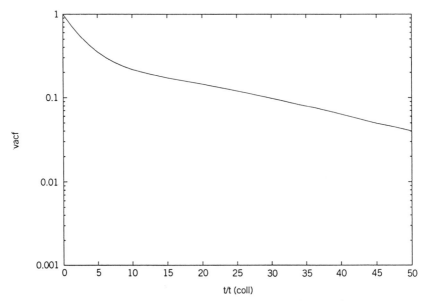

Figure 3.4. Translational velocity autocorrelation function (log scale) for prolate ellipsoids with $e = 10$ at a reduced density $\rho/\rho_{cp} = 0.2$. Time is measured in units of the mean time between collisions per particle.

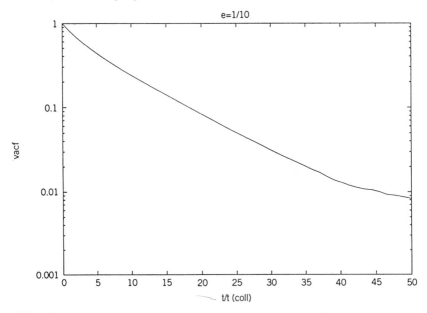

Figure 3.5. Translational velocity autocorrelation function (log scale) for oblate ellipsoids with $e = 1/10$ at a reduced density $\rho/\rho_{cp} = 0.2$. Time is measured in units of the mean time between collisions per particle.

D_T/D_T^E is much larger. Certainly, no evidence of algebraic decay is seen in the systems studied (in any case, it would be difficult to obtain direct evidence of this in the modest system sizes used). What is seen, especially in the prolate case, is a two-exponential decay in time (see Figs. 3.4, 3.5), indicating some coupling of the velocity with other degrees of freedom, possibly reorientation. A detailed theory of this phenomenon is still lacking, but related, and even more dramatic, effects are seen in isotropic fluids of thin hard needles [142–144] and in liquid crystalline phases of disklike molecules. [145] In the latter context (see Section VI), we briefly sketch a possible mechanism that can lead to such large deviations from the Enskog theory.

2. Pretransitional Collective Rotation

A key feature of the I→N transition is the appearance of pretransitional effects in molecular reorientation in the isotropic phase. Allen and Frenkel [146] compared single particle and collective rotation, in the isotropic phase near to the ordering transition, with a view to identifying these pretransitional fluctuations. The crucial theoretical prediction is the Kivelson–Keyes equation of Section (3.D.1) [131,147] where the single particle second-rank correlation time τ_2^s and its collective counterpart τ_2^c are calculated in the simulation as time integrals of appropriate time correlation functions, and the static second-rank Kirkwood factor g_2 is computed as a sum over pairs. In the simulations (for the $e = 3$ prolate ellipsoid system), collective rotation was seen to slow down dramatically near the I→N transition, but

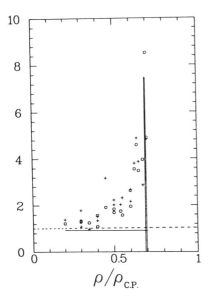

Figure 3.6. Density dependence of g_2 (open circles) and τ_2^c/τ_2^s (pluses) for ellipsoids with $e = 3$. The horizontal line is the best estimate for j_2, assuming the latter quantity to be independent of density. The dotted line marks a value of unity. The vertical lines mark the estimated boundaries of the $I - N$ coexistence region for this system size.

the effect is mostly explained by the divergence of the static factor g_2. There is little evidence for any dramatic variation in j_2, which seems to be just less than unity in agreement with (low density) kinetic theory (see Fig. 3.6) [35] (see Section III.D.1) and with the (scant) experimental evidence. [148,149] Indeed, in the absence of information to the contrary, j_2 had long been assumed to be essentially unity, and the simulations seem to provide some justification for this. However, for spherocylinders the situation is not so simple. [77] The low density theoretical prediction is again that j_2 is slightly less than unity, and this is confirmed by simulation. However, at higher densities, j_2 appears to rise to more than 1.5.

Throughout, one must remember that the I→N transition is weakly first order, rather than continuous, so actually it preempts the complete divergence of g_2 and τ_2^c/τ_2^s. Also, it is important to bear in mind the limitations of small system sizes and modest simulation run lengths, when discussing collective reorientational degrees of freedom near a phase transition.

PART TWO: THE LIQUID-CRYSTALLINE PHASE

IV. PHASE TRANSITIONS

A. Theory

In order to describe phase transitions in liquid-crystalline systems,[1] we need a statistical mechanical formalism capable of dealing with spontaneous symmetry breaking. The formalism based on the canonical partition function employed in the section on the isotropic phase is not up to this task. We therefore introduce the general framework of density functional theory (DFT), with which one can give a unified description of all phase transition phenomena in continuous systems. Next we give an overview of the diverse approximations that have been proposed to obtain workable functionals for the description of liquid-crystalline phase transitions in HCB fluids. Finally, we review the application of these ideas to specific transitions.

---

[1]In the compilation of this section, we were able to make use extensively of the in-depth review by Vroege and Lekkerkerker *Phase Transitions in Lyotropic Colloidal and Polymer Liquid Crystals* [150]. We are grateful to these authors for supplying a preprint version of their review.

1. Density Functional Formalism

The density functional formalism is in essence just a variational formulation of the classical statistical mechanics of many-particle systems. It has, however, several advantages with respect to other approaches. First, variational problems are well understood and there exists a large body of mathematical knowledge concerning their solution. Secondly, the quantity being varied is a function of the degrees of freedom of a single particle only, so that we are shielded from the complexities of n-particle distributions. In fact, it is a deep and surprising result of the theory that there is a one-to-one correspondence between the single particle distribution function and the full N-particle probability density. Finally, the formalism can deal with spontaneous symmetry breaking in a direct and natural way, making it an ideal tool for the study of symmetry related phase transitions. To our knowledge the first application of density functional techniques to liquid crystals is due Workman and Fixman. [151]

In order to state the the theory, we first need to define the one-particle distribution function (ODF)

$$\rho^{(1)}(\xi) = \left\langle \sum_{i=1}^{N} \delta(\xi - \xi_i) \right\rangle \tag{4.1}$$

where $\xi = (\mathbf{r}, \Omega)$ is shorthand for the degrees of freedom of a single particle, being, for the HCBs we are considering, the position of the center of mass $\mathbf{r}$ and the orientation Ω with respect to a fixed reference frame. The sum on the right hand side of Eq. (4.1) runs over all the N particles in the system and the angular brackets denote equilibrium averaging. Note that the ODF so defined is normalized to the number of particles, that is,

$$\int_V d\xi \rho^{(1)}(\xi) = N \tag{4.2}$$

where the integration runs over the finite volume V of our system.

The full statement of density functional theory for classical many particle systems [152] is the assertion that there exists a functional $\mathcal{W}[\rho^{(1)}]$ of the ODF with the following properties:

1. For any arbitrary ODF $\rho^{(1)}$, the following inequality holds: $\mathcal{W}[\rho^{(1)}] \geq \mathcal{W}[\rho_{eq}^{(1)}]$ where $\rho^{(1)}{}_{eq}$ is the equilibrium distribution.
2. $\mathcal{W}[\rho_{eq}^{(1)}] = \mathcal{W}_{eq}$ where $\mathcal{W}_{eq}$ is the thermodynamic equilibrium value of the grand canonical potential.

As such, this statement is entirely nonconstructive: We are told that the functional exists, but not how to construct it. As we shall see, the actual construction involves approximations, which in final analysis can only be tested through comparison with simulation data. Nevertheless, the general structure of the functional can be analyzed somewhat further. First, recalling the thermodynamic relation $W = F - \mu N$ relating the grand canonical potential W to the free energy F, the chemical potential μ and the number of particles N, we can write

$$\mathcal{W}[\rho^{(1)}] = \mathcal{F}[\rho^{(1)}] - \mu \int_V d\xi \rho^{(1)}(\xi) \tag{4.3}$$

which introduces the functional $\mathcal{F}[\rho^{(1)}]$ representing the Helmholtz free energy of the system. This free energy functional can be split into an ideal and an excess part, the latter containing all contributions due to the interactions between the particles.

$$\beta \mathcal{F}[\rho^{(1)}] = \int d\xi \rho^{(1)}(\xi)\{\ln \mathcal{V}_T \rho^{(1)}(\xi) - 1\} - \Phi[\rho^{(1)}] \tag{4.4}$$

where $\mathcal{V}_T$ is the thermal de Broglie volume discussed in Appendix A.A and we have introduced the functional $\Phi[\rho^{(1)}] = -\beta \mathcal{F}_{\text{excess}}[\rho^{(1)}]$ that serves as a generating functional for the n-particle direct correlation functions

$$c^{(n)}(\xi_1, \xi_2, ..., \xi_n; [\rho^{(1)}]) = \frac{\delta^n \Phi[\rho^{(1)}]}{\delta \rho^{(1)}(\xi_1) \delta \rho^{(1)}(\xi_2) \cdots \delta \rho^{(1)}(\xi_n)} \tag{4.5}$$

Historically, the first approach to the actual construction of the functional is in terms of a diagram expansion obtained by re-summing the diagrams in an expansion of the grand canonical potential in terms of an external field, which is eliminated in favor of the ODF through a topological reduction technique. [153] The result is the following identity:

$$\Phi[\rho^{(1)}] = \begin{cases} \text{Sum of all connected, irreducible diagrams with } \rho^{(1)} \text{ vertices and Mayer function bonds:} \\ f(\xi_i, \xi_j) = e^{-\beta v(\xi_i, \xi_j)} - 1 \end{cases} \tag{4.6}$$

For hard particles the Mayer function is just the negative of the characteristic function χ of the mutual excluded volume of the particles,

$$f(\xi_i, \xi_j) = -\chi(\xi_i, \xi_j) = \begin{cases} -1, & \text{if } i \text{ and } j \text{ overlap} \\ 0, & \text{if } i \text{ and } j \text{ do } not \text{ overlap} \end{cases} \tag{4.7}$$

One can also use the definition (4.5) to relate the excess part Φ to some properties of a reference state $\rho_0^{(1)}$ through a double functional integration. Consider an arbitrary one parameter family of ODFs $\rho_\alpha^{(1)}$ interpolating smoothly between a reference system at $\rho_0^{(1)}$ and a final state at $\rho_1^{(1)}$ as a parameter α goes from 0 to 1. Since the functionals are unique, the final result will *not* depend on the path of integration. [154] A first integration yields

$$
\begin{aligned}
\Phi[\rho_1^{(1)}] - \Phi[\rho_0^{(1)}] &= \int_0^1 d\alpha \frac{d}{d\alpha} \Phi[\rho_\alpha^{(1)}] \\
&= \int_0^1 d\alpha \int d\xi \frac{\delta\Phi[\rho_\alpha^{(1)}]}{\delta\rho^{(1)}(\xi)} \frac{\partial\rho_\alpha^{(1)}(\xi)}{\partial\alpha} \\
&= \int_0^1 d\alpha \int d\xi c^{(1)}(\xi, [\rho_\alpha^{(1)}]) \frac{\partial\rho_\alpha^{(1)}(\xi)}{\partial\alpha}
\end{aligned}
\tag{4.8}
$$

One can now repeat such an integration to eliminate the one-particle direct correlation function in favor of a description in terms of the two-particle direct correlation function.

$$
c^{(1)}(\xi, [\rho_\alpha^{(1)}]) = c^{(1)}(\xi, [\rho_0^{(1)}]) + \int_0^\alpha d\alpha' \int d\xi' c^{(2)}(\xi, \xi', [\rho_{\alpha'}^{(1)}]) \frac{\partial\rho_{\alpha'}^{(1)}(\xi')}{\partial\alpha'} \tag{4.9}
$$

Putting these two results together yields

$$
\begin{aligned}
\Phi[\rho_1^{(1)}] =& \\
& \Phi[\rho_0^{(1)}] + \int_0^1 d\alpha \int d\xi c^{(1)}(\xi, [\rho_0^{(1)}]) \frac{\partial\rho_\alpha^{(1)}(\xi)}{\partial\alpha} \\
& + \int_0^1 d\alpha \int_0^\alpha d\alpha' \int d\xi d\xi' c^{(2)}(\xi, \xi', [\rho_{\alpha'}^{(1)}]) \frac{\partial\rho_\alpha^{(1)}(\xi)}{\partial\alpha} \frac{\partial\rho_{\alpha'}^{(1)}(\xi')}{\partial\alpha'}
\end{aligned}
\tag{4.10}
$$

This expression, which can be used to relate the excess free energy of a system to the structural properties of a (possibly simpler) reference phase, is the starting point for many approximations considered.

In order to locate the equilibrium ODF of our system, we have to minimize the grand canonical functional. This leads first of all to the necessary (but not yet sufficient) criterion that the functional be stationary to variation in the equilibrium ODF,

$$
\frac{\delta\mathcal{W}[\rho^{(1)}]}{\delta\rho^{(1)}(\xi)} \bigg|_{\rho^{(1)}=\rho_{eq}^{(1)}} = 0 \tag{4.11}
$$

Inserting the explicit forms (4.3) and (4.4) and using the definition (4.5), we find the following equation for the ODF:

$$\rho^{(1)}(\xi) = \frac{1}{V_T} e^{\beta\mu} \exp c^{(1)}(\xi, [\rho^{(1)}]) \tag{4.12}$$

The chemical potential μ can be eliminated from this equation by using the normalisation condition (4.2). One can look on this equation as a non-linear self-consistency relation for the ODF, the one-particle direct correlation function $c^{(1)}$ playing the role of an effective field that acts on a particle due to interactions with the rest of the system.

To ensure that a stationary distribution satisfying the condition (4.11) is indeed a minimum, we need the following condition to be met for arbitrary variations $\Delta\rho^{(1)}$:

$$\int d\xi d\xi' \frac{\delta W[\rho^{(1)}]}{\delta\rho^{(1)}(\xi)\delta\rho^{(1)}(\xi')} \Delta\rho^{(1)}(\xi)\Delta\rho^{(1)}(\xi') > 0 \tag{4.13}$$

It should be stressed that this condition is necessary but not always sufficient to determine the stability of the phase in question. This becomes apparent when calculating phase coexistence at constant pressure where the Gibbs free energy is the relevant thermodynamic potential. This necessitates a separate calculation of the pressure of each potential phase.

A useful technique for analyzing nonlinear equations like the stationarity condition (4.11) in the neighborhood of phase transitions, where, as a rule, coexisting solutions to the equations occur, is bifurcation analysis. This technique probes new solutions to the equations that branch off from a known reference solution by making a parametric expansion around the branch or bifurcation point. In our case, a reference solution would be a locally stable phase $\rho_0^{(1)}$ that exists in a given range of densities. The procedure starts by assuming the following form for a solution close to the reference solution at the density n_0:

$$\rho_\epsilon^{(1)} = \rho_0^{(1)} + \epsilon\rho_1^{(1)} + \epsilon^2\rho_2^{(1)} + \cdots$$
$$n_\epsilon = n_0 + \epsilon n_1 + \epsilon^2 n_2 + \cdots \tag{4.14}$$

These expansions are inserted into the stationarity equation (4.11) and a solution is constructed order by order in the arbitrary parameter ϵ. The first order equation is usually called the bifurcation equation and provides information both about the location of the bifurcation, that is, the density at which it occurs, and the nature of the new solution (symmetry). It takes the form of a generalized eigenvalue problem

$$\rho_1^{(1)}(\xi) = \rho_0^{(1)}(\xi) \int d\xi' c^{(2)}(\xi, \xi', [\rho_0^{(1)}])\rho_1^{(1)}(\xi') \tag{4.15}$$

One can successively solve the higher order equations in the hierarchy generated by these expansions, thus constructing approximations to the bifurcating solution that are valid at ever increasing distances (in the solution space) from the reference solution.

2. Specific Approximations

a. The Onsager Approximation. Historically, the first and, undoubtably, the most influential approach to ordering phenomena in HCBs is that proposed by Onsager in 1949. Although derived from first principles using the theory of mixtures, it is in fact a density functional theory avant la lettre. In terms of the expansion (4.6), it consists of neglecting all higher order diagrams, keeping just the lowest order single bond diagram. The explicit expression for this diagram is given by

$$\Phi_2[\rho^{(1)}] = \frac{1}{2} \int d\xi d\xi' f(\xi, \xi') \rho^{(1)}(\xi) \rho^{(1)}(\xi') \tag{4.16}$$

Considering for a moment just spatially homogeneous phases for which the ODF factorizes $\rho^{(1)}(\mathbf{r}, \Omega) = n\psi(\Omega)$, where n is the number density and ψ a normed orientational distribution function, we can simplify to

$$\Phi_2[\psi] = -\frac{1}{2} N n \int d\Omega \, d\Omega' \mathcal{E}(\Omega, \Omega') \psi(\Omega) \psi(\Omega') \tag{4.17}$$

where we have introduced the mutual excluded volume of two HCBs at fixed orientations,

$$\mathcal{E}(\Omega, \Omega') = \int d\mathbf{r}' \chi(\mathbf{r} - \mathbf{r}', \Omega, \Omega') \tag{4.18}$$

Onsager justified his approximation by arguing that in the isotropic phase of a system of extremely elongated rods, this diagram dominates all higher order ones. More recently, it was shown that the same holds true for d-dimensional generalizations of HCB fluids in the limit that the dimensionality is very high. [155] Especially this latter observation seems to indicate that the Onsager approximation is the hard-particle analog of the mean field approach for long-range interactions, which is also exact in the infinite dimensionality limit. Moreover, the stationarity equations obtained within the approximation are also formally equivalent to the mean field one, the pair excluded volume playing the role of the potential and the number density that of the inverse temperature. An added advantage of the approximation is that the pair excluded volume at fixed relative orientation has been determined analytically for a number of convex bodies (ellipsoids of revolution, spherocylinders, spheroplatelets, right circular cylinders) allowing detailed calculations to be made.

A natural extension of the Onsager approximation is, of course, the consideration of higher order diagrams in the virial expansion. Here one meets with severe difficulties due to the complexity of the integrals involved. The only head on attempt to face this problem was made by Tjipto-Margo and Evans [41] who expanded the kernel of the third virial coefficient of hard ellipsoids into a set of invariant functions. The lower order coefficients in this expansion could be determined through a Monte Carlo procedure. These results were then used both in a straightforward extension of the original Onsager theory as well as a y-expansion approach (to be discussed below). The results are a definite improvement over the original second-virial coefficient based theories.

A related approach, originally proposed by Barboy and Gelbart, [49–51], is the so called y-expansion, where the diagram expansion for a homogeneous fluid is re-summed in terms of the reduced density variable

$$y = \frac{\eta}{1 - \eta} \tag{4.19}$$

where we introduce the packing fraction $\eta = nv_0$, with v_0 the proper volume of the HCB. The idea behind this expansion is to exploit similarity with the exact solution of the Percus–Yevick equation for hard spheres, which suggests that all relevant quantities can be expressed as short power series in the variable y. For a homogeneous phase, the free energy per particle can be expressed as an expansion in the number density n

$$\frac{\beta \mathcal{F}[\psi]}{N} = \int d\Omega \psi(\Omega) \ln \psi(\Omega) + \ln \mathcal{V}_T n + \sum_{k=2}^{\infty} \frac{B_k[\psi]}{k - 1} n^{k-1} \tag{4.20}$$

where the B_k are the generalized virial coefficients

$$B_k = -\frac{k - 1}{V} \sum \begin{pmatrix} \text{Connected diagrams with} \\ k \ \psi\text{-vertices and } f\text{-bonds} \end{pmatrix} \tag{4.21}$$

Eliminating the number density in favor of the variable y, we find

$$\frac{\beta \mathcal{F}[\psi]}{N} = \int d\Omega \psi(\Omega) \ln \psi(\Omega) + \ln \frac{\mathcal{V}_T}{v_0} + \ln y + \sum_{k=2}^{\infty} \frac{C_k[\psi]}{k - 1} y^{k-1} \tag{4.22}$$

where the coefficients $C_k[\psi]$ are expressed in terms of the generalized virial coefficients of order k and lower,

$$C_k[\psi] = (k-1) \sum_{j=2}^{k} \binom{-j+1}{k-j} \frac{B_j[\psi]v_0^{j-1}}{j-1} + (-)^{j-1} \qquad (4.23)$$

b. *Scaled Particle Theory.* Scaled particle theory (SPT), originally developed in the context of the hard sphere fluid, [42] was extended by several authors to apply to fluids of non-spherical particles. [156–161] SPT is based on the intriguing and original approach to calculate the amount of reversible work needed to add a scaled copy of the original particles to the fluid. The most elaborate version of the theory as applied to nonspherical particles is the one presented in Ref. [158] which considers hard spherocylinders and allows both the length and the breadth of the particles to be considered as independent scaling parameters. This approach, however, is effective only for spherocylinders, due to the special form of the pair excluded volume as a function of the particle's dimensions. [161] Here, for simplicity, we present the theory with a single scaling parameter. We do, however, follow Cotter [160] in her perscription for achieving consistent thermodynamics, which was lacking in the earlier applications of the theory.

Consider an HCB fluid to which we add a single scaled copy of the other particles. The size of the inserted particle is governed by the scaling parameter λ. When λ is taken to be zero, we are adding a point particle to the fluid, when $\lambda = 1$, the inserted particle is identical in size to the other particles, and when $\lambda \gg 1$, we are effectively creating a particle-shaped macroscopic cavity in the fluid. The main assumption of SPT is that the reversible work W done to insert the scaled particle smoothly interpolates between the limits $\lambda \ll 1$ and $\lambda \gg 1$. Denoting by $v_0(\lambda)$ the proper volume of the particle as a function of the scaling parameter, we can easily calculate the work for the case that the inserted particle is macroscopic in size,

$$W(\Omega|\lambda) \sim Pv_0(\lambda) = \lambda^3 Pv_0 \qquad (4.24)$$

where P is the equilibrium pressure. The fact that the particle volume, and hence this contribution to the work, scales as the third power in the scaling parameter suggests that if we can calculate the contribution of adding a very small particle up to second order in λ, we can by adding these two together obtain an expression that works well in both limits and so allow us to interpolate to $\lambda = 1$. For the case of a homogeneous but possibly orientationally ordered fluid, where the ODF has the form $\rho^{(1)}(\mathbf{r}, \Omega) = n\psi(\Omega)$ the expansion to second order in λ can be calculated from a generalization of the expansion, originally derived by Reiss et al. [42]

$$-\beta W(\Omega|\lambda) = \ln \sum_{k=0}^{\infty} \frac{(-)^k}{k!} n^k F^{(k)}(\Omega|\lambda) \tag{4.25}$$

where the $F^{(k)}$ are related to the equilibrium k-particle distribution functions through

$$F^{(k)}(\Omega|\lambda) = \prod_{j=1}^{k} \int d\Omega_j \psi(\Omega_j) \int_{\chi(\mathbf{r}_j,\Omega,\Omega_j|\lambda)=1} d\mathbf{r}_j g^{(k)}(\mathbf{r}_1\Omega_1,\dots,\mathbf{r}_k\Omega_k,[\psi]) \tag{4.26}$$

The spatial integrations run over all configurations where the fluid particles overlap with the scaled particle, which is inserted at the origin. Note that in an orientationally ordered phase, the work done depends on the orientation of the inserted particle with respect to a fixed reference frame. The k-particle distribution functions that appear are generalizations of the well-known two-particle radial distribution function and are defined through the relation

$$\rho^{(k)}(\xi_1,\dots,\xi_k,[\rho^{(1)}]) = \Big\langle \sum_{i_1 \neq i_2 \neq \dots \neq i_k}^{N} \delta(\xi_1 - \xi_{i_1})\dots\delta(\xi_k - \xi_{i_k}) \Big\rangle$$
$$\equiv \Big(\prod_{i=1}^{k} \rho^{(1)}(\xi_i)\Big) g^{(k)}(\xi_1,\dots,\xi_k,[\rho^{(1)}]) \tag{4.27}$$

Since the k-particle distribution functions are identically zero in any configuration where two or more particles overlap each other (the probability of these configurations being zero), we can write down the expansion (4.25) in the case where the scaled particle is so small that no three particles can overlap with the scaled particle without also overlapping which each other:

$$e^{-\beta W(\Omega|\lambda)} = 1 - n \int d\Omega_1 \psi(\Omega_1) \int_{\chi(\Omega,\Omega_1|\lambda)=1} d\mathbf{r}_1$$
$$+ \frac{1}{2} n^2 \int d\Omega_1 \psi(\Omega_1) \int d\Omega_2 \psi(\Omega_2)$$
$$\times \int_{\chi(\Omega,\Omega_1|\lambda)=1} d\mathbf{r}_1 \int_{\chi(\Omega,\Omega_2|\lambda)=1} d\mathbf{r}_2 g^{(2)}(\mathbf{r}_1\Omega_1,\mathbf{r}_2\Omega_2,[\psi]) \tag{4.28}$$

One can even reduce this expression further by using exact constraints on the pair distribution function to show that the second term does not contribute to second order in λ. One is thus left with

$$e^{-\beta W(\Omega|\lambda)} = 1 - n \int d\Omega_1 \psi(\Omega_1) \mathcal{E}(\Omega, \Omega_1|\lambda) \tag{4.29}$$

where we have introduced the excluded volume at fixed orientations between the scaled particle and one of the system particles. Recalling Eqs. (2.15) and (2.32) from Section II and the results from Appendix A.D, we can express this excluded volume explicitly as

$$\mathcal{E}(\Omega, \Omega_1|\lambda) = v_0(1 + \lambda^3) + \lambda(1 + \lambda)\left(\frac{1}{2}\mathcal{E}(\Omega, \Omega_1) - v_0\right) \tag{4.30}$$

Performing the expansion to second order in λ and adding the cubic term, we arrive at our final approximation for the average reversible work needed to add the scaled particle:

$$-\beta\bar{W} = \ln(1 - nv_0) - yW_2[\psi]\lambda - \frac{1}{2}\left(2W_2[\psi]y + W_3[\psi]y^2\right)\lambda^2 - \beta Pv_0\lambda^3 \tag{4.31}$$

where the parameter $y = nv_0/(1 - nv_0)$ is the one also introduced in the previous subsection and the functionals W_k are defined by

$$W_2[\psi] = \int d\Omega \int d\Omega' \psi(\Omega)\psi(\Omega')\mathcal{C}(\Omega, \Omega') \tag{4.32}$$

$$W_3[\psi] = \int d\Omega \int d\Omega' \int d\Omega'' \psi(\Omega)\psi(\Omega')\psi(\Omega'')\mathcal{C}(\Omega, \Omega')\mathcal{C}(\Omega, \Omega'') \tag{4.33}$$

with the kernel $\mathcal{C}$ is given by

$$\mathcal{C}(\Omega, \Omega') = \left(\frac{1}{2v_0}\mathcal{E}(\Omega, \Omega') - 1\right) \tag{4.34}$$

All the necessary ingredients have now been collected. The contact with thermodynamics is made through the obeservation that the excess chemical potential of the fluid is related to the average reversible work needed to create a cavity of the same size as the original particles, hence

$$\mu_{\text{ex}} = \int d\Omega \psi(\Omega) W(\Omega|\lambda = 1) \tag{4.35}$$

The equation of state can now be found by integrating the Gibbs–Duhem relation

$$\frac{\partial P}{\partial n} = 1 + n\frac{\partial\mu_{\text{ex}}}{\partial n} \tag{4.36}$$

The free energy is now easily calculated by using the relation $G = \mu N = F + PV$, yielding as our final result

$$\beta \mathcal{F}[\psi]/N = \ln \mathcal{V}_T v_0 + \ln y + \int d\Omega \psi(\Omega) \left(\ln \psi(\Omega) - 1\right) + W_2[\psi]y + \frac{1}{6}W_3[\psi]y^2 \quad (4.37)$$

This result looks identitical in structure to the y-equation result (Eq. 4.20) if one keeps terms up to second order in y. Indeed upon closer inspection, one finds that W_2 is in fact identical to C_2. The coefficient W_3, however, differs from the corresponding C_3, [161] except in the case of hard spheres where both approaches lead to the same result, which is also identitical to the one obtained from the exact solution to the Percus–Yevick equation [1] using the compressibility route for the equation of state.

It should be pointed out that the method of implementing SPT outlined above and followed, with minor variations, by all the papers cited, is not the optimal way of doing it. Tully-Smith and Reiss [58] have devised a more general scheme, based on the same ideas, that removes the limitation of expanding the work function just to third order in the scaling parameter and replacing it by an asymptotic expansion of the kernel of the work function in inverse powers of the scaling parameter. The coefficients in this expansion are then fixed by maximally exploiting a number of exact constraints.

c. *Direct Correlation Function Approaches.* In this section, we discuss the approaches to the problem of constructing a free energy functional for the nonisotropic HCB fluid that take the expansion about a reference fluid (4.10) as a starting point. We will take the reference fluid to be both isotropic and homogeneous, that is, $\rho_0^{(1)} = n_0/8\pi^2$ and at the same number density as the system, that is, $n_1 = n_0$. Since the free energy is a true functional, the path of the functional integration can be freely chosen to be of the simple form $\rho_\lambda^{(1)} = \rho_0^{(1)} + \lambda \Delta \rho^{(1)}$ where $\Delta \rho^{(1)} = \rho_1^{(1)} - \rho_0^{(1)}$. Using the fact that the one-particle direct correlation in the reference phase does not depend on the particle's coordinates, we arrive at

$$\Phi[\rho^{(1)}] = \Phi[\rho_0^{(1)}] + \frac{1}{2}\int d\xi d\xi' c^{(2)}(\xi, \xi, [\rho_0^{(1)}])\Delta\rho^{(1)}(\xi)\Delta\rho^{(1)}(\xi') \quad (4.38)$$

We now need to specify the two-particle direct correlation function in the reference phase. The various approaches discussed below distinguish themselves by the way in which they deal with this problem.

First, and as an aside, it is instructive to see that the previously described

Onsager approximation is recovered, if we replace the direct correlation by the first term in its low density expansion

$$c^{(2)}(\xi, \xi', [\rho_0^{(1)}]) = f(\xi, \xi'), \quad n_0 \ll 1 \tag{4.39}$$

where f is the Mayer function.

The most direct route to obtaining $c^{(2)}$ in the isotropic phase was followed by Patey and co-workers [23,25,162] who used results from the numerical solution of the various liquid state integral equations for the isotropic HCB fluid. In this scheme, one obtains a set of density dependent coefficients in an invariant expansion of the direct correlation function which can be used as input to the functional expansion. This is of necessity a rather involved procedure. Many authors therefore take the simplifying step of mapping the problem onto the direct correlation function for hard spheres, for which we possess an approximation in analytical form through the solution to the Percus–Yevick equation as well as several semi-empirical extensions thereof. Given the scarcity of hard information, these approximations are all, to a greater or a lesser degree, inspired guesses.

A first proposal, originally suggested by Pynn, [163,164] is to model the direct correlation function by the hard-sphere one, artificially made orientation dependent by scaling the interparticle distance with the distance of closest approach between the two HCBs $\sigma(\hat{\mathbf{r}}, \Omega, \Omega')$, where $\hat{\mathbf{r}}$ is the unit vector along the line connecting the centers of the two particles,

$$c^{(2)}(\mathbf{r}, \Omega, \Omega', [\rho_0^{(1)}]) = c_{HS}^{(2)}(r/\sigma(\hat{\mathbf{r}}, \Omega, \Omega'), n_0) \tag{4.40}$$

This expression of course reduces to the correct expression for hard spheres. Since in this approach the influence of relative orientation is decoupled from the influence of interparticle distance, the name decoupling approximation is appropriate. It has the defect, first noticed by Lado, [165] that for $r \to 0$ the direct correlation function is predicted to be isotropic, contrary to reality.

Another proposal, due to Baus et al. [166] who formulated it with ellipsoids of revolution in mind, is the following:

$$c^{(2)}(\mathbf{r}, \Omega, \Omega', [\rho_0^{(1)}]) = \frac{\mathcal{E}(\Omega, \Omega')}{v_0} c_{HS}^{(2)}(r/\sigma_{\text{eff}}, n_{\text{eff}}) \tag{4.41}$$

where in the simplest case the effective hard-sphere diameter is determined by the equal volume rule $4\pi\sigma_{\text{eff}}^3/3 = v_0$ and the effective density n_{eff} is taken equal to the reference density n_0. By construction, this expression does not have the deficiency of being isotropic at short distances.

One can go one step further by assuming that one can use the expression (4.41) to describe an approximate direct correlation of an orientationally ordered phase. This new reference phase will already contain some of the effects of orientational order but still posses a rotationally invariant direct correlation function. To correct for the fact that there is on average less interaction between the partially aligned particles at a given density as compared to the isotropic fluid of the same density, the pseudonematic reference state should have a lower effective density. This effective density can be fixed by requiring the same value at contact of the direct correlation function in the purely isotropic case and in the effective nematic case, where the average contact distance in the latter case should be reduced in order to take into account the ordering of the particles. The perscription suggested by Colot et al. is the following

$$c_{HS}^{(2)}(r/\sigma_{eff} = 1, n_0) = c_{HS}^{(2)}(r/\sigma_{eff} = x, n_{eff}) \tag{4.42}$$

where x is a factor that accounts for the reduced average contact distance. For ellipsoids, Colot et al. [167] unfortunately chose $x = b/a$ where a and b are the lengths of the major and minor axes, respectively. This implies an average contact distance *smaller* than the minor axis of the ellipsoids, which is of course unphysical.

d. Weighted Density Approximations. Next, we briefly discuss a class of methods for the construction of density functionals for HCB fluids, the so-called weighted density approximations (WDA), all traceable to the ideas originally suggested by Tarazona and Evans [168,169] and Curtin and Ashcroft [170,171] in their work on the inhomogeneous hard-sphere fluid. Consider our general expression for the density functional (4.10) in the case that the reference state is the zero density state, that is, $\Delta\rho^{(1)}(\xi) = \rho^{(1)}(\xi)$. In that case, the excess free energy functional can be written as

$$\Phi[\rho^{(1)}] = \int d\xi \rho^{(1)}(\xi)\Delta\psi(\xi, [\rho^{(1)}]) \tag{4.43}$$

where $\Delta\psi(\xi, [\rho^{(1)}])$ is a nonlocal functional of the ODF. The idea of the WDA is to replace this nonlocal functional by a *function* $\Delta\phi(\bar{\rho}^{(1)}(\xi))$ of a suitably chosen *local* density $\bar{\rho}^{(1)}(\xi)$. This local density is then taken to be some kind of weighted average of the true ODF over a neigborhood surrounding the point in question. In practice, the effective density is taken to depend only on the position and the orientational effects are included in an effective way through the averaging procedure. The usual form of this averaging is therefore

$$\bar{n}(\mathbf{r}) = \int d\mathbf{r}' \, d\Omega \, d\Omega' w(\mathbf{r} - \mathbf{r}', \Omega, \Omega')\rho^{(1)}(\mathbf{r}', \Omega') \tag{4.44}$$

The approximated excess free energy functional will then take the form

$$\Phi[\rho^{(1)}] = \int d\mathbf{r} n(\mathbf{r}) \Delta\phi\left(\bar{n}(\mathbf{r})\right) \tag{4.45}$$

where we have introduced

$$n(\mathbf{r}) = \int d\Omega \rho^{(1)}(\mathbf{r}, \Omega) \tag{4.46}$$

The two inputs of a specific WDA are thus the form of the weight function $w(\mathbf{r} - \mathbf{r}', \Omega, \Omega')$ and the explicit form of the function $\Delta\phi$.

The proposal put forward by Poniewierski and Hołyst [172,173] is to take the following definition for the averaged local density:

$$\bar{n}(\mathbf{r}) = \int d\Omega \int d\mathbf{r}' \, d\Omega' \frac{1}{2B_2^{\text{iso}}} \chi(\mathbf{r} - \mathbf{r}', \Omega, \Omega') f(\mathbf{r}, \Omega) \rho^{(1)}(\mathbf{r}', \Omega') \tag{4.47}$$

where the orientational distribution function $f(\mathbf{r}, \Omega)$ is defined as

$$f(\mathbf{r}, \Omega) = \rho^{(1)}(\mathbf{r}, \Omega) / \int d\Omega' \rho^{(1)}(\mathbf{r}, \Omega') \tag{4.48}$$

and the isotropic second virial coefficient is introduced to obtain a proper normalization:

$$\int d\mathbf{r} \, d\Omega \, d\Omega' \chi(\mathbf{r} - \mathbf{r}', \Omega, \Omega') = \int d\Omega \, d\Omega' \mathcal{E}(\Omega, \Omega') \equiv 2(8\pi^2)^2 B_2^{\text{iso}} \tag{4.49}$$

Given the fact that the effective local density is independent of orientation so that the system is effectively sphericalized, it is natural to select an effective excess free energy density related to the hard sphere system. Poniewierski and Hołyst chose

$$-\Delta\phi(n) = nB_2 + \left(\frac{\beta F_{\text{CS}}^{\text{excess}}}{N}(\eta) - 4\eta\right) \tag{4.50}$$

which is the Carnahan–Starling expression for the excess free energy per particle of the hard-sphere fluid in terms of the packing fraction η (see Ref. [1]) corrected to give the proper second virial coefficient for the non-spherical particles in the isotropic phase.

$$-\frac{\Delta\phi^{\text{CS}}(\eta)}{N} \equiv \frac{\beta F_{\text{CS}}^{\text{excess}}}{N}(\eta) = \frac{\eta(4 - 3\eta)}{(1 - \eta)^2} \tag{4.51}$$

A related approach is the one introduced by Somoza and Tarazona [174, 175]. They base their analysis on a mapping onto the system of parallel hard ellipsoids (PHE) rather than onto the the hard-sphere fluid. In this way, they hope to include more of the anisotropic structure of the full HCB fluid into the reference system. Moreover, in the case of PHEs, one is able to make use of the fact that this system can be related to the hard sphere fluid using an affine scale changing transformation. [176] Their procedure is as follows. First select the equivalent PHE system through the rule that the volume of the PHE is equal to that of the HCB in question and that the the average moments of inertia of the HCB are proportional to those of the PHE, that is,

$$\langle \mathbf{I}_{HCB} \rangle = \int d\Omega f(\mathbf{r}, \Omega) \mathbf{I}_{HCB}(\Omega) \propto \mathbf{I}_{PHE} \tag{4.52}$$

Then use a scaled version of the weighting function applied to the inhomogeneous hard-sphere fluid, [177] the scaling given by the mapping between the HS and PHE, to obtain the effective local density $\bar{n}(\mathbf{r})$ of the PHE. Finally, construct the WDA excess free energy functional for the PHE with an extra factor that corrects for the second virial coefficient of the HCB in a manner similar to that of the phenomenological Parsons approach.

$$\Phi[\rho^{(1)}] =$$
$$\int d\mathbf{r}\, d\Omega \rho^{(1)}(\mathbf{r}, \Omega) \Delta\phi_{HPE}(\bar{n}(\mathbf{r})) \frac{\int d\mathbf{r}'\, d\Omega' \rho^{(1)}(\mathbf{r}', \Omega') \chi_{HCB}(\mathbf{r} - \mathbf{r}', \Omega, \Omega')}{\int d\mathbf{r}'\, d\Omega' \rho^{(1)}(\mathbf{r}', \Omega') \chi_{PHE}(\mathbf{r} - \mathbf{r}')}$$
$$\tag{4.53}$$

e. Empirical Approaches Finally, we discuss some of the more empirical approaches to the construction of theories for phase transitions in HCB fluids. In all cases an attempt is made to somehow incorporate nonsphericity into known results on the hard sphere system.

One of the first such theories is due to Flapper and Vertogen [178,179] who argued that the packing fraction η for hard spheres of diameter σ can also be interpreted as

$$\eta = \frac{\pi}{6}\sigma^3 n = \frac{1}{8} V_{excl} n \tag{4.54}$$

where V_{excl} is the excluded volume of two spheres. They therefore proposed to identify the effective packing fraction for nonspherical particles (in a homogeneous, e.g., nematic phase) as

$$\eta_{eff}[\psi] = \frac{1}{8} n \langle \mathcal{E}[\psi] \rangle = \frac{1}{8} n \int d\Omega\, d\Omega' \psi(\Omega) \psi(\Omega') \mathcal{E}(\Omega, \Omega') \tag{4.55}$$

The equation of state for the HCB fluid is then postulated to be the one for the HS fluid, but with the above effective packing fraction. Choosing for instance the Carnahan–Starling relation [53] to model the HS behavior, we can construct the approximate free energy functional

$$\frac{\beta \mathcal{F}[\psi]}{N} = \int d\Omega \psi(\Omega) \ln \psi(\Omega) + \ln \mathcal{V}_T n + \frac{\eta[\psi](4 - 3\eta[\psi])}{(1 - \eta[\psi])^2} \qquad (4.56)$$

In a similar vein we have the "scaling" approach introduced by Lee. [180,181] Here the excess free energy of the Carnahan–Starling HS equation is rather arbitrarily multiplied by a scaling function that contains the dependence on the orientational distribution

$$\beta \mathcal{F}^{\text{excess}}[\psi] = \frac{\langle \mathcal{E}[\psi] \rangle}{8v_0} \left(\frac{\eta(4 - 3\eta)}{(1 - \eta)^2} \right) \qquad (4.57)$$

As was pointed out by Vroege and Lekkerkerker, [150] this result coincides with that of the so called decoupling approximations, both the direct correlation version discussed in Section. IV.A.2 above as well as one based on the radial distribution function discussed by Parsons. [182]

3. Specific Transitions

In this section we presents the results of the theoretical approaches described above as they are applied to the various liquid crystalline phase transitions thatoccur in HCBs.

a. *Isotropic–Nematic Transition.* The most studied transition occurring in HCB fluids is, of course, the isotropic–nematic transition. As is observed in the computer simulations, HCBs that are sufficiently nonspherical undergo a weakly first order transition from the low density isotropic phase to the uniaxially symmetric nematic phase. The first order nature of the transition is correctly captured by all theories. They differ, however, in their predictions about the strength and the location of the transition. Key quantities to compare are therefore the packing fractions η_{iso} and η_{nem} of the coexisting phases, the coexistence pressure P_{coex} conveniently expressed in dimensionless units as $P^* = \beta P v_0$ and the jump in the nematic order parameter $< P_2 >_c$ at the transition.

For ellipsoids of revolution, the most detailed comparisons can be made for particles with a length-to-width ratio $x = 3$. The results are summarized in Table I.

From these results it is clear that most theories *under*estimate the coexistence densities and probably *over*estimate the jump in the order parameter.[2] The scaling approach of Lee[181] seems incredibily accurate

---

[2]No reliable results are quoted for this quantity in the literature, but it is generally believed to be of the order of $< P_2 > \approx 0.35$

TABLE I

Comparisons of Results for the Isotropic–Nematic Transition of Several Theories with MC Data for the Case of Hard Ellipsoids at $x = 3$

Method	Source Reference	η_{iso}	η_{mem}	P^*_{coex}	$<P_2>_c$
Monte Carlo	[183]	0.507	0.517	9.79	–
y-expansion true B_3	[41]	0.465	0.481	8.107	0.641
y-expansion approx. B_3	[160]	0.420	0.438	5.31	0.568
Integral equations	[25]	0.418	0.436	—	0.657
Decoupling approximation	[184]	0.493	0.494	–	0.017
Decoupling + partial $c^{(3)}$	[185]	0.314	0.335	–	0.547
Structural mapping	[167]	0.472	0.494	7.76	0.561
Structural mapping + decoupling	[186]	0.475	0.494	–	0.547
Weighted density	[173]	0.454	0.474	4.68	0.485
Scaling approach	[181]	0.508	0.517	10.00	0.533

as far as the densities and the pressure are concerned but this might be fortuitous since it does less well at length-to-breadth ratio $x = 2.75$ (although it might be argued that that is in fact a much harder case since we are close to the isotropic-nematic-solid triple point). In the light of the recent results from more extended simulations with larger number of particles at $x = 3$, [187] which indicate that the true transition densities are probably even higher than the ones quoted here, we should be careful to draw definite conclusions about the relative merit of these theories based on the numbers given.

We now turn to spherocylinders where we make a comparison at length-to-breadth ratio $x = 5$ in Table II The same general picture already apparent from the results on the ellipsoids is also evident here. The phenomenology of the transition is well captured, the numbers seem to be of the right order of magnitude but it is unclear whether they represent a *systematic* approximation to the the actual values. We return to these questions in the concluding remarks of this section.

The question of what happens to the isotropic–nematic transition if the particles are no longer uniaxially symmetric has a history dating back to Freiser's observation [189] that one could even imagine the formation

TABLE II

Comparisons of Results for the Isotropic–Nematic Transition of Several Theories with MC Data for the Case of Hard Spherocylinders at $x = 5$

Method	Source Reference	η_{iso}	η_{mem}	P^*_{coex}	$<P_2>_c$
Monte Carlo	[188]	0.40	0.40	4.9	–
Integral equations	[25]	0.325	0.338	–	0.635
Weighted density	[173]	0.38	0.41	2.9	0.59
Scaling	[180]	0.400	0.418	5.36	0.667

of a biaxial nematic phase, that is, a phase where both a major and the minor axes of the molecule are orientationally ordered. In the context of hard particle fluids, the theoretical approaches were at first confined to systems of rectangular slabs that were only allowed a discrete set of orthogonal orientations. [193,190–193]. More progress was made with the analytical solution of the excluded volume problem at arbitrary relative orientations of a biaxial generalization of the spherocylinder, the so-called spheroplatelet, which allows, for example, the Onsager theory to applied to such a system. [55] Using the methods of bifurcation analysis, Mulder [194] was later able to clarify the distinct features of the phasediagram in the Onsager approximation for all hard particles having the same symmetry as rectangular slabs (symmetry group D_{2h}). The most important feature in this phase diagram, depicted schematically in Fig. 4.1, is the occurrence of a so-called Landau bicritical point as one continuously deforms a prolate particle into an oblate particle. This bicritical point is the end point of two

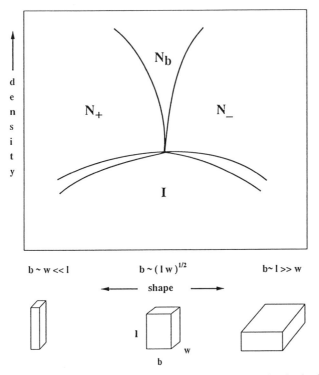

Figure 4.1. Schematic phasediagram for hard biaxial particles. On the horizontal axis the shape of the particle is varied by changing one of its characteristic lengths so that particles ranging from elongated rods to flat disks are obtained

first order transition lines to the rod-like (N_+) and plate-like (N_-) nematic phases formed by the rather prolate particles and rather oblate particles, respectively. The intermediate case is a particle which is neither distinctly rod-like nor platelike and undergoes a second order phase transition to a biaxial nematic (N_b) phase. The theory of Ref. [194] gives an analytical expression for determining this intermediate particle shape in terms of moments of the pair-excluded volume. If we characterize the particle by three distinct dimensions *length* ≤ *breadth* ≤ *width* then the "bicritical" shape is given approximately by the intuitively reasonable result

$$b_{\text{bi-cr}} = \sqrt{lw} \qquad (4.58)$$

that is, the middle dimension is the geometric mean of the outer two. Hołyst and Poniewierski [195] considered the same type of system using their version of the weighted-density approximation as well as an adapted version of the scaling theory of Lee. [180] Although this involves more elaborate free energy functionals than the Onsager approximation, the results for the bicritical shape were nevertheless the same as in Ref. [194], a fact which can be explained by an application of formal methods [196] showing how relatively *in*sensitive results concerning the symmetry properties of phase diagrams are to the type of approximations employed. In fact the predictions of the theory are well borne out by the results of simulations performed by Allen [197] (see Section IV.B.3) on a system of hard biaxial spheroids.

A very recent result concerns the isotropic–cholesteric transition in a HCB fluid. By applying a torsional deformation to a biaxial spheroid, Evans [198] was able to construct a chiral HCB (this construction is explained in Appendix A.C). Employing a version of the Parsons–Lee [180] scaling approach, he was able to determine the pitch of the cholesteric helix which,for a realistic choice of molecular size and chirality, turns out to be a visible wavelength as is also observed in experiments.

b. Nematic-Smectic Transition. The fact that particles with purely repulsive interactions would be able to form a smectic would have been actively dismissed by most workers in the field until just a couple of years ago. Nevertheless, the possibility had been clearly demonstrated by Hosino et al. [199] as early as 1979. They considered a fluid of hard rods with their long axis constrained to point in a chosen preferential direction thus forming a "nematic" by design. Using an approximate method to locate a possible instability of such a phase to a density wave along the preferred direction, they were able to give predictions in the Onsager approximation for the packing fraction at which a smectic phase would set in and the initial wavelength of the density modulation. Later on they showed [200] that the smectic phase even occurred in their model when the rods were allowed to point in three mutually orthogonal directions.

Since the first observation of the actual formation of a smectic phase in a system of parallel hard cylinders by Stroobants et al. [201] most theoretical work has concentrated on sytems of aligned rods. Interestingly enough, one of the simplest HCBs, the ellipsoid of revolution, does not show a smectic phase for the perfectly aligned system. This fact is easily understood on the basis of the observation made by Lebowitz and Perram [176] that a system of aligned ellipsoids can be mapped onto the hard sphere fluid by a global change of scale. Since hard-sphere fluids do not possess a smectic phase neither do the aligned ellipsoids. Theories have therefore concentrated on hard right circular cylinders and spherocylinders. We can compare their results by looking at their predictions for the critical packing fraction η_c and the wavelength λ_c of the smectic density modulation at the transition. In Table III, we collect the results on parallel cylinders which undergo a continuous transition, as is correctly predicted by all theoretical approaches. For hard spherocylinders the situation is somewhat more complex since the results depend on the length-to-width ratio $x = L/D$ of the particles. In the limit that $x \to \infty$, the results should coincide with those on the cylinders since in that limit the influence of the hemi-spherical caps becomes negligible. In Table IV we compare some results for $x = 3$.

By considering oblique cylinders or parallelepipeds, a number of authors have also studied smectic phases where the density wave is not parallel to the direction of molecular alignment thereby modelling the so-called smectic-C phase. [199,208]

A number of authors have also considered the possibility of the formation of a columnar phase in these systems [205,204,207,209] but since

TABLE III

Comparisons of Results for the Nematic–Smectic Transition of Several Theories with MC Data for the Case of Parallel Hard Cylinders. (Note That the Critical Wavelength Scales with the Length L of the Cylinder

Method	Source Reference	η_c	λ_c/L
Monte Carlo	[201]	0.35	1.27
Onsager	[202]	0.58	1.40
Onsager+B_3	[202]	0.36	1.35
Onsager+B_3+B_4	[202]	0.37	1.34
Effective free energy	[203]	0.35	1.28
Weighted density	[204]	0.28	1.40

TABLE IV

Comparisons of Results for the Nematic–Smectic Transition of Several Theories with MC Data for the Case of Parallel Hard Spherocylinders with Length-to-Breadth Ratio $x = 3$

Method	Source Reference	η_c	λ_c/L
Monte Carlo	[201]	0.43	≈ 1.6
Cell model	[205] [1]	≈ 0.5	≈ 1.6
Integral equation	[206]	≈ 0.38	≈ 1.5
Decoupling	[207]	0.36	1.65
Weighted density	[208]	0.42	1.6

[1]This theory predicts a first order N–S transition

the recent simulation results indicate that the stability of this phase is questionable [210] (see Section IV.B.4), we will not discuss this point in detail.

A final result on the nematic–smectic transition in aligned sytems worth mentioning is the recent work by Hołyst. [211] He considers parallel particles built up from a thinner and a thicker cylindrical segment, that is, particles that are not up-down symmetric. In a fluid where half of the particles points up and the other half down, he finds in the Onsager approximation a transition to a smectic phase with a double modulation of the density. The smallest wavelength is due to the tendency of the fluid to form closely coupled alternating layers with the molecules pointing up and down, respectively. These bilayers themselves are at a typical distance from each other given by the second, larger wavelength. This phase is commonly known as a smectic A_d phase.

The problem of the nematic–smectic transition for the case of freely rotating particles has also been considered by several authors: Poniewierski and Hołyst [172,212,213] using the weighted density approximation (see Eq. (4.47)) and Somoza and Tarazona [175,208,214] using their density functional theory with the aligned ellipsoids of revolution fluid as a reference system (see Eq. (4.53)). In both cases, the particles considered were sphero-cylinders. Unfortunately there exists only a single set of simulation data on this system for the case $x = 5$ making quantitative comparisons next to impossible. Nevertheless one notes major differences in the results of the two approaches. A key factor determining these differences is their respective predictions for the location of the tricritical point on the nematic–smectic transition line. For very elongated rods the theory predicts a continuous transition which is consistent with the idea that these systems at the typical N–S transition densities are almost fully orientationally ordered and thus should behave like their perfectly aligned counter-parts. [3] At a finite length-to-width ratio x_{tri} the transition is predicted to become first order. In the Poniewierski–Hołyst theory this critical aspect ratio is predicted to be $x_{tri} = 5.9$ [213] while the Somoza–Tarazona result is $x_{tri} = 50$. The nature of this large discrepancy is at present not under-stood. Moreover, there is no indications from the simulation data at $x = 5$ that the transition is first order. Both theories do agree on the location of the isotropic-nematic-smectic triple point at $x \approx 3$.

Taylor and Herzfeld [216] applied scaled particle theory to a system of hard spheroplatelets with a view on determining the role of particle bi-axiality on the nature and location of the smectic phase. They restricted themselves, however, to discrete orientations for the molecules. New fea-tures in their work are the appearance of a biaxial smectic phase and a near complete suppression of the biaxial nematic phase in favor of the smectic phase.

Finally, Evans [217] studied the question of how the shape of a particle influences the propensity to form smectic phases. He introduces a particle intermediate between the ellipsoid of revolution and the spherocylinder, the "ellipocylinder" which allows one to study a one-parameter family of shapes ranging from the ellipsoid of revolution, which does not form a smectic phase, to a spherocylinders that does. Using the Parsons–Lee "scaling" approach, he determines the phase diagrams of the different sytems. The main result is that the freely rotating "ellipocylinders" do form a smectic phase indicating that the ellipsoids of revolution are indeed a singular case.

c. *Mixtures.* Theoretical work on mixtures of hard particles has been very limited, although most of the theoretical approaches discussed here can in principle be generalized to treat multicomponent systems. Lekkerkerker and co-workers [218,219] considered binary mixtures of rods of different length in the Onsager approximation. The results show a host of new phe-nomena of which we mention (i) fractionation effects; the concentration of the longer rods is enhanced in the nematic phase; (ii) widening of the co-existence region; the density of the isotropic phase at coexistence may be substantially smaller than that of the nematic phase; (iii) re-entrancy; on

---

[3] Recent work by Poniewierski [215] who analyzed the behavior of spherocylinders in the limit $L/D \rightarrow \infty$ shows that this is a rather subtle question. In fact there are non-zero contributions to the free energy due to the rotational freedom of the particles even in the limit. This shows that one can not simply identify the infinitely long rods with the aligned system as had been tacitly assumed by most workers

the increace of density some mixtures can undergo a re-entrant transition from a nematic phase to an isotropic phase.

Another system of interest is a mixture of rod-like and plate-like particles. Alben [220] predicted on the basis of a lattice model that such a mixture might form a biaxial nematic phase at certain compositions. In fact the phase diagram of this mixture as a function of composition should be isomorphic to the one shown in Fig. 4.1. Rabin et al. [221] and Stroobants and Lekkerkerker [219] later considered this question for hard particles in the Onsager approximation and essentially confirmed Alben's predictions. A later paper by Palffy-Muhoray et al. [222], albeit dealing with a mean field description, raised the question whether phase separation between a rod-like and plate-like nematic phase would not be thermodynamically more favourable. To our knowledge, this is still an open question.

d. Two-Dimensional Nematics. Two-dimensional liquid crystals are very different from their three-dimensional counterparts. This can be clearly demonstrated by considering the isotropic–nematic transition in two dimensions. On basis of the Landau theory of phase transitions, we expect that the isotropic nematic transition in two dimensions should be of second order. Density functional theory [223] predicts the same. However, the actual situation is much more subtle than that. The point is that two-dimensional nematics are very similar to the two-dimensional Heisenberg system ("2D-xy model") and hence there is a possibility that topological defects have a pronounced effect on the nature of the phase transitions. [224] In order to see this, we should consider the change in free energy of a two-dimensional nematic, due to distortions of the director field. In fact, in Section V.A.4, we discuss the so-called Frank distortion free energy of a 3D nematic liquid crystal in some detail. Here we simply "borrow" the expression for the distortion free energy of a *3D* nematic (Eq. 5.8) and adapt it to the 2D case under consideration. We choose the average director along the y-axis. We denote the angle between the average director and the instantaneous local director by θ. By analogy to the three-dimensional case, [2] the expression for the deformation free energy is of the form

$$
\begin{aligned}
F_D &= \frac{1}{2}K_1(\partial_x n_x)^2 + \frac{1}{2}K_3(\partial_y n_x)^2 \\
&= \frac{1}{2}K_1(\partial_x \theta)^2 + \frac{1}{2}K_3(\partial_y \theta)^2
\end{aligned}
\tag{4.59}
$$

where K_1 and K_3 are the "splay" and "bend" elastic constants discussed in Section V.A.4. In the last line of Eq. (4.59), we have assumed that θ

is small. We shall simply postulate that the deformation free energy of a 2D nematic is given by Eq. (4.59). Moreover, we shall assume for the sake of convenience that $K_1 = K_3$. In that case, we obtain a very simple expression for the deformation free energy density;

$$F_D = \frac{1}{2} K (\nabla \theta)^2 \qquad (4.60)$$

Using this expression, it is easy to compute the elastic contribution to the free energy of a single π-disclination in a 2D nematic. Consider a circular path (circumference $2\pi r$) around the disclination core. Along this path, the director rotates over an angle π. Hence $(\nabla \theta)^2 = (1/2r)^2$. If we insert this expression in Eq. (4.60) and integrate from the disclination core (radius a_0, say) to L (the linear dimension of the system), then we find that the elastic energy of an isolated disclination is

$$F_{el} = \frac{1}{2} K \int_{a_0}^{L} \frac{2\pi r}{4r^2}\, dr = \frac{\pi K}{4} \log(L/a_0) \qquad (4.61)$$

Clearly, $F_{el} \to \infty$ if $L \to \infty$. This would seem to suggest that no free disclinations are possible in a $2D$ nematic. However, we should also consider the "configurational entropy" of a single disclination, that is, the entropy $k \log \Omega$ associated with the number of distinct ways in which we can place a disclination in a two-dimensional area L^2. If we use a_0 as our unit of length, then the configurational entropy is given by $k \log(L/a_0)^2$ (where we have neglected an additional constant, independent of system size). Combining this expression for the configurational entropy with our expression for the elastic free energy (Eq. 4.61), we obtain the following expression for the total free energy of a single disclination in a 2D nematic:

$$F_{\text{total}} = \left(\frac{\pi K}{4} - 2kT \right) \log(L/a_0) \qquad (4.62)$$

Equation (4.62) suggests that if $kT < (\pi K/8)$, no free disclinations are possible, whereas for $kT > (\pi K/8)$, spontaneous generation of free disclinations may take place. However, if a nematic contains a finite concentration of free disclinations, orientational correlations are destroyed over distances longer than the characteristic separation of the free defects and the resulting phase is an isotropic fluid. This simple version of the Kosterlitz–Thouless scenario for defect mediated phase transitions predicts that the nematic phase cannot be stable above a critical temperature $kT^* = (\pi K/8)$. At that temperature, there is a continuous phase transition (of "infinite" order) from the nematic to the isotropic phase. However, there is an alternative possibility: namely that the I–N transition is simply first order. But if the I–N transition is of first order, then this transition must occur *before* the nematic phase has reached the point where it becomes unstable with respect to the formation of free disclinations; that is, at a first order I–N transition, the following inequality *must* hold:

$$K > \frac{8kT}{\pi}$$

This condition also follows from the more rigorous version of the KT-theory.

Note that our discussion of the disclination-mediated I–N transition was based on the assumed from of the Frank free energy (Eq. 4.59). It should be stressed that this form of the deformation free energy has quite drastic consequences for the nature of orientational order in 2D nematics. In particular, it implies that there exists no true long-ranged orientational order in such systems. We define the ℓth orientational correlation function as

$$g_\ell(r) \equiv < \cos 2\ell(\theta(0) - \theta(r)) >= \text{Re} < \exp[-2i\ell(\theta(0) - \theta(r))] > \quad (4.63)$$

Using the fact that the free energy (Eq. 4.60) is quadratic in $\theta(\mathbf{k})$, it is easy to show that $g_\ell(r)$ has the following form:

$$g_\ell(r) = \left(\frac{r}{a_0}\right)^{-2\ell^2 kT/\pi K} \equiv \left(\frac{r}{a_0}\right)^{-\eta_\ell} \quad (4.64)$$

where the last term on the right-hand side of Eq. (4.64) defines the exponent η_ℓ. Note that this equation implies that, provided that Eq. (4.59) is valid, there is no true long-range orientational order in a 2D nematic, but algebraic or "quasi long-range" order. Similarly, it can be shown that the order parameter $< \cos 2\theta >$ also vanishes algebraically with increasing system size:

$$< \cos 2\theta >\propto \left(\frac{L}{a_0}\right)^{-kT/\pi K} \quad (4.65)$$

Now recall that a 2D nematic is only expected to be stable against the spontaneous generation of free disclinations, when K is larger than $\pi/(8kT)$. Hence, at the K–T transition, the orientational correlation functions and the nematic order parameter must satisfy the following relations:

$$g_\ell(r) = \left(\frac{r}{a_0}\right)^{-\ell^2/4} \quad (4.66)$$

$$< \cos 2\theta >\propto \left(\frac{L}{a_0}\right)^{-1/8} \tag{4.67}$$

Two points should be stressed: (1) if the I–N transition is first order, then at the transition the exponents of g_ℓ and $< \cos 2\theta >$ must be less than the critical values given by Eq. (4.66); (2) the above arguments rest on the assumption that the deformation free energy is of the form given by Eq. (4.60). If this expression is valid, 2D nematics can only have algebraic orientational order. However, it has thus far only been possible to prove the absence of true long-range orientational order for a certain class of short-ranged potentials called *separable*. [225] A pair-potential is called separable if the interaction energy of two molecules at fixed center-of-mass separation $\mathbf{r}_{ij}$ depends only on the relative orientation of the two molecular axes $\mathbf{u}_i \cdot \mathbf{u}_j$, but not on $\mathbf{r}_{ij} \cdot \mathbf{u}_i$ or $\mathbf{r}_{ij} \cdot \mathbf{u}_j$. Realistic pair-potentials are hardly ever separable. We come back to this point in Section IV.B.5.

4. Critical Discussion

After having reviewed the various approaches to the theory of phase transitions in HCB fluids it seems fitting to comment on what has been achieved so far. On the one hand we observe that, more or less, all theories are able to reproduce the phenomenology of the phase transitions in question. It is worth pointing out that one does not need an elaborate density functional to achieve this goal. In fact most of the transitions that occur can be understood on the basis of the Onsager approximation alone. For the class of functionals that do not explicitly contain a three particle contribution, this observation can even be given a formal justification, [196] at least as far as transitions from the isotropic phase are concerned.

As far as the quantitative predictions of the various theories are concerned the situation is less clear. In many cases, reasonable to good results (as compared to simulations) are obtained. However, one is at a loss to understand why a specific choice of functional performs well or not. Moreover it is not evident that one can systematically improve on the proposals given here. The techniques based on series expansions of the free energy functional (generalized Onsager and y-expansion) are liable to suffer from convergence problems even if the difficulties of obtaining higher order contributions was surmounted. Scaled particle theory should in principle be extensible but its quantitative predictions at the level where we are able to use it are not encouraging. The direct correlation function approaches are even harder to generalize. In the case of the integral equations, one first of all has to deal with the unknown influence of the choice of the closure equation. Next, in order to deal with true breaking of symmetry, one should really solve a coupled set of equations involving both the ODF and a correlation function, rather than study the instabilities of the isotropic phase. This has not been attempted so far. The approaches based on the idea of a local weighted density are burdened by the fact that the choice of weighting function is essentially arbitrary and that it is unclear whether the hard-sphere equation of state is indeed the proper reference system for the thermodynamics of nonspherical particles.

At a more general level, one should keep in mind that constructing a density functional also implies certain approximations for the correlation functions. Since density functionals are usually employed to study the thermodynamics of the systems involved, this is not often discussed. It remains to be seen whether the approximate correlation functions bear any resemblance to the real ones for these systems. Probably the theory of HCB fluids still has a long way to go before it achieves the level of simultaneous prediction of both the structure and the equation of state that is obtained in the current theories of the hard-sphere fluid.

B. Simulations

1. Phase Transitions and Free Energy

In order to map the "phase diagram" of a hard-core model system by computer simulation, we must be able to accurately locate all phase transitions. In this section, we section we discuss the special techniques that are required to locate first-order phase transitions.

The most direct way to study first order phase coexistence in a computer simulation would be to simply change the temperature or pressure of the system under study until a phase transformation occurs. In the real world, it is often (although by no means always) possible to ensure that such a phase change takes place reversibly. The coexistence point is defined as the point where the reversible phase transformation occurs. At coexistence, the temperature and pressure of the coexisting phases are equal. In addition, the chemical potential of every individual species α must have the same value in every phase.

$$P_{\mathrm{I}} = P_{\mathrm{II}}, \qquad T_{\mathrm{I}} = T_{\mathrm{II}}, \qquad \mu_{\mathrm{I}}^{\alpha} = \mu_{\mathrm{II}}^{\alpha} \qquad (4.68)$$

Although it is usually not possible to locate a phase transition in a simulation by direct observation of the coexistence of two phases, much progress has been made during the past few years in the simulation of phase coexistence of moderately dense fluid phases. The so-called Gibbs-ensemble method of Ref. [226] relies on the fact that it is possible to satisfy the conditions for coexistence between two bulk phases (or, to be more precise, homogeneous phases with periodic boundary conditions) by allowing them to exchange both volume and molecules. Unfortunately, such a direct simulation method is of limited value in computer simulations of transitions involving dense phases that have some translational order. The reason why the Gibbs-ensemble method breaks down under those circumstances is twofold. First, pronounced hysteresis effects are usually observed in computer simulations of a strong first order phase transition, such as melting. This implies that it is difficult for the molecules in the system to spontaneously rearrange from a configuration belonging to the "old" phase, to one that corresponds to the "new" phase. But even if the two different phases have somehow been prepared, it is usually impossible to exchange particles between them. As a consequence, we cannot ensure the equality of the chemical potential in the two phases.

Under those circumstances, it is still possible to locate the point where the two phases coexist. But in order to do so, we must explicitly compute the chemical potential of the homogeneous phases at the same temperature and pressure and find the point where the two μ's are equal. Below, we describe several techniques that can be used to compute the chemical potential (or, equivalently, the Helmholtz free energy) of particles in dense phases.

a. The Natural Way. When discussing techniques to measure free energies, it is useful to recall how such quantities are measured experimentally. In the real world, free energies cannot be obtained from a single measurement either. What can be measured, however, is the derivative of the free energy with respect to volume V and temperature T:

$$\left(\frac{\partial F}{\partial V}\right)_{NT} = -P \tag{4.69}$$

and

$$\left(\frac{\partial F/T}{\partial 1/T}\right)_{NV} = E \tag{4.70}$$

Here P is the pressure and E the energy of the system. The trick is now to find a reversible path that links the state under consideration to a state of known free energy. The change in F along that path can then simply be evaluated by integration of Eqs. (4.69) and (4.70). In the real world, the free energy of a substance can only be evaluated directly for a very limited number of thermodynamic states. One such state is the ideal gas phase, the other is the perfectly ordered ground state at $T = 0K$. In computer simulations, the situation is quite similar. In order to compute the free energy of a dense liquid, one may construct a reversible path to the very dilute gas phase. It is not really necessary to go all the way to the ideal gas. But at least one should reach a state that is sufficiently dilute that the free energy can be computed accurately, either from knowledge of the first few terms in the virial expansion of the compressibility factor PV/Nk_BT, or that the chemical potential can be computed by other means (see Ref. [227] and Section IV.B.2 below). For the solid, the ideal gas reference state is less useful (although techniques have been developed to construct a reversible path from a dense solid to a dilute (lattice-) gas [228]). The obvious reference state for solids is the harmonic lattice. Computing the absolute free energy of a harmonic solid is relatively straightforward, at least for atomic and simple molecular solids. However, for hard-core models, the crystalline phase is never harmonic. Hence, other techniques are required to compute the free energy of hard-core solids.

b. *Artificial Reversible Paths.* Fortunately, in computer simulations we do not have to rely on the presence of a "natural" reversible path between the phase under study and a reference state of known free energy. If such a path does not exist, we can construct an artificial path (see e.g. Ref. [1]). It works as follows: consider a case where we need to know the free energy $F(V, T)$ of a system with a potential energy function U_1, where U_1 is such that no "natural" reversible path exists to a state of known free energy. Suppose now that we can find another model system with a potential energy function U_0 for which the free energy *can* be computed exactly. Now let us define a generalized potential energy function $U(\lambda)$, such that $U(\lambda = 0) = U_0$ and $U(\lambda = 1) = U_1$. The free energy of a system with this generalized potential is denoted by $F(\lambda)$. Although $F(\lambda)$ itself cannot be measured directly in a simulation, we can measure its derivative with respect to λ:

$$\left(\frac{\partial F}{\partial \lambda}\right)_{NVT\lambda} = \left\langle \frac{\partial U(\lambda)}{\partial \lambda} \right\rangle_{NVT\lambda} \tag{4.71}$$

If the path from $\lambda = 0$ to $\lambda = 1$ is reversible, we can use Eq. (4.71) to compute the desired $F(V, T)$. We simply measure $(< \partial U/\partial \lambda >$ for a number of values of λ between 0 and 1. Typically, 10 quadrature points will be sufficient to get the absolute free energy per particle accurate to within $0.01 k_B T$. However, it is important to select a reasonable reference system. For solids, one of the safest approaches is to choose as a reference system an Einstein crystal with the same structure as the phase under study. [229] This choice of reference system makes it improbable that the path connecting $\lambda = 0$ and $\lambda = 1$ will cross an (irreversible) first order phase transition from the initial structure to another, only to go back to its original structure for still larger values of λ. Nevertheless, it is important that the parametrization of $U(\lambda)$ be chosen carefully. Usually, a linear parametrization (i.e., $U(\lambda) = \lambda U_1 + (1 - \lambda)U_0$) is quite satisfactory. But for hard particles, such a λ-parametrization leads to problems because one cannot continuously switch off a hard-core potential using a *linear* parametrization. In Appendix B.A, we briefly sketch how to compute the Helmholtz free energy of the crystalline state of a system of non-spherical hard particles. More details about such free energy computations can be found in Refs. [81,227,230,231].

2. *Phase Transitions in Liquid Crystals*

Let us consider how the techniques sketched in the previous section can be applied to first order phase transitions in liquid crystals. At the outset, it should be stressed that it is often not trivial to construct a reversible path that will link a liquid-crystalline phase to a state of known free energy. Usually, the liquid-crystalline phase of interest will be separated by first order phase transitions from both the dilute gas and the low temperature (harmonic) solid. In the case of the nematic phase, this problem has been resolved by switching on a strong ordering field. In the presence of such a field, the first order isotropic–nematic transition is suppressed and a reversible expansion to the dilute gas becomes possible. [75] However, this approach has one obvious disadvantage: in order to compute the (very small) difference in the free energy of the isotropic and the nematic phase, we subtract two large numbers. The first is the change in (excess) free energy of the isotropic phase upon compression from the dilute gas to the I–N transition. The second is the change in free energy associated with (**a**) the alignment of the dilute gas in a strong field, (**b**) the compression of this aligned fluid to the density of the nematic phase and (**c**) the switching off of the aligning field at this density. Usually, the free energy change in step (**a**) is evaluated analytically. However, the other steps all require numerical integration. As a consequence, our estimate of difference in free energy of the isotropic and nematic phases is usually rather inaccurate. In this respect, the method is similar to the one used to locate the liquid–vapour transition by integrating around the critical point. It also suggests that the solution of this problem may be similar; rather than constructing a long "physical" integration path, it may be advantageous to construct a short "artificial" path. For instance, it is in principle straightforward to compute the reversible work needed to transform the isotropic phase directly into the nematic phase along a "reaction path" where we constrain the nematic order parameter to take on values intermediate between isotropic and nematic. In a different context, this approach has been explored by van Duijneveldt and Frenkel. [232]

a. Particle Insertion Method. An alternative method that can be used to compute the free energy of a fluid phase (including the nematic phase) is the so-called particle insertion method of Widom. [16] The statistical mechanics that is the basis for this method is quite simple. Consider the definition of the chemical potential μ_α of a species α. From thermodynamics, we know that μ is defined as

$$\mu = \left(\frac{\partial F}{\partial N}\right)_{VT} \tag{4.72}$$

where F is the Helmholtz free energy of the N-particle system. If we express the Helmholtz free energy of an N-particle system in terms of the partition function Q_N (Eq. 2.104), then it is obvious from Eq. (4.72) that, for sufficiently large N the chemical potential is given by $\mu = -kT \ln(Q_{N+1}/Q_N)$. If we use the explicit form (Eq. 2.104) for Q_N, we find

$$\begin{aligned}
\mu &= -kT \ln(Q_{N+1}/Q_N) \\
&= -kT \ln\left(\frac{q(T)V}{N+1}\right) - kT \ln\left(\frac{\int ds^{N+1} \exp(-\beta U(s^{N+1})))}{\int ds^N \exp(-\beta U(s^N))}\right) \\
&= \mu_{\text{id}}(V) + \mu_{\text{ex}}.
\end{aligned} \tag{4.73}$$

In the last line of Eq. (4.73), we have indicated the separation in the ideal gas contribution to the chemical potential, and the excess part. As $\mu_{\text{id}}(V)$ can be evaluated analytically, we focus on μ_{ex}. We now separate the potential energy of the $(N+1)$-particle system into the potential energy function of the N-particle system, $U(s^N)$, and the interaction energy of the $(N+1)$th particle with the rest: $\Delta U \equiv U(s^{N+1}) - U(s^N)$. Using this separation, we can write μ_{ex} as

$$\mu_{\text{ex}} = -kT \ln \left< \int ds_{N+1} \exp(-\beta \Delta U) \right>_N \tag{4.74}$$

where $< \cdots >_N$ denotes canonical ensemble averaging over the configuration space of the N-particle system. The important point to note is that Eq. (4.74) expresses μ_{ex} as an ensemble average that can be sampled by the conventional Metropolis scheme. This last integral can be sampled as follows: we carry out a constant NVT Monte Carlo simulation on the system of N particles. At frequent intervals during this simulation we randomly generate a coordinate s_{N+1}, uniformly over the unit cube. With this value of s_{N+1}, we then compute $\exp(-\beta \Delta U)$. By averaging the latter quantity over all generated trial positions, we obtain the average that appears in Eq. (4.74). So, in effect, we are computing the average of the Boltzmann factor associated with the random insertion of an additional

particle in an N-particle system, but we never accept any such trial inser-
tions, because then we would no longer be sampling the average needed in
Eq. (4.74). The Widom method is a very powerful method to compute the
chemical potential of (not too dense) atomic and simple molecular liquids.
Its main advantage is its great simplicity, and the fact that it can be added
onto an existing constant NVT MC program, without any modifications to
the original sampling scheme; we are simply computing one more thermal
average.

The Widom method was first applied to the evaluation of the free energy
of the isotropic and nematic phases of infinitely thin platelets by Eppenga
and Frenkel. [67] The particle insertion scheme is well suited for the latter
system, as the method works best for strongly anisometric molecules at low
density. In fact, in Ref. [67] it is shown that, in the case of nematics, the
scheme can be made more efficient by inserting particles that are aligned
with the nematic direction. For details, we refer the reader to Ref. [67].

b. Free Energies of Smectic and Columnar Phases. For the calculation of
free energies of smectic and columnar phases, other techniques have to be
used. The situation is simplest if the transition from the nematic to smectic phase is continuous (or, at least, free of hysteresis). In that case, the
"natural" thermodynamic integration of Eq. (4.69) may be used to compute the free energy of the smectic phase, assuming that we know the free
energy of the nematic phase. This approach was, for instance, followed in
Refs. [201,233]. Often, however, the smectic phase cannot be expanded or
compressed reversibly into a phase of known free energy. In such cases,
we should construct an artificial path to a phase of known free energy.
For instance, in the study of the phase behavior of freely rotating spherocylinders, a smectic phase was observed that was separated by first order
phase transitions from both the isotropic and the solid phase. [234] However, in the corresponding system of *parallel* spherocylinders, the smectic
phase could be expanded reversibly to the dilute gas reference state. In
this case, it was possible to compute the free energy of the smectic phase
of freely rotating molecules, by computing the (excess) reversible work
required to align the spherocylinders.

Even if such a convenient reference system is not at hand, there exist
fairly robust schemes to compute the free energy of an arbitrary smectic or,
for that matter, columnar phase. In the case of a smectic, we can exploit
the fact that a smectic phase can be considered as a one-dimensional solid
stacking of 2D fluid (S_A) *or* solid (S_B) layers. Similarly, a columnar phase
resembles a 2D crystal of 1D fluid columns. In Ref. [230], a technique
is described that makes it possible to reversibly decompose such smectic
(columnar) phases into isolated fluid layers (columns). However, this ap-

proach has, to our knowledge, not yet been applied to such liquid crystals. It should be stressed that such absolute free energy calculations need not be repeated for *every* model that we may care to study. For instance, if we have computed the absolute free energy of one state point in the smectic phase of rod-like molecules with an aspect ratio of 5 (say), then we can compute the free energy of the smectic phase of similar molecules with another aspect ratio simply by computing the reversible work needed to change the shape of our model particles from the initial aspect ratio to the desired value. [235] Such an approach has been followed by Allen in his study of biaxial ellipsoids. [197]

c. Alternatives to Free Energy Calculations. Up to this point, we have discussed various techniques that allow us to locate a first order phase transition in a computer simulaion. Sometimes, however, such calculations may result in a less accurate estimate of the phase transition than can be obtained from other criteria. This is of particular relevance in the case of weak first order phase transitions, such as the one separating the isotropic fluid from the nematic phase. In the latter case, the coexistence point is bracketed by a rather narrow pressure (or temperature) range where hysteresis occurs. Clearly, if our free energy calculations result in an inaccuracy in the coexistence pressure that exceeds the range where hysteresis occurs, we might as well have estimated the location of the phase transition directly from the equation of state data. Another, less well founded but quite convenient, criterion to locate the I–N transition is to use the equivalent of the Lindemann rule for the I–N transition. Experimentally, it is known that, at the isotropic–nematic transition, the order–parameter in the nematic phase has a value of 0.35 ± 0.15. It would seem that such an ill-defined "melting rule" cannot possibly provide us with a very accurate estimate of the I–N transition. However, in the vicinity of the I–N transition, the nematic order parameter varies quite steeply. As a consequence, the above rule of thumb usually yields an estimate of the density of the I–N transition that is as least as good as the one obtained by free energy calculations.

3. Results: Spheroids

Computer simulations of hard-core models for two-dimensional liquid crystals were pioneered by Vieillard-Baron in the early 1970s. [5] Vieillard-Baron also made much progress towards the study of three-dimensional model systems, [6] but did not observe spontaneous nematic ordering in 3D. The first systematic simulation study of a three-dimensional hard-core nematogen was performed by Frenkel and Mulder [75] who studied a system of hard ellipsoids of revolution for a number of length-to-width ratios between 1/3 and 3.

   The shape of hard ellipsoids of revolution is characterized by a single parameter, x, the ratio of the length of the major axis $(2a)$ to that of the minor axis $(2b)$: $x = a/b$ (in the literature, both e and x are used to denote the ratio a/b). Prior to the simulations reported in Ref. [75], the phase behavior of hard spheroids was only known for a few special values of x, viz. $x = 1$: hard spheres, which freeze at 66% of close packing. [236] $x \to \infty$: thin hard needles, because this limit is equivalent to the Onsager's model. The latter system has a transition to the nematic phase at vanishing volume fraction. And $x \to 0$: thin hard platelets, which also form a low density nematic. [67] The simulations of Ref. [75] were performed on a system containing $\sim 10^2$ particles and for values of x between 3 and 1/3. In order to locate all phase transitions, the absolute free energy of all phases was computed. Figure 4.2 shows how the stability of the different phases of hard ellipsoids depend on their length-to-width ratio. Four distinct phases can be identified, namely the low density isotropic fluid, an intermediate density nematic liquid-crystalline phase, which is only stable if the length-to-width ratio of the ellipsoids is larger than 2.5 or less than 0.4, and a high density orientationally ordered solid phase. In the case of weakly anisometric ellipsoids, an orientationally disordered solid

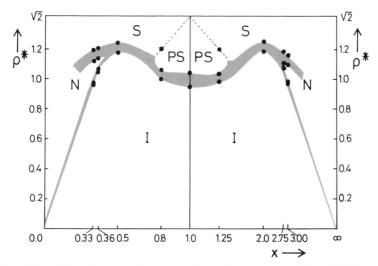

Figure 4.2. "Phase diagram" of a system of hard ellipsoids of revolution. [75] The ratio of the length of the semi-major to the semi-minor axis is denoted by x. The reduced density ρ^* is defined such that the density of regular close packing is equal to $\sqrt{2}$ for all x. The shaded areas indicate two-phase regions associated with a first order phase transition. The following phases can be distinguished: **I**; isotropic fluid, **S**; orientationally ordered crystalline solid. **PS**; orientationally disordered ("plastic") crystal, and **N**; nematic liquid-crystalline phase. The densities of coexisting phases at a first order phase transition (black dots) were computed in a free-energy calculation. Note that no stable nematic is possible for $0.4 < x < 2.5$.

phase was also observed. One thing to note about the phase transitions in the hard-ellipsoid system is that for particles with $3 \geq x \geq (1/3)$ the relative density-jump at the I–N transition is much smaller than for the Onsager model. Typically, the density changes only by some 2% at the I–N transition. Hence the very large density discontinuity at the I–N transition in the Onsager model (more than 20%) is peculiar to long rods and not to hard-core models in general. Recent simulations of Zarragoicoechea et al. [187] indicate that, although the nematic phase is stable in a system of $\mathcal{O}(100)$ ellipsoids with an axial ratio $x = 3$, the nematic phase becomes unstable for systems of 256 ellipsoids. This result is surprising in view of the fact that Allen and Frenkel found an (apparently) stable nematic phase for a system consisting of up to 216 particles. It seems likely that the very weakly first order isotropic nematic transition is more sensitive to finite-size effects than a strong first order transition, such as freezing. In any event, the findings of Ref. [187] suggest that the isotropic–nematic transition line may have a slightly larger slope than indicated in Fig. 4.2. It seems unlikely that this will affect the qualitative features of the phase diagram. However, when we speak about "the" phase diagram of hard ellipsoids of revolution, the reader should bear in mind that the results that we discuss were obtained for system sizes of the order of 100 particles.

Perhaps the most striking feature of the phase diagram in Fig. 4.2 is the near symmetry between the behavior of oblate and prolate ellipsoids with inverse length-to-width ratios. Prolate-oblate ($x \rightarrow 1/x$) symmetry of ellipsoids is to be expected at *low* densities because the second virial coefficient $B_2(x)$ equals $B_2(1/x)$. However, no such relation holds between the third and higher virial coefficients. To give a specific example: in the limit $x \rightarrow \infty$ (the Onsager limit), $B_3/B_2^2 \rightarrow 0$, whereas for $x \rightarrow 0$ (hard platelets [67]) $B_3/B_2^2 \rightarrow 0.4447(3)$. Hence there is no reason to expect any exact symmetry in the phase diagram of hard ellipsoids of revolution. For larger anisometries than studied in the simulations of Ref. [75] one should expect to see asymmetric behavior in the location of the isotropic–nematic transition. In fact, Allen has performed simulations of ellipsoids with aspect ratios 5, 10, 0.2 and 0.1. [237] These simulations show that, as the molecular anisometry increases, the isotropic–nematic transition continues to shift to lower densities. This is to be expected in view of the known limiting behavior of infinitely thin hard platelets and infinitely thin hard rods (see above). However, in Ref. [237] the exact location of the isotropic–nematic transition is not computed.

Even though we expect to see appreciable prolate-oblate asymmetry in the location of the isotropic–nematic transition for highly anisometric spheroids, it is doubtful if the near symmetry of the melting line will be much affected. Strongly aligned rods and platelets follow the same

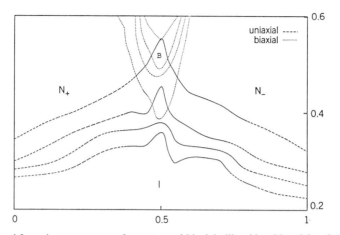

Figure 4.3. $\rho/\rho_{\text{close-packed}}$ of a system of biaxial ellipsoids with axial ratio $c/a = 10$, while b/a varies between 1 (prolate limit) and 10 (oblate limit). In this figure, b/a is represented on a logarithmic scale with base 10. The dashed curves separate state points belonging to the different phases: I denotes the isotropic phase, N_+ the "rod-like" nematic phase, N_- the "platelike" nematic phase and B the biaxial phase. The drawn curves connect the measured state-point data within one phase at a given reduced density . Note that the biaxial phase ends in a Landau bi-critical point at an aspect ratio $a/b \approx \sqrt{10}$. This behavior is in agreement with the theoretical predictions of Mulder. [194]

equation of state $(P = 3\rho)$ and a simple estimate of the melting point of very anisometric ellipsoids [238] suggests that in the limit $x \to \infty$, the symmetry between oblate and prolate ellipsoids is still present.

More recently, Allen has studied the effect of molecular biaxiality on the mesogenic properties of hard ellipsoids. [197] In particular, Allen studied the nature of the liquid-crystalline phase as a prolate spheroid was made increasingly biaxial and was finally transformed into an oblate spheroid. In this case it was found that the rod-like and plate-like nematic phases are separated by a biaxial phase. Figure 4.3 shows how the stability of the different liquid-crystalline phases depends on the molecular biaxiality.

4. Beyond Nematics

The existence of a nematic phase in a system of hard-core molecules is not surprising. In fact, the earliest analysis of any statistical-mechanical model for a liquid-crystalline phase, that is, Onsager's study of a system of thin hard rods, shows that this simple hard-core system *must* form a nematic phase at sufficiently high density. It would, of course, be nice if there existed something like the Onsager model for smectics: an exactly solvable model system that exhibits a transition to the smectic-A phase. Unfortunately, no such model is known. Hence the only way to test approximate "molecular" theories of the smectic phase is to compare with computer simulations. In the spirit of Section IV.B.3 we look for the simplest possible model that will form a smectic phase. In the case of nematics, convex hard-core models were the natural candidates because these constituted the natural generalization of the Onsager model. However, for smectics it is not obvious that hard-core models will work. In fact, in the existing textbooks on liquid-crystal physics the possibility of a hard-core smectic is not even discussed. The only presimulation article discussing the possibility of hard-core smectics is a paper by Hosino et al. [239] The "traditional" approach was to ascribe smectic ordering to attractive interactions between the molecular cores or, alternatively, to the change in packing entropy of the flexible tails of the mesogenic units. [240–243]

a. Parallel Molecules. Whereas essentially *any* fluid of sufficiently nonspherical convex hard bodies will form a nematic phase, nonsphericity alone is not enough to from a smectic phase. This is best demonstrated by the following simple example. We know from experiment that in many smectic phases, the orientational order parameter $S \approx 1$. Let us therefore first consider the possibility of forming a smectic phase in a fluid of *perfectly aligned* molecules ($S = 1$). We know that sufficiently nonspherical hard ellipsoids can form a nematic phase (see Section IV.B.3). It is natural to ask whether a perfectly aligned nematic of hard ellipsoids can transform into a smectic phase. The answer to this question is *no*. The reason is quite simple. Consider a fluid of ellipsoids with length-to-width ratio a/b, all aligned along the z-axis (say). Now we perform an affine transformation that transforms all z coordinates into coordinates z', such that $z' = (b/a)z$. At the same time we transform to new momenta in the z-direction: $p'_z = (a/b)p_z$. Clearly, this transformation does not change the partition function of the system, and hence all thermodynamic properties of the system are unchanged. However, the effect of this affine transformation is to change a fluid of parallel ellipsoids into a system of hard spheres. But, as far as we know, hard spheres can only exist in two phases; fluid and crystal. Hence parallel ellipsoids can only occur in the (nematic) fluid phase and in the crystalline solid phase. In particular, no smectic phase is possible. This makes it extremely improbable that a fluid of nonparallel ellipsoids will form a stable smectic. Such a phase is only expected in the unlikely case that the orientational fluctuations would *stabilize* smectic order. This example demonstrates that we should be careful in selecting possible models for a hard-core smectics. Surprisingly (and luckily) it turned out that another very simple hard-core model system, namely a system of parallel hard spherocylinders, does form a smectic phase. [201,233] A stable smectic phase is possible for length-to-width ratios $L/D \geq 0.5$. In addition, we find that another phase appears at high densities and larger L/D values. In small systems, the phase appeared to be columnar, [201] but in larger systems the range of stability of the columnar phase shrinks (see Fig. 4.4) and is largely replaced by a hexagonal solid phase. In order to tell whether the latter phase is indeed truly solid or, for example, smectic-B, would require simulations on systems that contain many more particles than the 1000–2000 that we were thus far able to study systematically. It should be noted that, although the present evidence suggests that there may not be a columnar phase in a system of *pure* parallel spherocylinders, very recent work of Stroobants [244] indicates that, in a binary mixture of parallel spherocylinders of unequal length, the columnar phase reappears.

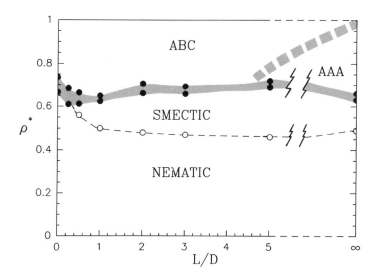

Figure 4.4. Schematic "phase diagram" of a system of parallel hard spherocylinders as obtained by computer simulation. [210] The abcissa indicates the length-to-width ratio L/D. The ordinate measures the density referred to the density at regular close packing.The dashed area indicates the two-phase region at the first order freezing transition. For $L/D < 5$, the solid consists of "ABC"-stacked triangular planes. For larger values of L/D, we find evidence for a hexagonal ("AAA") stacking of the molecules (i.e., triangular lattices stacked on top of one another. At very high densities and large L/D values we find a pocket where the system appears to form a columnar phase. However, the range of stability of this phase is strongly dependent on the size of the system studied. Although we still observed this columnar phase for a system of 1080 particles, it is conceivable that this phase will disappear altogether in the thermodynamic limit. The dashed curve indicates the nematic–smectic transition.

b. The Effect of Rotation. Of course, a model system consisting of *parallel* spherocylinders is rather unphysical. It is therefore of considerable interest to know if a system of freely rotating hard-core molecules can form a smectic phase. This question is of some practical interest, in view of the experimental evidence that smectic [245] and columnar [246] ordering may take place in concentrated solutions of rod-like DNA molecules.

Simulations of a system of freely rotating spherocylinders with length-to-width ratio $L/D = 5$ [188,247] revealed the presence of a stable smectic phase, in addition to the expected isotropic, nematic and solid phases. This work was recently extended to other aspect ratios by Veerman and Frenkel. [234] These authors show that the smectic phase disappears at $L/D = 3$. At this aspect ratio, the nematic phase is no longer (meta)-stable. Figure 4.5 shows a tentative phase diagram of freely rotating hard spherocylinders.

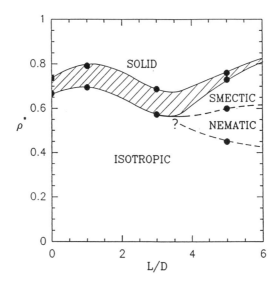

Figure 4.5. Phase diagram of a system of freely rotating spherocylinders as a function of the ratio between the length of the cylindrical part (L) and the diameter (D). The ordinate ρ^* measures the density divided by the density at regular close packing. The grey (dashed) area is the two-phase region separating the densities of the coexisting solid and fluid phases. The black dots indicate computed phase-coexistence points. The nematic-smectic transition is indicated by a dashed curve, as is the isotropic-nematic transition. Although the latter transition is expected to be of first order, the resolution of the current simulations was insufficient to determine the density discontinuity at this transition. The location of the isotropic-nematic-smectic triple point can only be estimated approximately and is indicated in the figure by a question mark.

c. *Columnar Phases* If hard-core models exhibit smectic phases, one may wonder if excluded volume effects can also induce the formation of the even more ordered, columnar phase. In this case, it is natural to look for a convex, plate-like molecule. Oblate ellipsoids are not expected to form columnar phases. The argument why this should be so is essentially the same as the one that "explains" why prolate ellipsoids should not form smectic phases. Rather, we should look for the oblate equivalent of the spherocylinder.

For ellipsoids, the transition from prolate to oblate shapes is controlled by a single parameter (the axial ratio a/b). In contrast, spherocylinders cannot be changed into oblate particles simply by changing L/D (unless we allow for the possibility of negative L/D). It turns out that a particularly convenient "oblate spherocylinder" model is the so-called *truncated sphere* [248] discussed in Section II.D.1

At high densities, truncated spheres can be stacked in a regular close-packed lattice. The volume fraction at regular close packing is

$$\eta_{cp} = (\pi/6)\sqrt{3 - (L/D)^2}$$

Note that for $L/D = 1$ (hard spheres), this reduces to the well-known hard-sphere result $\eta_{cp} = \pi/\sqrt{18}$. For $L/D \to 0$ (flat, cylindrical platelets), we obtain the 2D hard-disk value $\eta_{cp} = \pi/\sqrt{12}$.

Veerman and Frenkel [81] have reported Monte Carlo simulations on a system of truncated spheres with L/D=0.1, 0.2 and 0.3, over a range of densities between dilute gas and crystalline solid [81,248]. Surprisingly, it turned out that all three model systems behaved completely differently.

For the system with $L/D = 0.1$, it was observed that the system spontaneously ordered to form a nematic phase at a reduced density of 0.335 (i.e., at 33.5% of regular close packing). At a density corresponding to 49% of regular close packing, this nematic phase undergoes a strong first-order transition to a columnar phase (at a reduced density $\rho^* = 0.534$. The columnar–crystalline transition occurs at much higher density ($\rho^* > 0.80$). Surprisingly, the columnar phase is not present in systems of *aligned* truncated spheres. [249]

Next, we turn to the system of platelets with $L/D = 0.2$. At first sight, the behavior of this system looks quite similar to that observed for the thinner platelets. In particular, the equation of state for truncated spheres with $L/D = 0.2$ looks similar to the one corresponding to $L/D = 0.1$. However, we do not find a nematic phase in the dense fluid, close to freezing. In particular, if we measure the orientational correlation function $g_2(r) \equiv\ < P_2(\mathbf{u}(0) \cdot \mathbf{u}(r)) >$ of the dense fluid, we find that it decays to zero within one molecular diameter, even at the highest densities of the "fluid" branch (see Fig. 4.6).   In a nematic phase, $g_2(r)$ should tend to a finite limit: $g_2(r) \to S^2$ as $r \to \infty$, where S is the nematic order parameter. It should be stressed that the absence of nematic order in the $L/D = 0.2$ system is not a consequence of the way in which the system was prepared. Even if we started with a configuration at a reduced density $\rho/\rho_{cp} = 0.50$ with all the molecules initially aligned, the nematic order would rapidly dissapear. In other words, at that density the nematic phase is *mechanically* unstable. If we had to base our analysis exclusively on Fig. 4.6, we would have concluded that truncated spheres with $L/D = 0.2$ freeze from the *isotropic* phase.

The surprise comes when we consider the higher order orientational correlation function $g_4(r) \equiv < P_4(\mathbf{u}(0) \cdot \mathbf{u}(r)) >$. Usually, when $g_2(r)$ is short-ranged, the same holds a fortiori for $g_4(r)$. However, Fig. 4.7 shows that for densities $\rho/\rho_{cp} > 0.55$, $g_2(r)$ is much longer ranged than $g_2(r)$. This suggests that the system has a strong tendency towards orientational order with *cubic* symmetry ("cubatic", not to be confused with cubic, which refers to a system that also has *translational* order).   In computer simulations one should always be very suspicious of any spontaneous order-

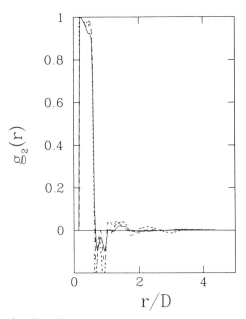

Figure 4.6. Density dependence of the orientational correlation function $g_2(r) \equiv < P_2(\mathbf{u}(0) \cdot \mathbf{u}(r)) >$ in a system of hard truncated spheres with a length-to-width ratio $L/D = 0.2$. Drawn curve; $\rho/\rho_{cp} = 0.51$; long dashes; $\rho/\rho_{cp} = 0.57$, short dashes; $\rho/\rho_{cp} = 0.63$. Note that even at the highest densities studied, $g_2(r)$ is short ranged.

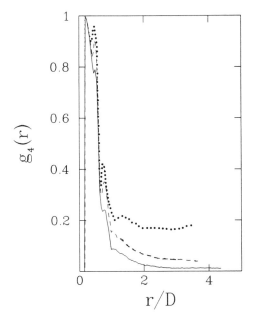

Figure 4.7. Density dependence of the orientational correlation function $g_4(r)$ $\equiv<$ $P_4(\mathbf{u}(0)\cdot\mathbf{u}(r))>$ in a system of hard truncated spheres with a length-to-width ratio $L/D=0.2$. Symbols as in Fig. 4.6. Note that, unlike $g_2(r)$, $g_4(r)$ appears to be long-ranged at high densities.

ing with cubic symmetry, because such ordering could be induced by the (cubic) periodic boundary conditions. In order to test if the boundary conditions were responsible for the cubatic order, we did a number of long simulations with systems of up to 2048 particles. These simulations strongly suggest that the onset of cubatic orientational order is not an artifact of the boundary conditions. Another indication that the boundary conditions are not the cause of the observed ordering is that still higher order correlations (g_6 and g_8) that could also be induced by the periodic boundaries, are in fact rapidly decaying functions of r. If we make a log–log plot of $g_4(r)$ in the large system for several densities between $\rho = 0.51$ and $\rho = 0.63$ (see Fig. 4.8), it appears that the cubatic order is not truly long-ranged but quasi-long-ranged, that is, $g_4(r) \sim r^{-\eta}$, where η depends on the density ρ. This observation should, however, be taken with a large grain of salt, as the range over which linear behavior in the log–log plot is observed corresponds to less than one decade in r.

Finally, for truncated spheres with $L/D=0.3$, both the nematic and the "cubatic" phase are absent. We have summarized our knowledge of the phase behavior of the truncated-sphere system [81] in a tentative phase

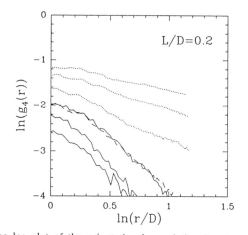

Figure 4.8. Log–log plot of the orientational correlation function $g_4(r) \equiv< P_4(\mathbf{u}(0) \cdot \mathbf{u}(r)) >$ in a system of hard truncated spheres with a length-to-width ratio $L/D=0.2$ as a function of density. With increasing density, the amplitude of this correlation function goes up. The lowest curve corresponds to $\rho^* = 0.51$. The following drawn curves to $\rho^* = 0.54$ and $\rho^* = 0.56$. The long-dashed curve corresponds to $\rho^* = 0.57$. At higher densities $\rho^*=0.58$, 0.60 and 0.63 (long-dashed curves), $g_4(r)$ appears to decay algebraically over the narrow range of distances $(1 < r/D < 3.2)$ where we could observe monotonic decay of $g_4(r)$.

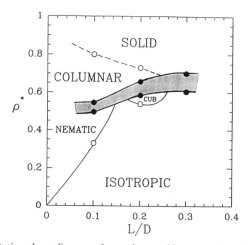

Figure 4.9. Tentative phase diagram of cut spheres with an aspect ratio L/D between 0.1 and 0.3. The high density solid/columnar phase is separated by a first order phase transition from the low density fluid phase. The densities of the coexisting phases are indicated by black dots. The open circles indicate the approximate location of phase transitions that have been estimated using techniques other than free energy calculations. The isotropic–nematic transition is weakly first order. The same appears to be the case with the isotropic–cubatic transition. In our simulations, the transition between the solid and the columnar phases appeared to be continuous.

diagram (Fig. 4.9). In fact, we know a few features of this phase diagram for both larger and smaller values of L/D. In particular, we know that for $L/D=1$ (hard spheres), the two-phase region is located between $\rho^*=0.67$ and $\rho^*=0.74$. For $L/D \rightarrow 0$, the isotropic–nematic transition occurs at $\rho^*=0$.

5. Simulation of 2D Nematics

In Section IV.A.3, we discussed the peculiar nature of the nematic phase in two-dimensional systems. The discussion in that section suggests that there are two obvious questions about 2D nematics that one could try to answer by simulation. (1) If the pair potential is nonseparable, do we find algebraic or true long-range order? (2) If we find algebraic order, do we observe a first order I–N transition or a continuous one of the Kosterlitz–Thouless type. For the first question, a good starting point would be to choose a pair potential that is as nonseparable as possible. An obvious

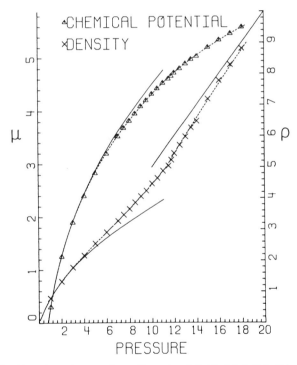

Figure 4.10. Equation of state of two-dimensional fluid of infinitely thin hard needles of length 1. Note that in this figure the reduced pressure is the independent variable. The reduced density (ρL^2) is indicated by crosses, the chemical potential μ by triangles. The drawn curves at low pressure were computed using a 5-term virial series.

candidate is a two-dimensional model of infinitely thin hard needles, [250] that is, a two-dimensional hard-body fluid. This pair potential is very non-separable in the sense that, at fixed $|r_{ij}|$ and fixed $\mathbf{u}_i.\mathbf{u}_j$ the pair potential is *not* constant, but may vary between 0 and ∞. The equation of state of this system is shown in Fig. 4.10.   According to the bifurcation analysis of the corresponding Onsager limit, [251] a second order isotropic–nematic transition is expected at a density $\rho L^2 = (3\pi/2) = 4.712\cdots$ and a pressure$PL^2 = 11.78\cdots$. At first sight this seems to be quite a reasonable estimate of the I–N transition, because very close to this point the equation of state appears to exhibit a change of slope. However, analysis of the long-range behavior of the orientational correlation functions and of the system-size dependence of the order parameter $< \cos 2\theta >$ indicate that the higher density phase is not a stable nematic. The orientational correlation functions decay either exponentially (see Fig. 4.11) or with an

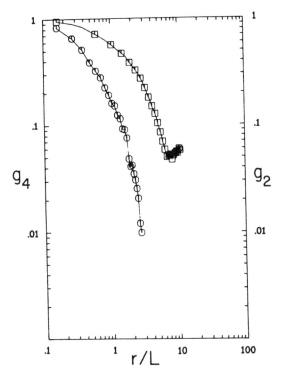

Figure 4.11. Orientational correlation functions $g_2(r) \equiv < \cos 2(\phi(0) - \phi(r)) >$ and $g_4(r) \equiv < \cos 4(\phi(0) - \phi(r)) >$ for a two-dimensional system of hard needles of length L=1. This figure shows that at a reduced density ρL^2=6.75, the orientational order decays exponentially. In other words: the phase is isotropic.

apparent algebraic exponent that is larger than the critical value given in Eq. (4.66). Only at a density that is almost twice the Onsager transition point does the observed behavior conform to what is expected for a stable nematic with algebraic order (see Fig. 4.12).   However, at this density, the equation of state is completely featureless. Such behavior is to be expected of the I–N transition and is in fact of the K–T type.

Subsequently, Cuesta and Frenkel [223] have studied the isotropic to nematic transition in a system of 2D hard ellipses with aspect ratios 2, 4 and 6. It is found that in all cases where a stable nematic phase is found (aspect ratios 4 and 6), this phase exhibits algebraic orientational order. However, whereas the I–N transition appears to be of the K–T type for aspect ratio 6 (and larger), the transition is found to be of first order

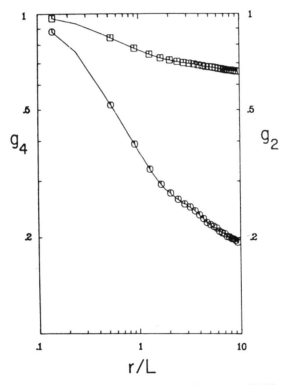

Figure 4.12. Orientational correlation functions $g_2(r) \equiv < \cos 2(\phi(0) - \phi(r)) >$ and $g_4(r) \equiv < \cos 4(\phi(0) - \phi(r)) >$ for a two-dimensional system of hard needles of length $L=1$. This figure shows that at a reduced density $\rho L^2 = 8.75$, the orientational order decays algebraically. From the values of the algebraic exponents η_2 and η_4 the effective Frank elastic constant can be computed. At $\rho = 8.75$, this Frank constant is large enough to make the 2D nematic stable with respect to disclination unbinding.

for aspect ratio 4. This implies that in the latter case, the 2D nematic undergoes a first-order transition *before* it has reached the point where it becomes absolutely unstable with respect to diclination unbinding. A very puzzling feature is the nature of the high density phase of 2D ellipses. A snapshot of such a phase is shown in Fig. 4.13.   This phase does not appear to have true crystalline order, nor for that matter, smectic order.

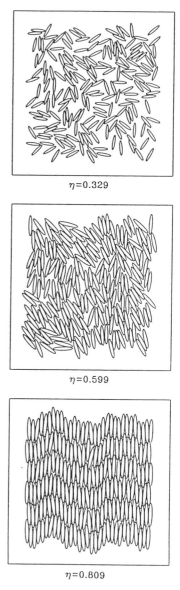

$\eta=0.329$

$\eta=0.599$

Figure 4.13. Snapshots of typical configurations of a system of hard ellipses in the isotropic phase ($\eta = 0.329$), near the estimated estimated isotropic–nematic transition ($\eta=0.599$) and in the high density phase ($\eta = 0.809$). Although the latter phase exhibits local solid-like ordering, it is not a true (two-dimensional) solid.

$\eta=0.809$

Yet it does clearly have a large amount of local order. The precise nature of this high density phase is currently under investigation.

V. STATIC PROPERTIES

A. Theory

This sections deals with the static properties of HCBs in the liquid-crystalline phase. Most of the scalar properties like the equation of state and the "vector" properties like the various correlation functions, that were discussed in Section I, are of course also relevant in this case. Only their definitions will have to adjusted to reflect the presence of long-range orientational and (partial) positional order that occurs. There are, however, also a few static properties that are unique to the liquid crystalline state, that is, the order parameters that describe the nature and degree of order, and the elastic constants that describe the response of the system to long wavelength distortions of the initially uniform field of preferred local orientations. Since it is basic to the whole discussion, we start by discussing the structure of the ODF in the LC phase, which leads us to the definitions of the order parameters. The other properties are then dealt with in turn.

1. Order Parameters

The ODF (one particle distribution function) $\rho^{(1)}$ was introduced in Section III, but its explicit form was left undetermined. This makes sense in the context of density functional theory where it plays the role of the basic variable as a single unit. In practice, however, we need to describe the ODF in more detail, even if only to be able to approximate it numerically. The most natural way to obtain such a description is to expand it into a complete set of functions that are in fact the irreducible representations of the original, unbroken symmetry group of the low density phase. In our case this is the group $G_0 = T \otimes O_3$ of the translation group T and the full three-dimensional orthogonal group O_3. The basis functions in this expansion are just products of plane waves and irreducible rotation-group matrix elements. [252] A side product of this approach is the fact that the expansion coefficients will turn out to be order parameters not only in the intuitive sense that they will be zero in the disordered phase and nonzero in the ordered one but also in the formal sense of the Landau theory of phase transitions, that is, they themselves transform as irreducible basis sets of the unbroken symmetry group. [253] In the most general case, we can thus write

$$\rho(\mathbf{r}, \Omega) = \frac{1}{V} \sum_{\mathbf{k}}{}' \sum_{\ell,m,n} \frac{(2\ell + 1)}{8\pi^2} a_{\mathbf{k},\ell,m,n} e^{i\mathbf{k}\cdot\mathbf{r}} D^\ell_{m,n}(\Omega) \qquad (5.1)$$

where the prime on the first summation indicates a restriction to **k**-vectors in the first Brillioun zone.

For the typical system studied this expansion expansion reduces to a much simpler one. Consider, for example, a nematic phase formed by uniaxially symmetric molecules with inversion symmetry. In this case, we have no dependence on position and dependence on the orientation only through the angle θ the major molecular axis makes with the z-axis of an arbitrary reference frame. A systematic way of deducing the correct expansion is then to average Eq. (5.1) over the extra symmetries introduced. For the example of the nematic, we thus average over the system volume (no positional order) , the azimuthal angle ϕ of the lab frame (uniaxial symmetry of the phase), the angle ψ around the molecular symmetry axis and the inversion operation applied to the molecular frame. The result is the expansion

$$\rho(\mathbf{r}, \Omega)_{\text{nem}} \equiv nf(\theta) = n \sum_{\ell=\text{even}} \frac{(2\ell + 1)}{2} a_\ell P_\ell(\cos \theta) \tag{5.2}$$

The second expansion coefficient

$$a_2 = \langle P_2 \rangle = \int d \cos \theta f(\theta) P_2(\theta) \tag{5.3}$$

is, of course, the well known Maier–Saupe order parameter. [254]

In some cases it is preferable to work with the basis set generated by the so-called irreducible Cartesian tensors [255] rather than the rotation matrices. The uniaxial component of the second rank irreducible tensor is known as the de Gennes order parameter

$$\mathbf{S} = \frac{1}{2} (3\hat{\mathbf{n}} \otimes \hat{\mathbf{n}} - \mathbf{1}) \tag{5.4}$$

The unit vector $\hat{\mathbf{n}}$ is the so-called nematic director, which is actually defined up to a sign, reflecting the inversion symmetry of the phase. In a frame where **S** is diagonal, that is, the director is along the z-axis of the lab frame, the component S_{zz} is actually equal to the Maier–Saupe order parameter.

2. Correlation Functions

The usual method of the defining equilibrium distribution functions in a disordered fluid phase uses the explicit representation of the full N-particle distribution defined in terms of the system Hamiltonian. In a phase with any kind of long-range order, this method breaks down, since it cannot deal with the occurrence of spontaneous order which is an effect associated with the thermodynamic limit, $N \to \infty$, where the finite N

representation is meaningless. As mentioned in the previous chapter, these difficulties can be bypassed using the density functional formalism. The various correlation functions are then defined through their relation to the direct correlation functions which are obtained directly from the density functional itself. Recalling Eq. (4.5) from Section IV, we find for the two-particle direct correlation function

$$c^{(2)}(\xi_1, \xi_2) = \frac{\delta^2 \Phi[\rho^{(1)}]}{\delta\rho^{(1)}(\xi_1)\delta\rho^{(1)}(\xi_2)} \tag{5.5}$$

where we have used the shorthand $\xi = (\mathbf{r}, \Omega)$. From here, we can go to the usual two-particle distribution function $\rho^{(2)}$ via the generalized Ornstein–Zernike equation [152]

$$\rho^{(2)}(\xi_1, \xi_2) - \rho^{(1)}(\xi_1)\rho^{(1)}(\xi_2) = \rho^{(1)}(\xi_1)\rho^{(1)}(\xi_2)c^{(2)}(\xi_1, \xi_2) \tag{5.6}$$

$$+\rho^{(1)}(\xi_2) \int d\xi_3 \left(\rho^{(2)}(\xi_1, \xi_3) - \rho^{(1)}(\xi_1)\rho^{(1)}(\xi_3) \right) \rho^{(1)}(\xi_3)c^{(2)}(\xi_3, \xi_2)$$

3. Equation of State

In the context of density functional theory, the equation of state is probably best obtained directly from its definition in terms of the equilibrium free energy.

$$P = -\left(\frac{\partial F}{\partial V}\right)_{N,T}, \qquad F = \mathcal{F}[\rho^{(1)}{}_{eq}] \tag{5.7}$$

Lacking any substantial theory for the correlation function themselves, the traditional virial (cf. Eq. 2.10) and compressibility relations (cf. Eq. 2.27) are of limited use in an ordered phase.

4. Frank Elastic Constants

We consider here the effect of long wavelength distortions of the local preferred direction in a nematic liquid. Such a distortion is most conveniently described by introducing the director field $\hat{\mathbf{n}}(\mathbf{r})$ specifying the *direction* of the local preferred molecular orientation with respect to a fixed lab frame. The derivation of the macroscopic free energy of distortion is then an exercise in the construction of the relevant second order rotational invariants constructed from $\hat{\mathbf{n}}(\mathbf{r})$ and its derivatives. We quote here the well known result[2]

$$F_d = \tfrac{1}{2}K_1(\nabla \cdot \hat{\mathbf{n}})^2 + \tfrac{1}{2}K_2(\hat{\mathbf{n}} \cdot \nabla \wedge \hat{\mathbf{n}})^2 + \tfrac{1}{2}K_3(\hat{\mathbf{n}} \wedge (\nabla \wedge \hat{\mathbf{n}}))^2 \tag{5.8}$$

which introduces the Frank elastic constants: K_1 (splay), K_2 (twist) and K_3 (bend). The distortions of the director field associated with each of the three constants are depicted in Fig. 5.1 It is now the task of statistical mechanics to come up with microscopic expressions for the Frank constants. For hard particles this question was first addressed by Priest [256] and Straley.[257] The first to derive an essentially exact formulation on the basis of density functional theory were Poniewierski and Stecki. [258–260] Here we present a slightly adapted version of their results. [261]

Consider a nematic phase of uniaxially symmetric molecules. The undistorted ODF has the form $\rho^{(1)}(\mathbf{r}, \hat{\mathbf{u}}) = nf(\hat{\mathbf{n}} \cdot \hat{\mathbf{u}})$, where n is the number density, $\hat{\mathbf{n}}$ the (constant) director and $\hat{\mathbf{u}}$ a unit vector along the molecular symmetry axis. We now allow the director to vary spatially with typical wavelengths much larger than the molecular scale. In this case, we can assume that the local orientational distribution function retains the same functional form as in the undistorted bulk, that is, the only thing that varies is the direction of the preferred orientation. In other words, if the orientational distribution in the undistorted bulk is given by $f_0(\hat{\mathbf{n}} \cdot \hat{\mathbf{u}})$, then the position dependent orientational distribution of the distorted system is given by $f(\mathbf{r}, \hat{\mathbf{u}}) = f_0(\hat{\mathbf{n}}(\mathbf{r}) \cdot \hat{\mathbf{u}})$, where $\hat{\mathbf{n}}(\mathbf{r})$ is the director at $\mathbf{r}$. If the distortions are not too large we can functionally expand the free energy functional in terms of the difference $\Delta f(\mathbf{r}, \hat{\mathbf{u}}) = f(\mathbf{r}, \hat{\mathbf{u}}) - f_0(\hat{\mathbf{n}} \cdot \hat{\mathbf{u}})$, yielding to second order

$$\beta \Delta \mathcal{F}[\Delta f] = -\tfrac{1}{2}n^2 \int d\mathbf{r}\, d\hat{\mathbf{u}} \int d\mathbf{r}'\, d\hat{\mathbf{u}}' \Delta f(\mathbf{r}, \hat{\mathbf{u}}) \Delta f(\mathbf{r}', \hat{\mathbf{u}}') c^{(2)}(\mathbf{r}, \hat{\mathbf{u}}, \mathbf{r}', \hat{\mathbf{u}}', [f_0])$$

(5.9)

There are no first order terms in this expansion because the undistorted phase is stable, so that its first order variation with respect to arbitrary changes in the ODF vanishes (cf. Eq. 4.11). As we assumed only long wavelength distortions, we will make a gradient expansion of the difference

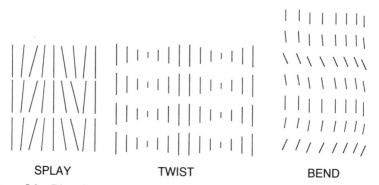

SPLAY   TWIST   BEND

Figure 5.1. Distortions of the director in the nematic phase: from left to right: splay, twist and bend.

in the orientation distribution functions. To this end we first choose a more convenient reference system for the two-particle integration, defining the center-of-mass coordinates $\mathbf{R} = \frac{1}{2}(\mathbf{r} + \mathbf{r}')$ and the relative separation $\Delta\mathbf{r} = \mathbf{r} - \mathbf{r}'$. The distribution function at $\mathbf{r}$ and $\mathbf{r}$ can then be related to the one in the center of mass through the Taylor expansion

$$f(\hat{\mathbf{n}}(\mathbf{R} \pm \tfrac{1}{2}\Delta\mathbf{r}) = f_0(\hat{\mathbf{n}}(\mathbf{R}) \cdot \hat{\mathbf{u}}) \pm \frac{(\Delta\mathbf{r} \cdot \nabla_{\mathbf{R}})}{2} f_0(\hat{\mathbf{n}}(\mathbf{R}) \cdot \hat{\mathbf{u}}) \qquad (5.10)$$

$$+ \frac{1}{2} \frac{(\Delta\mathbf{r} \cdot \nabla_{\mathbf{R}})}{2} \frac{(\Delta\mathbf{r} \cdot \nabla_{\mathbf{R}})}{2} f_0(\hat{\mathbf{n}}(\mathbf{R}) \cdot \hat{\mathbf{u}}) \pm \mathcal{O}(\Delta^3\mathbf{r})$$

Inserting this into Eq. (5.9) and performing a partial integration, we find

$$\beta\Delta\mathcal{F}[\Delta f] = \frac{1}{4}n^2 \int d\mathbf{R} \, d\Delta\mathbf{r} \, d\hat{\mathbf{u}} \, d\hat{\mathbf{u}}' \dot{f}_0(\hat{\mathbf{n}}(\mathbf{R}) \cdot \hat{\mathbf{u}}) \dot{f}_0(\hat{\mathbf{n}}(\mathbf{R}) \cdot \hat{\mathbf{u}}')$$

$$\times \left[(\Delta\mathbf{r} \cdot \nabla_{\mathbf{R}})\hat{\mathbf{n}}(\mathbf{R}) \cdot \hat{\mathbf{u}} \right] \left[(\Delta\mathbf{r} \cdot \nabla_{\mathbf{R}})\hat{\mathbf{n}}(\mathbf{R}) \cdot \hat{\mathbf{u}}' \right] c^{(2)}(\Delta\mathbf{r}, \hat{\mathbf{u}}, \hat{\mathbf{u}}', [f_0]) \quad (5.11)$$

where the dot on the distribution functions denotes differentiation with respect to their single argument, that is, $\hat{\mathbf{n}} \cdot \hat{\mathbf{u}}$.

The elastic constants can now be deduced by imposing infinitesimal distortion patterns corresponding to the three fundamental "modes" discussed earlier

$$\hat{\mathbf{n}}(\mathbf{R}) = \hat{\mathbf{e}}_z + \epsilon X \hat{\mathbf{e}}_x \quad \text{(splay)}$$

$$\hat{\mathbf{n}}(\mathbf{R}) = \hat{\mathbf{e}}_z + \epsilon X \hat{\mathbf{e}}_y \quad \text{(twist)}$$

$$\hat{\mathbf{n}}(\mathbf{R}) = \hat{\mathbf{e}}_z + \epsilon Z \hat{\mathbf{e}}_x \quad \text{(bend)} \qquad (5.12)$$

Inserting these one by one into Eq. (5.11) one can simply read off the relevant constants

$$\beta K_1 = \tfrac{1}{2}Nn \int d\Delta r \, d\hat{\mathbf{u}} \, d\hat{\mathbf{u}}' \dot{f}_0(\hat{\mathbf{n}} \cdot \hat{\mathbf{u}}) \dot{f}_0(\hat{\mathbf{n}} \cdot \hat{\mathbf{u}}')(\Delta x)^2 u_x u'_x c^{(2)}(\Delta\mathbf{r}, \hat{\mathbf{u}}, \hat{\mathbf{u}}', [f_0])$$

$$\beta K_2 = \tfrac{1}{2}Nn \int d\Delta r \, d\hat{\mathbf{u}} \, d\hat{\mathbf{u}}' \dot{f}_0(\hat{\mathbf{n}} \cdot \hat{\mathbf{u}}) \dot{f}_0(\hat{\mathbf{n}} \cdot \hat{\mathbf{u}}')(\Delta x)^2 u_y u'_y c^{(2)}(\Delta\mathbf{r}, \hat{\mathbf{u}}, \hat{\mathbf{u}}', [f_0])$$

$$\beta K_3 = \tfrac{1}{2}Nn \int d\Delta r \, d\hat{\mathbf{u}} \, d\hat{\mathbf{u}}' \dot{f}_0(\hat{\mathbf{n}} \cdot \hat{\mathbf{u}}) \dot{f}_0(\hat{\mathbf{n}} \cdot \hat{\mathbf{u}}')(\Delta z)^2 u_x u'_x c^{(2)}(\Delta\mathbf{r}, \hat{\mathbf{u}}, \hat{\mathbf{u}}', [f_0])$$

$$(5.13)$$

Of course, the major obstacle in the application of such formulas is, yet again, the lack of any hard facts about the direct correlation function of the nematic phase. Explicit predictions have therefore been given mainly for the Onsager approximation, where the direct correlation is at least known (cf. Eq. 4.39). The results of this approximation, which assumes that the direct correlation is independent of both the density and the bulk phase of the system, can at best be of a rather qualitative nature.

Poniewierski and Stecki [260] have also shown how microscopic expressions for the Frank constants can also be derived from hydrodynamic fluctuation theory. [262] This formulation, although no more convenient from the purely theoretical point of view, has the distinct advantage of relating the elastic constants to properties more easily measurable in simulations. In order to give their results we need to define the Ursell function $U^{(2)}$

$$U^{(2)}(\mathbf{r}, \hat{\mathbf{u}}, \mathbf{r}', \hat{\mathbf{u}}') \equiv \rho^{(2)}(\mathbf{r}, \hat{\mathbf{u}}, \mathbf{r}', \hat{\mathbf{u}}') - \rho^{(1)}(\mathbf{r}, \hat{\mathbf{u}})\rho^{(1)}(\mathbf{r}', \hat{\mathbf{u}}') \quad (5.14)$$

Note that in the nematic phase this quantity is translationally invariant. The Frank constants are then given by

$$\frac{n < P_2 >}{\beta K_1} = \lim_{q_x \to 0} \lim_{q_z \to 0} q_x^2 \int d\hat{\mathbf{u}}\, d\hat{\mathbf{u}}'\, u_x u_x' u_z u_z' \hat{U}^{(2)}(\mathbf{q}, \hat{\mathbf{u}}, \hat{\mathbf{u}}')$$

$$\frac{n < P_2 >}{\beta K_2} = \lim_{q_y \to 0} \lim_{q_z \to 0} q_y^2 \int d\hat{\mathbf{u}}\, d\hat{\mathbf{u}}'\, u_y u_y' u_z u_z' \hat{U}^{(2)}(\mathbf{q}, \hat{\mathbf{u}}, \hat{\mathbf{u}}')$$

$$\frac{n < P_2 >}{\beta K_3} = \lim_{q_z \to 0} \lim_{q_x \to 0} q_z^2 \int d\hat{\mathbf{u}}\, d\hat{\mathbf{u}}'\, u_x u_x' u_z u_z' \hat{U}^{(2)}(\mathbf{q}, \hat{\mathbf{u}}, \hat{\mathbf{u}}') \quad (5.15)$$

B. Simulations

The detection of different kinds of orientational and translational order in a computer simulation requires special care. For instance, the structural information that is contained in the familiar radial distribution function, $g(r)$, is insufficient to distinguish between different kinds of (liquid-) crystalline ordering. Hence, other functions that probe the relevant forms of translational and orientational order must be introduced. Below, we describe the different structural probes that can be used to probe (liquid-) crystalline order in computer simulations.

1. Orientational Order Parameters

In an isotropic molecular liquid, the one-particle distribution function ρ_1 is a constant. In an ordered system, such as a liquid crystal, ρ_1 depends on the molecular orientation Ω and possibly also on the center of mass coordinate $\mathbf{r}$. In a nematic, ρ_1 is a function of Ω only.

Although knowledge of ρ_1 suffices to determine the nature and degree of ordering in a liquid crystal, it is often convenient to be able to quantify the liquid-crystalline order with a few numbers rather than with a continuous function. The quantities a_ℓ with $\ell \geq 2$, defined in Section V.A.1 can be used as a measure of the nematic order. Often the quantity $a_2 \equiv < P_2(\cos \theta) >$ is referred to as *the* nematic order parameter, and is denoted by S. Similarly, it is possible to define order parameters that quantify smectic and columnar order by expanding the spatial variation of the single particle density function ρ_1 in Fourier components. The amplitude of the Fourier component with the lowest nonzero wave vector that is commensurate with the periodicity of the density modulation is a measure for the smectic (columnar) order.

From a computational point of view, the definition of the nematic order parameters in Eq. (5.2) is not entirely satisfactory for the following reason: a_ℓ is defined as the average of $P_\ell(\cos\theta)$, where θ is the angle between the molecular orientation vector $\hat{\mathbf{u}}$ (for convenience we consider axially symmetric molecules) and the nematic director $\hat{\mathbf{n}}$. But the director is defined as the average alignment direction of the molecules in a nematic. In the absence of external forces the direction of $\hat{\mathbf{n}}$ is not known a priori. Hence Eq. (5.2) suggests that in order to measure the order parameter, we should first determine the preferred alignment (i.e., we should already know if the sample is in the nematic phase) and only then can we measure the degree of nematic order.

Fortunately, it is possible to give a definition of S that does not presuppose knowledge of $\hat{\mathbf{n}}$. To see this, consider the expression for $< P_2(\hat{\mathbf{u}} \cdot \hat{\mathbf{e}}) >$, where $\hat{\mathbf{e}}$ is an arbitrary unit vector:

$$< P_2(\hat{\mathbf{u}} \cdot \hat{\mathbf{e}}) > = \frac{1}{N} \sum_{i=1}^{N} \hat{\mathbf{e}} \cdot \left(\frac{3\hat{\mathbf{u}}^i \hat{\mathbf{u}}^i - \mathbf{I}}{2}\right) \cdot \hat{\mathbf{e}} \qquad (5.16)$$

$$\equiv \hat{\mathbf{e}} \cdot \mathbf{Q} \cdot \hat{\mathbf{e}}$$

where $\mathbf{I}$ is the second-rank unit tensor and the last line of Eq. (5.16) defines the tensor order parameter $\mathbf{Q}$. $\mathbf{Q}$ is a traceless, symmetric second-rank tensor. Its eigenvalues correspond to the expectation values of $S_\alpha \equiv < P_2(\hat{\mathbf{u}} \cdot \hat{\mathbf{e}}_\alpha) >$ for the three orthonormal eigenvectors $\hat{\mathbf{e}}_\alpha$. We now *define* the director $\hat{\mathbf{n}}$ to be the eigenvector of $\mathbf{Q}$ that has the largest eigenvalue S. In a uniaxial nematic, the other two eigenvalues must be equal. If we combine this with the fact that $\mathbf{Q}$ is traceless, it follows that the latter eigenvalues must be equal to $-S/2$. In the nematic phase, we can therefore rewrite the tensor order parameter $\mathbf{Q}$ as

$$\mathbf{Q} = S\frac{3\hat{\mathbf{n}}\hat{\mathbf{n}} - \mathbf{I}}{2} \qquad (5.17)$$

In a numerical simulation, it is quite simple to determine the eigenvalues and eigenvectors of $\mathbf{Q}$. The nematic order parameter is then simply the largest eigenvalue of this second-rank tensor. Although this definition of S is indeed convenient to measure order *in the nematic phase*, it is less convenient to detect the isotropic–nematic transition. The reason is that the largest eigenvalue of $\mathbf{Q}$ is, by construction, positive. Hence, even in the isotropic phase it does not fluctuate around zero, but remains positive. In Appendix B.B, we show that, in the isotropic phase, the largest eigenvalue of $\mathbf{Q}$ only vanishes as $1/\sqrt{N}$. If, in contrast we define the nematic order parameter as -2 times the *middle* eigenvalue of $\mathbf{Q}$, then we find that this quantity *does* fluctuate around zero in the isotropic phase. Its average value has a much weaker system size dependence, viz. as $1/N$. For details, see Appendix B.B.

2. Correlation Functions

a. Orientational Correlation Functions. A convenient probe, both of the long-range orientational order in liquid crystalline phases, and of the *local* orientational order in the isotropic phase, is provided by the orientational correlation functions $g_\ell(r)$, defined as

$$g_\ell(r) = \langle P_\ell(\hat{\mathbf{u}}(0) \cdot \hat{\mathbf{u}}(r)) \rangle \qquad (5.18)$$

where $P_\ell(x)$ is the ℓth Legendre polynomial and $\hat{\mathbf{u}}(r)$ is a unit vector along the axis of the molecule at distance r from the reference molecule. In the nematic phase, all $g_\ell(r)$ with even ℓ are long-ranged. In the isotropic phase, they typically decay to zero within one molecular diameter, except very close to the isotropic–nematic transition. It is straightforward to verify that, in a phase with long-ranged orientational order, the following equality must hold

$$\lim_{r \to \infty} g_\ell(r) = S_\ell^2$$

where a_ℓ ($= S_\ell$) is the ℓth orientational order parameter, defined in Eq. (5.2). In a simulation, we can only study orientational correlations for interparticle separations less than $L/2$, where L is the diameter of the periodic simulation box. Hence, in practice, we will use the approximate estimate

$$S_\ell \approx \sqrt{g_\ell(r = L/2)}$$

For $\ell=2$, the above estimate of the nematic order agrees well with the value obtained using Eq. (5.16).

b. Translational Correlation Functions. In order to characterize (liquid-) crystalline phases with partial translational order, it is convenient to define correlation functions that probe this kind of ordering. Simplest among these functions is the longitudinal density correlation function denoted by $g_\parallel(r)$. This function measures the amplitude of density correlations in the direction of the alignment of the molecules. We can also define correlation functions that probe density correlations in the directions perpendicular to the molecular alignment axis. The simplest function to measure such correlations is denoted by $g(r_\perp)$. Here, $r_\perp$ is defined as the component of the distance between two particles perpendicular to the molecular alignment axis. $g(r_\perp)$ can, for instance, be used to distinguish a columnar phase, where the ordering of columns in a two-dimensional crystal lattice creates a strong modulation of $g(r_\perp)$, from a nematic phase, where $g(r_\perp)$ is rather featureless. A detailed description of these, and other probes of partial translational order in liquid crystals can be found in Ref. [81]

3. Frank Elastic Constants

As discussed earlier, the Frank elastic constants determine the response of the system to any external perturbation causing an orientational deformation. In the simplest case of the nematic phase, the free energy of deformation may be written as in Eq. (5.8), which essentially defines the elastic constants K_1, K_2 and K_3.

In simulations, the elastic quantities are most conveniently computed from equilibrium orientational fluctuations. The necessary equations have been summarized by Forster. [262] We assume, for simplicity, that the molecules are axially symmetric, with a unit vector $\hat{\mathbf{e}}_i = (e_{ix}, e_{iy}, e_{iz})$ along the axis of each molecule i. The center of mass of molecule i is at position $\mathbf{r}_i$. Then the ordering tensor in reciprocal space, for N molecules in volume V, is defined as

$$\hat{Q}_{\alpha\beta}(\mathbf{k}) = \frac{V}{N} \sum_{i=1}^{N} \frac{3}{2} \left(e_{i\alpha} e_{i\beta} - \frac{1}{3} \delta_{\alpha\beta} \right) \exp(i\mathbf{k} \cdot \mathbf{r}_i) \qquad (5.19)$$

Here $\delta_{\alpha\beta}$ is the Kronecker delta, $\alpha, \beta = x, y, z,$ and $\mathbf{k}$ is the wave vector. This is the Fourier transform of the real space orientation density

$$Q_{\alpha\beta}(\mathbf{r}) \equiv \frac{1}{V} \sum_{\mathbf{k}} \hat{Q}_{\alpha\beta}(\mathbf{k}) \exp(-i\mathbf{k} \cdot \mathbf{r}) \qquad (5.20)$$

In an unperturbed system, the orientation density is independent of position:

$$\langle \mathbf{Q}(\mathbf{r}) \rangle_0 = \langle \mathbf{Q} \rangle_0 = \langle \hat{\mathbf{Q}}(0) \rangle_0 / V = \text{constant} \qquad (5.21)$$

As discussed earlier, the order parameter $\overline{P}_2$ is the highest eigenvalue of $\langle \mathbf{Q} \rangle_0$, and $\hat{\mathbf{n}}$ is the corresponding eigenvector. If we choose $\hat{\mathbf{n}} = (0, 0, 1)$, then $\langle \mathbf{Q} \rangle_0$ is diagonal with

$$\langle Q_{xx} \rangle_0 = \langle Q_{yy} \rangle_0 = -\frac{1}{2}\overline{P}_2 \qquad (5.22)$$

$$\langle Q_{zz} \rangle_0 = \overline{P_2} \tag{5.23}$$

Fluctuations of $\hat{\mathbf{Q}}$ are related to the elastic constants as follows:

$$\left\langle \hat{Q}_{xz}(\mathbf{k}) \hat{Q}_{xz}(-\mathbf{k}) \right\rangle_0 = \frac{9}{4} \left(\frac{\overline{P_2}^2 V k_B T}{K_1 k_x^2 + K_3 k_z^2} \right) \tag{5.24}$$

$$\left\langle \hat{Q}_{yz}(\mathbf{k}) \hat{Q}_{yz}(-\mathbf{k}) \right\rangle_0 = \frac{9}{4} \left(\frac{\overline{P_2}^2 V k_B T}{K_2 k_x^2 + K_3 k_z^2} \right) \tag{5.25}$$

where the wave vector $\mathbf{k} = (k_x, 0, k_z)$ is chosen in the xz plane. Just as the elastic constants are defined for long wavelength director fluctuations, so the above equations are valid only in the limit of small k. In practical applications, it is necessary to extrapolate to $k = 0$. Particular attention must be paid to this, since in a small simulation box, there will be only a limited range of wave vectors accessible.

It should be emphasized that these fluctuation expressions are exactly equivalent to the forms involving the direct correlation function $c^{(2)}$, as discussed in Section V.A. This correspondence is made clear in the work of Poniewierski and Stecki, [260] Somoza and Tarazona [263] and others. It is also worth stressing that the Frank constants for hard-particle liquid crystals exist and are well behaved. The free energy for these systems is entirely entropic in origin: a deformation of the director field changes the entropy through its effect on molecular freedom and packing. The appropriate thermodynamic derivative is the Frank elastic constant. In a similar way, the hard-sphere crystal elastic constants are well known and well behaved (except in the limit of close packing).

Attempts to calculate elastic constants have been made for hard ellipsoids and spherocylinders. [264,265] Typical system sizes are in the range $125 \leq N \leq 600$. For these system sizes, the accessible range of k is limited, and deviations from the ideal equations (5.24,5.25) can be seen. Nonetheless, the $k \to 0$ extrapolation is practicable, and results reliable to about 15% can be obtained. These are shown for two different ellipsoid shapes in Figs. 5.2, 5.3. The density functional theories of Section V.A [63] are also depicted in the figure, and it can be seen that theory generally underestimates the simulation results. A similar comparison has been carried out for spherocylinders by Somoza and Tarazona, [208] using the results of Allen and Frenkel. [264,265] Once more, the theory generally underestimates the results of simulation. Recall that these same theories are comparatively successful in reproducing the transition density and order parameter variation. This discrepancy is presumably due to the sensitivity of the elastic constants to the variation of the direct correlation function $c^{(2)}$ at larger r_{ij}, outside the hard core overlap region. A more detailed

e=5

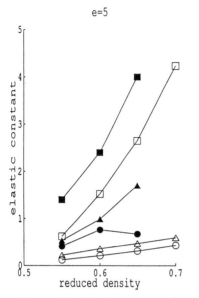

e=1/5

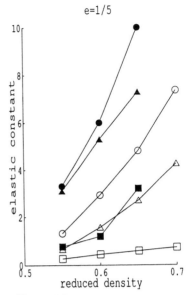

Figure 5.2. Elastic constants for prolate hard ellipsoids with $e = 5$ as a function of reduced density ρ/ρ_{cp}. We show K_1 ($\triangle$), K_2 ($\bigcirc$) and K_3 ($\square$), with simulation results denoted by filled symbols, and the theory of Tjipto-Margo and Evans [63] by open symbols.

Figure 5.3. Elastic constants for oblate hard ellipsoids with $e = 1/5$ as a function of reduced density ρ/ρ_{cp}. Notation as for Fig. 5.2.

knowledge of the form of $c^{(2)}$ for both isotropic and nematic molecular liquids would be helpful.

Very recently, a more direct approach has been made [266] to calculate the twist elastic constant K_2, by directly measuring the torque density in a system of molecules in twisted periodic boundary conditions. For a cuboidal simulation box of dimensions $L_x = L_y \neq L_z$, periodic replicas in the $\pm z$ direction are rotated by, respectively, $\pm\pi/2$ about the z axis relative to the original. This rotation is applied to center of mass coordinates as well as molecular orientations, but for a nematic fluid the distortion of the positional degrees of freedom is inconsequential. A uniformly twisted nematic director field, with the director everywhere perpendicular to the z axis, of wave vector $k_z = \pi/2L_z$, is stabilized in these boundaries, as shown in Fig. 5.4. The torque density associated with this deformation is [2]

$$\tau/V \equiv -V^{-1}\partial F_d/\partial k_z = -K_2 k_z \qquad (5.26)$$

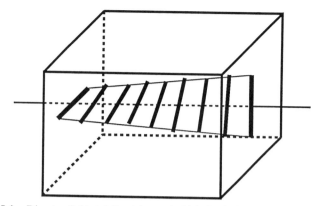

Figure 5.4. Director field stabilized in a cuboidal simulation box with twisted periodic boundary conditions applied. Neighboring boxes in the direction of the helix axis have all position and orientation vectors rotated by $\pm \pi/2$ about the axis.

where V is the volume of the box. The torque is measured as a sum of pairwise contributions

$$\tau/V = \frac{1}{2V} \left\langle \sum_i \sum_{j>i} (\tau_{ij}^z - \tau_{ji}^z) z_{ij} \right\rangle \tag{5.27}$$

where τ_{ij}^z is the z component of the torque on i exerted by j, and τ_{ji}^z the corresponding torque exerted on j by i. Further work is needed to establish the range of applicability of this technique, and to assess its value as a cross-check of the fluctuation expression approach discussed above.

VI. DYNAMIC PROPERTIES

In the nematic phase, two separate diffusion coefficients $D_\parallel$ and $D_\perp$ describe translation parallel and perpendicular to the director, respectively. Each is the time integral

$$D_\parallel = \int_0^\infty dt \, c_\parallel(t) \tag{6.1}$$

$$D_\perp = \int_0^\infty dt \, c_\perp(t) \tag{6.2}$$

of an appropriate component of the center of mass velocity autocorrelation function

$$c_\parallel(t) = \langle v_\parallel(0) v_\parallel(t) \rangle \tag{6.3}$$

$$c_\perp(t) = \langle v_\perp(0)v_\perp(t)\rangle \tag{6.4}$$

Here $v_\perp$ is either of the two Cartesian components perpendicular to the director (say, v_x, if the director is parallel to the z-axis) while $v_\parallel$ lies along it. Note that the chosen units of temperature and mass imply that $c_\perp(0) = c_\parallel(0) = 1$, that is, the correlation functions are normalized.

Diffusion in the nematic phase has been studied for ellipsoids of revolution. [267] For highly elongated prolate ellipsoids an unusual increase with density is seen for $D_\parallel$ just above the I→N transition; for very flat oblate ellipsoids a similar effect is observed for $D_\perp$. Examples are shown in Figs. 6.1 and 6.2. This seems to be associated with the rapid variation in order parameter close to the transition; physically the increasing order of the environment surrounding a molecule promotes slow decay of the velocity correlations, offsetting the general damping influence of an increasing collision rate. The correlations persist for many tens of collision times; these effects are illustrated in Figs. 6.3 and 6.4.

The most striking feature of these velocity correlations is the very slow long-time decay, extending to many tens of single particle collision times, of the more persistent component ($c_\parallel(t)$ for the prolate case, $c_\perp(t)$ for the oblate). We tentatively attribute this to coupling of the velocity with slow molecular reorientation, *not* to coupling with the hydrodynamic vortex

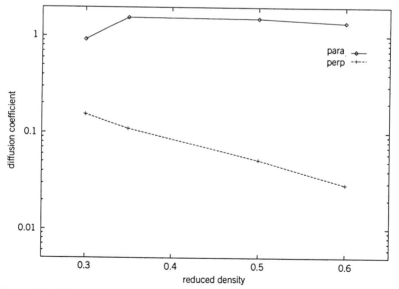

Figure 6.1.   Diffusion coefficients parallel and perpendicular to the director (log scale) as functions of reduced density ρ/ρ_{cp} for prolate ellipsoids with $e = 10$.

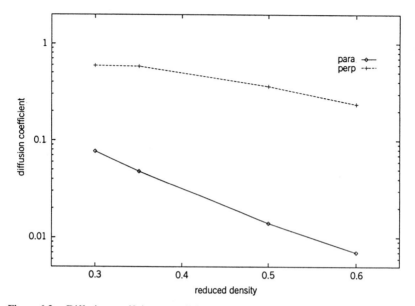

Figure 6.2.   Diffusion coefficients parallel and perpendicular to the director (log scale) as functions of reduced density ρ/ρ_{cp} for oblate ellipsoids with $e = 1/10$.

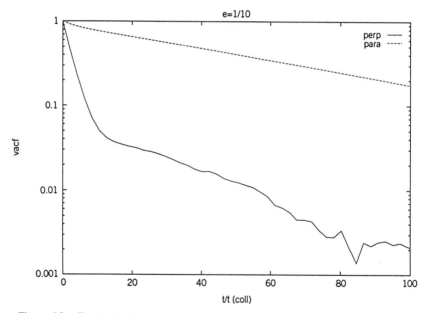

Figure 6.3.   Translational velocity autocorrelation function (log scale), parallel and perpendicular to the director, for prolate ellipsoids with $e = 10$ at a reduced density $\rho/\rho_{cp} = 0.35$. Time is measured in units of the mean time between collisions per particle.

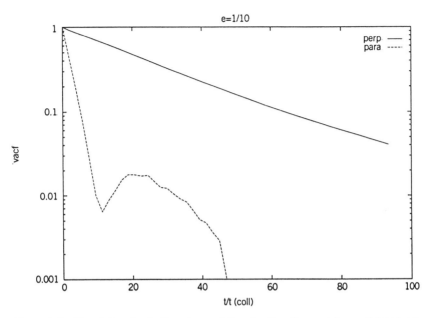

Figure 6.4. Translational velocity autocorrelation function (log scale), parallel and perpendicular to the director, for oblate ellipsoids with $e = 1/10$ at a reduced density $\rho/\rho_{cp} = 0.35$. Time is measured in units of the mean time between collisions per particle.

mode which gives rise to algebraic $t^{-3/2}$ long-time tails. [139–141] The long-time decay seems to be exponential, not algebraic. Yet, although we see no evidence for algebraic decay, we have no reason to doubt that the latter effect should be present at sufficiently long times. For shorter times, it is apparently masked by other effects.

Hess [268] has developed a theory based on relating the highly ordered system of ellipsoids to a reference system of hard spheres, by affine transformation. For perfect alignment (order parameter $S = 1$) a scaling of all the coordinates and ellipsoid shapes by a factor $1/e$ converts each ellipsoid configuration into an equivalent configuration of hard spheres. However, the scaling cannot simply be applied to the velocities as well, so there is not a one-to-one correspondence for dynamical properties. Nonetheless, it is possible to apply the scaling to the diffusion equation, and moreover to generalize the result to cater for imperfect orientational ordering ($S < 1$). The principal result of this analysis is an equation relating the anisotropy of diffusion to the order parameter S and the elongation e:

$$\Delta \equiv \frac{D_\parallel - D_\perp}{D_\parallel + 2D_\perp} = S\left(\frac{e^2 - 1}{e^2 + 2}\right) \qquad (6.5)$$

In the limit of high ordering the anisotropy Δ is predicted to be substantial: for example, as $S \to 1$, $\Delta \to 0.97$ for $e = 10$, and $\Delta \to 0.89$ for $e = 5$. A previous theory, due to Chu and Moroi [269], when applied to ellipsoids yields a slightly different formula:

$$\Delta \equiv \frac{D_\parallel - D_\perp}{D_\parallel + 2D_\perp} = S\left(\frac{e-1}{e+2}\right). \tag{6.6}$$

Predicted values of Δ are rather lower than for Eq. (6.5): as $S \to 1$, $\Delta \to 0.75$ for $e = 10$, and $\Delta \to 0.57$ for $e = 5$.

Remarkably, the simulation results agree very well (see Fig. 6.5) with the universal affine-transformation prediction, Eq. (6.5), for all the shapes studied. Much higher values of Δ are observed than can be explained by the original Chu and Moroi theory (Fig. 6.6). However, the relationships between $D_\parallel$, $D_\perp$ and the "equivalent" hard-sphere values are not explained so well by the affine transformation model.

The increase of $D_\parallel$ with density for rod-like molecules is reminiscent of the divergence of the longitudinal diffusion coefficient predicted by Doi and Edwards [270–273] and tested by simulation of the *isotropic* hard needle fluid [142–144]. The Doi–Edwards theory applies in the semi-dilute regime, and their idea of "tube dilation" associated with orientational ordering may be valid here, although a detailed comparison for the orientationally ordered fluid in the prolate ellipsoid case is not valid.

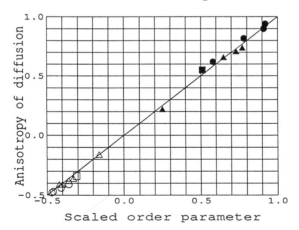

Figure 6.5. Test of the Hess theory for diffusion in the nematic phase. We plot the anisotropy of diffusion Δ versus the shape-scaled order parameter according to Eq. (6.5). Results for $e = 10$ ($\bullet$), $e = 5$ ($\blacktriangle$), $e = 3$ ($\blacksquare$), $e = \frac{1}{3}$ ($\square$), $e = \frac{1}{5}$ ($\triangle$), and $e = \frac{1}{10}$ ($\bigcirc$) are shown.

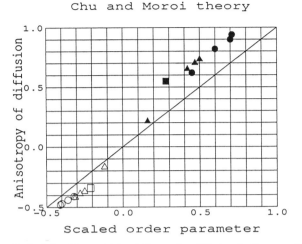

Figure 6.6. Test of the Chu and Moroi theory for diffusion in the nematic phase. We plot the anisotropy of diffusion Δ versus the shape-scaled order parameter according to Eq. (6.6). Notation as for Fig. 6.5.

Very recently, Tang and Evans [S. Tang and G.T. Evans, J. Chem. Phys., to appear] have formulated the kinetic theory in a way that explains the two-exponential decay very well. Using a projection operator formalism, they separate the decay rates of $c_\parallel(t)$ and $c_\perp(t)$ into components for different relative orientations of molecular axis, director, and direction of motion. Without any adjustable parameters, and taking only the density and order parameter as input, the measured diffusion coefficients and correlation functions are predicted quite accurately. This approach also explains two-exponential decay in the isotropic phase (see Section III.E.1), in terms of an Enskog-like theory.

Finally, we discuss an extreme example of anistropic diffusion in the nematic phase, namely the diffusion of infinitely thin hard platelets. [145] The latter model can be considered as oblate ellipsoids, in the limit $a/b \to 0$.

This system orders into a nematic phase at a density $\rho B_2 \approx 4$ ($B_2 = d^3 \pi^2/16$), [67] where d denotes the diameter of a platelet. For such platelets, this nematic phase is stable at all finite densities with $\rho B_2 \geq 4$ (and therefore its nematic order parameter can be made arbitrarily close to one), in contrast to molecules with finite proper volume which freeze at high enough densities.

For this rather extreme model system, we can use a simple scaling argument to predict that, in the nematic phase, the transverse diffusion constant *increases* with increasing density. Consider an assembly of smooth hard platelets of diameter d at a density ρ in which the system is in the nematic phase. The transverse diffusion coefficient for such a system can be estimated from a knowledge of the initial slope of the velocity autocorrelation function (assuming that this function decays exponentially with time). This slope is given by

$$1/\tau_\perp = -\frac{\langle \mathbf{v}_\perp^i . \Delta \mathbf{v} . \Gamma \rangle}{\langle \mathbf{v}_\perp^i . \mathbf{v}_\perp^i \rangle} \tag{6.7}$$

where Γ is the rate at which molecule i suffers collisions and $\Delta \mathbf{v}$ is the velocity change per collision. Assume that the normal to the plane of molecule i is inclined at an angle θ from the nematic director. A simple geometrical construction shows that [67]

$$\theta \sim \frac{1}{\pi d^3 \rho} \tag{6.8}$$

Γ the collision frequency goes as

$$\Gamma \sim \rho v_c d^2 \tag{6.9}$$

where v_c is the average relative velocity of the platelets at contact. v_c contains contributions from the relative translations and rotations of the platelets. In the nematic phase, v_c is dominated by $v_\parallel$ and rotations, and is only weakly dependent on density. Also $\Delta v_\perp \sim -\theta^2 v_\perp$, since the impulse imparted at a rim-platelet collision is always perpendicular to the plane of the platelet which suffers the collision. (Note that $v_\perp$ is the velocity *parallel* to the plane of the platelet). In other words

$$\langle v_\perp(0)\dot{v}_\perp(0^+)\rangle \sim -\langle v_\perp^2(0)\rangle \frac{1}{\rho^{*2}} \frac{\rho^*}{d} \tag{6.10}$$

where the reduced density $\rho^* = \rho d^3$ has been substituted in (7), and some numerical factors have been omitted. Hence integrating the velocity autocorrelation function over all time, we obtain

$$D_\perp \sim \rho^* \tag{6.11}$$

Thus we arrive at the remarkable conclusion that the diffusion coefficient diverges with density. Physically this divergence can be understood simply as a consequence of the fact that as the density increases, the efficiency with which collisions can transfer momentum decreases (owing to the increased nematic order), so that the platelets are able to slide past each other with greater and greater ease, while diffusion perpendicular to the molecular plane is effectively suppressed. Incidentally, a scaling argument similar to the one sketched above has been applied to diffusion in an isotropic system of infinitely thin needles. In the latter case, the scaling theory predicts a divergence of D with $\sqrt{\rho^*}$. [142]

To test whether such peculiar behavior of the diffusion constant is indeed observed, Alavi and Frenkel [145] performed molecular dynamics calculations of hard platelets in the density range $5 \leq \rho B_2 \leq 12$, which corresponds to a regime with pronounced nematic order $(0.92 < S < 0.99)$ The diffusion coefficients $D_\perp$ and $D_\parallel$ were determined by examining the long-time limit of the mean-square displacement curves.

In Figure 6.7, we have plotted the diffusion coefficient $D_\perp$ against ρB_2.

From this figure, it is clear that $D_\perp$ does indeed increase with density, as had been anticipated earlier by the scaling arguments. In contrast, $D_\parallel$ drops with increasing density. This feature highlights the fact that the overall increase in diffusion is associated with the greater ease with which platelets move parallel to their planes as the density (and hence the nematic order) increases. At densities beyond $\rho B_2 = 8$, $D_\perp$ apparently increases approximately linearly with ρB_2, a result also anticipated by the scaling argument.

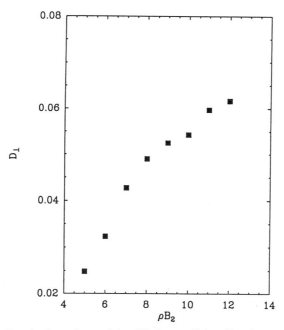

Figure 6.7.   Density dependence of the diffusion coefficient $D_\perp$ of a system of infinitely thin, hard platelets in the nematic phase. The diffusion constant is expressed in reduced units $d\,(k_B T/M)^{1/2}$

VII. OUTLOOK

In this review, we have tried to summarize our current theoretical understanding of hard-convex body fluids. In view of the vast literature on this
subject, this review is inevitably incomplete In fact, an up to date compilation of computer simulations on all kind of kind of models for molecular
fluids can be found in the excellent review of Levesque and Weis. [274]
It is clearly even more hazardous to make predictions about the future
developments in this area. However, it seems clear that certain problems
have received much more attention in the past than others. In particular,
a vast amount of work has been reported on equations of state of pure
HCB fluids in the isotropic phase. Much less is known, at present, about
mixtures of HCBs. Yet, this seems to be an interesting topic because we
may expect a much richer phase behavior in mixtures of nonspherical objects than in hard-sphere mixtures. If we then recall that, even for hard
sphere mixtures, the presence or absence of a fluid–fluid phase separation
is still a matter of debate, [275] it seems fair to say that the study of mixtures (binary, and, *a fortiori* polydisperse) of HCBs is a wide open area. It
should also be stressed that there is also an "experimental" reason to try
to understand the effect of polydispersity of the properties of hard-body
fluids: the best experimental realizations of hard-body fluids are suspensions of nonspherical colloidal particles. These systems are hardly ever
truly monodisperse and it seems likely that this polydispersity can effect
the phase behavior of these colloidal suspensions in a nontrivial way. In
fact, the recent numerical work of Stroobants on mixtures of hard parallel
spherocylinders [244] is a case in point.

Next, let us consider transport properties. It is clear that our knowledge of the dynamics of HCB fluids is still quite limited. Most of what
we know is limited to single particle properties ((rotational) diffusion) in
the isotropic and, to a lesser extent, the nematic phase. Our knowledge of
collective properties (viscosity, heat conductivity, collective orientational
fluctuations) is quite limited for isotropic fluids and virtually nonexistent
for the liquid-crystalline phases (see, however, Ref. [276]). As a consequence, the theories that describe the transport properties in isotropic
molecular liquids remain largely untested. In the case of transport in
liquid crystals, the situation is even more extreme as the corresponding
"Enskog" expressions for the transport coefficients have not even been
derived yet.

Finally, it is clear that, even if we stick to hard-core models, hard convex
bodies are but a subclass of all model systems that we could consider. First
of all, we could consider *rigid* nonconvex hard bodies (see, e.g., Refs. [65],
[277,278]), and secondly, *flexible* hard molecules (see, e.g., Ref. [279]).
Of course, very long-chain molecules have been studied extensively in

the context of polymer physics, but much less is known about short-chain molecules and the semi-flexible molecules ("worm-like chains"). Yet, flexibility is known to have a pronounced effect on the structure and phase behavior of hard-body fluids and, in particular, liquid crystals. Again, this area of research, and the study of dynamics in particular, appears wide open.

Of course, the study of highly idealized hard-core models cannot, and should not, replace the study of realistic models for molecular liquids and liquid crystals. Yet, for a fundamental understanding of real molecular systems, it is crucial to know how much of the behavior of such realistic models is already contained in simple hard-core models and how much of it is *truly* caused by other intermolecular interactions, such as dispersion forces, dipolar interactions or even many-body forces.

ACKNOWLEDGMENTS

We gratefully acknowledge the contributions of a large number of colleagues and collaborators. Special thanks are due to Dan Kivelson and Julian Talbot for their contribution to much of the work reported here. This research has been supported by the National Science Foundation (NSF), by the Petroleum Research Fund through grant PRF 18122-AC6 and by the Science and Engineering Research Council (SERC). The work of the FOM Institute is part of the research program of FOM and is supported by the Nederlandse Organisatie voor Wetenschappelijk Onderzoek (NWO). Computational facilities were provided at the John von Neumann Center for Scientific Computing at Princeton, NJ, at the University of Manchester Regional Computer Centre, and at the Amsterdam Computer Center (SARA), funded by NSF, SERC and WGS, respectively. International travel was made possible by a travel award from NATO.

A. APPENDICES TO PART I

A. The Ideal Free Energy of Rigid Bodies

In this appendix, we briefly discuss the calculation of the kinetic contribution to the free energy for a system of rigid bodies.

Consider a particle whose degrees of freedom are described by a set of n generalized coordinates, denoted by the n-dimensional vector $\mathbf{q}$. The Lagrangian of the particle in the absence of external force fields is just given by the kinetic energy

$$L = T = \frac{1}{2}\dot{\mathbf{q}} \cdot \mathbf{K}(\mathbf{q}) \cdot \dot{\mathbf{q}} \tag{A.1}$$

The generalized momenta conjugate to $\mathbf{q}$ are given by

$$\mathbf{p} = \frac{\partial L}{\partial \dot{\mathbf{q}}} = \mathbf{K}(\mathbf{q}) \cdot \dot{\mathbf{q}} \tag{A.2}$$

Using this to eliminate the generalized velocities $\dot{\mathbf{q}}$ from the kinetic energy, we find

$$T = \mathbf{p} \cdot \mathbf{K}(\mathbf{q})^{-1} \cdot \mathbf{p} \tag{A.3}$$

The canonical partition function of an interacting N particle system is then given by

$$Z_N = \frac{1}{h^n N!} \int \prod_{i=1}^{N} d\mathbf{p}_i d\mathbf{q}_i \exp[-\beta \sum_{i=1}^{N} \mathbf{p}_i \cdot \mathbf{K}(\mathbf{q}_i)^{-1} \cdot \mathbf{p}_i] \exp -\beta V(\{\mathbf{q}_k\}) \tag{A.4}$$

where $V(\{\mathbf{q}_k\})$ is the configurational energy of the system. We now perform the integration over the canonical momenta using the well-known generalized Gaussian integral identity

$$\int d\mathbf{x} e^{-\frac{1}{2}\mathbf{x} \cdot \mathbf{A} \cdot \mathbf{x}} = \sqrt{\frac{(2\pi)^n}{\det(\mathbf{A})}} \tag{A.5}$$

where $\mathbf{A}$ is a real, symmetric and positive definite $n \times n$ matrix. The result is

$$Z_N = \frac{1}{N!} \int \left\{ \prod_{i=1}^{N} d\mathbf{q}_i \sqrt{\frac{(2\pi)^n \det(\mathbf{K}(\mathbf{q}_i))}{h^{2n} \beta^n}} \right\} e^{-\beta V(\{\mathbf{q}_k\})} \tag{A.6}$$

where we have used the fact that $\det(\mathbf{A}^{-1}) = (\det(\mathbf{A}))^{-1}$.

For three-dimensional rigid bodies, the generalized coordinates are taken to be $\mathbf{q} = (x, y, x, \phi, \theta, \psi)$, that is, the coordinates of the center of mass in a Cartesian reference frame and the Euler angles specifying the orientation of a preferred frame (one in which the moment of inertia tensor is diagonal) of the particle with respect to this frame. The kinetic energy of such a particle is given by

$$T = \frac{1}{2} \left(m\dot{x}^2 + m\dot{y}^2 + m\dot{z}^2 + I_x \omega_x^2 + I_y \omega_y^2 + I_z \omega_z^2 \right) \tag{A.7}$$

where m is the mass of the particle and I_x, I_y, I_z the principal moments of inertia. The instantaneous angular velocities in the particle-fixed frame are given by [280]

$$\omega_x = \dot{\phi} \sin \theta \sin \psi + \dot{\theta} \cos \psi \tag{A.8}$$

$$\omega_y = \dot{\phi} \sin \theta \cos \psi - \dot{\theta} \sin \psi \tag{A.9}$$

$$\omega_z = \dot{\phi} \cos \theta + \dot{\psi} \tag{A.10}$$

Evaluating the determinant of the kinetic tensor $\mathbf{K}(\mathbf{q})$, we find

$$\det(\mathbf{K}(\mathbf{q})) = m^3 I_x I_y I_z \sin^2 \theta \tag{A.11}$$

Introducing this into the expression for the partition function, we arrive at

$$Z_N = \frac{1}{N! \mathcal{V}_T^N} \int \prod_{i=1}^{N} d\mathbf{r}_i d\Omega_i e^{-\beta V(\{\mathbf{r}_k, \Omega_k\})} \tag{A.12}$$

where we have introduced de orientational volume element $d\Omega = \sin \theta \, d\phi \, d\theta \, d\psi$ and the thermal volume

$$\mathcal{V}_T = \sqrt{\frac{(2\pi)^6 \hbar^{12} \beta^6}{m^3 I_x I_y I_z}} = \lambda^3 \gamma_x \gamma_y \gamma_z \tag{A.13}$$

which can be interpreted as the product of the translational thermal wavelength (one factor for each translational degree of freedom)

$$\lambda = \sqrt{\frac{2\pi \hbar^2 \beta}{m}} \tag{A.14}$$

and the orientational thermal angular spreads

$$\gamma_k = \sqrt{\frac{2\pi \hbar^2 \beta}{I_k}}, \quad k = x, y, z \tag{A.15}$$

Finally, we calculate the free energy per particle of a noninteracting gas of rigid particles in a volume V in the thermodynamic limit $N, V \to \infty$ at constant density $N/V = \rho$.

$$\beta F_N^{\text{id}}/N = \log \rho \mathcal{V}_T - 1 - \log 8\pi^2 \tag{A.16}$$

B. Center to Center Vectors

Consider two convex bodies whose mass centers are located at space points $\mathbf{r}_1$ and $\mathbf{r}_2$. The relative position of body 2 with respect to body 1 is $\mathbf{r}$,

$$\mathbf{r}_{12}(s) = \mathbf{r}_2 - \mathbf{r}_1 = \xi(\hat{\mathbf{s}}_1) + s\hat{\mathbf{s}}_1 - \xi(\hat{\mathbf{s}}_2) \tag{A.17}$$

Here $\xi(\hat{\mathbf{s}}_i)$ are the "radius" vectors for body i, emanating from its mass center and terminating at the point on its surface with surface normal $\hat{\mathbf{s}}_i$. Since the surface-to-surface distance, s, is measured along a common surface normal, which we take to point out of surface 1, then

$$\hat{\mathbf{s}}_1 = -\hat{\mathbf{s}}_2 = \hat{\mathbf{s}} \tag{A.18}$$

and so

$$\mathbf{r}_{12}(\mathbf{s}) = \xi(\hat{\mathbf{s}}) + s\hat{\mathbf{s}} - \xi(-\hat{\mathbf{s}}) \tag{A.19}$$

The radius vectors can be specified in terms of the support function of the convex body

$$h_i = \hat{\mathbf{s}}_i \cdot \xi(\hat{\mathbf{s}}_i) \tag{A.20}$$

We derive h by taking the derivative of h with respect to $\hat{\mathbf{s}}$,

$$\nabla_{\hat{\mathbf{s}}} h = (I - \hat{\mathbf{s}}\hat{\mathbf{s}}) \cdot \xi(\hat{\mathbf{s}}) + \hat{\mathbf{s}}\nabla_{\hat{\mathbf{s}}}\xi(\hat{\mathbf{s}}) \tag{A.21}$$

and as $\nabla_{\hat{\mathbf{s}}}\xi(\hat{\mathbf{s}})$ is perpendicular to $\hat{\mathbf{s}}$, then

$$\xi(\hat{\mathbf{s}}) = \hat{\mathbf{s}}h + \nabla_{\hat{\mathbf{s}}} h = \hat{\mathbf{s}}h + \hat{\mathbf{s}}_\theta \partial_\theta h + (1/\sin\theta)\hat{\mathbf{s}}_\phi \partial_\phi h \tag{A.22}$$

Given the support function, one can then determine the radius vector which traces out the exterior of the hard body.

C. Support Functions

Ideally one would like to have a support function that could represent convex shapes ranging from ellipsoids to spherocylinders. Although there is no unique choice of function that provides this feature, one function with some of these characteristics is a composite body formed by the addition of a sphere to a ellipsoid

$$h_i = (1/2)\sigma + h_i^e \tag{A.23}$$

where σ is the sphere diameter and h_i^e the ellipsoid support function of body i.

The equation for the surface of a biaxial ellipsoid with (semi-) axis lengths of a, b and c is given by

$$\mathbf{R}^T \cdot [\hat{\mathbf{u}}_x\hat{\mathbf{u}}_x a^{-2} + \hat{\mathbf{u}}_y\hat{\mathbf{u}}_y c^{-2} + \hat{\mathbf{u}}_z\hat{\mathbf{u}}_z b^{-2}] \cdot \mathbf{R} = 1 \tag{A.24}$$

where $\mathbf{R}$ is a Cartesian vector with components (R_x, R_y, R_z) in the body fixed principal axes. This surface has a support function

$$h^e = \hat{\mathbf{s}} \cdot \mathbf{R} = [(ax)^2 + (cy)^2 + (bz)^2]^{1/2} \tag{A.25}$$

where x, y, z are the projections of $\hat{\mathbf{s}}$ along the principal body axes,

$$(x, y, z) = \hat{\mathbf{s}}y(\hat{\mathbf{u}}_x, \hat{\mathbf{u}}_y, \hat{\mathbf{u}}_z) \tag{A.26}$$

A uniaxial body arises when $a = c$,

$$h^e = [a^2 + (b^2 - a^2)z^2]^{1/2} \tag{A.27}$$

and a sphere when $a = b = c$, wherein

$$h^e = a \qquad (A.28)$$

A chiral convex body may be formed by rotating the $\hat{\mathbf{u}}_x, \hat{\mathbf{u}}_y$ of the biaxial ellipsoid by an angle α which depends on R_z. The resulting surface is still a homogeneous quadratic form except that the principal axes, described by the unit vectors $\hat{\mathbf{u}}_x(\alpha), \hat{\mathbf{u}}_y(\alpha)$, depend on R_z

$$\hat{\mathbf{u}}_x(\alpha) = \hat{\mathbf{u}}_x \cos \alpha + \hat{\mathbf{u}}_y \sin \alpha$$
$$\hat{\mathbf{u}}_y(\alpha) = -\hat{\mathbf{u}}_x \sin \alpha + \hat{\mathbf{u}}_y \cos \alpha \qquad (A.29)$$

with $\alpha = cR_z/d$ and d being the intrinsic chiral period. For this twisted biaxial ellipsoid, $\mathbf{R}$ is given parametrically in terms of θ_p and ϕ_p by

$$R_x = \sin \theta_p [a \cos \phi_p - c \sin \phi_p \cos \alpha]$$
$$R_y = \sin \theta_p [c \cos \phi_p \cos \alpha + a \sin \phi_p \sin \alpha]$$
$$R_z = b \cos \theta_p \qquad (A.30)$$

and the surface by

$$\mathbf{R}^T \cdot [\hat{\mathbf{u}}_x(\alpha)\hat{\mathbf{u}}_x(\alpha)a^{-2} + \hat{\mathbf{u}}_y(\alpha)\hat{\mathbf{u}}_y(\alpha)c^{-2} + \hat{\mathbf{u}}_z\hat{\mathbf{u}}_z b^{-2}] \cdot \mathbf{R} = 1 \qquad (A.31)$$

A representative surface is shown in Fig A.1. The surface normal for the twisted biaxial ellipsoid is proportional to the gradient of the surface and this together with $\mathbf{R}$ suffices to determine h. In the limit of small twists, the twisted body is still convex. For large twists, the chiral body becomes a twisted ribbon and is no longer convex. By means of a perturbation analysis of the chiral correction to the support function, we find

$$h_1 = h_1^e + \delta h_1 \qquad (A.32)$$

where h_1^e is the support function of the convex biaxial ellipsoid and δh_1 is the first order chiral correction to body 1

$$\delta h_1 = \ell x_1 y_1 z_1 (3(c/h_1^e)^2 - (1 + (c/a)^2 + (c/b)^2)) \qquad (A.33)$$

and ℓ is a length

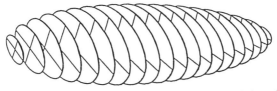

Figure A.1. The surface of a chiral particle as constructed by twisting a biaxial spheroid along its major axis.

$$\ell = (a^2 - c^2)b^2\pi/(c^2 d) \tag{A.34}$$

characterizing the extension of the surface due to chiral bulge.

At this point we can construct a variety of convex bodies: a sphere, a uniaxial ellipsoid, a biaxial ellipsoid and a twisted biaxial ellipsoidal body. By rolling a sphere over the exterior of these chosen bodies, we can flatten the sides of these bodies (which the exception, of course, of the sphere) to give shapes that approach spherocylinders, spheroplatelets, and twisted spheroplatelets but unlike the spherocylindric class of bodies, the composite bodies (sphere + ellipsoid) have a support function which is not defined piecewise. As an example, for two chiral bodies the support function for the pair is

$$h_{12} = h_1^e + h_2^e + \delta h_1 - \delta h_2 \tag{A.35}$$

whereas for a chiral body and a sphere of radius a,

$$h_{12} = a + h_1^e + \delta h_1 \tag{A.36}$$

If we choose to flatten either of these two shapes, we merely add to h_{12} a constant corresponding to the radius of the external sphere.

D. Jacobians

Typically, we shall require the Jacobian of the transformation from center to center coordinates to $\mathbf{s}$

$$d\mathbf{r}_{12} = |\partial\mathbf{r}_{12}/\partial\mathbf{s}|d\mathbf{s} \equiv S^{12}(\mathbf{s})d\mathbf{s} \tag{A.37}$$

This determinant simplifies as $d\mathbf{r}_{12}/d\mathbf{s} = \hat{\mathbf{s}}$ and so

$$S^{12}(\mathbf{s}) = |\hat{\mathbf{s}} \cdot \partial\mathbf{r}_{12}/\partial\theta \times \partial\mathbf{r}_{12}/\partial\phi| \tag{A.38}$$

When we take the θ and ϕ derivatives of $\mathbf{r}_{12}$, we obtain

$$\partial\mathbf{r}_{12}/\partial\theta = \hat{\mathbf{s}}_\theta[h + \partial_\theta^2 h] + \hat{\mathbf{s}}_\phi\partial_\theta[(1/\sin\theta)\partial_\phi h] \tag{A.39}$$

$$\partial\mathbf{r}_{12}/\partial\phi = \hat{\mathbf{s}}_\theta\partial_\theta[(1/\sin\theta)\partial_\phi h]\hat{\mathbf{s}}_\phi[h + \cot\theta\partial_\theta h + \partial_\theta^2 h] \tag{A.40}$$

and hence

$$S^{12} = (h + \partial_\theta^2 h)(h + \cot\theta\partial_\theta h + \partial_\theta^2 h) - (\partial_\theta[(1/\sin\theta)\partial_\phi h])^2 \tag{A.41}$$

In deriving these results, we have made use of the conventions that $\hat{\mathbf{s}}$ makes an anle θ and ϕ with respect to a space fixed (u_x, u_y, u_z) coordinate system. Further we have defined $\hat{\mathbf{s}}_\theta$ and $\hat{\mathbf{s}}_\phi$, the spherical polar unit vectors orthogonal to $\hat{\mathbf{s}}$.

$$\hat{\mathbf{s}} = \hat{\mathbf{u}}_z \cos\theta + \sin\theta[\hat{\mathbf{u}}_x \cos\phi + \hat{\mathbf{u}}_y \sin\phi]$$
$$= \hat{\mathbf{s}}_\theta \times \hat{\mathbf{s}}_\phi = -(\partial\hat{\mathbf{s}}_\theta/\partial\theta)$$
$$\hat{\mathbf{s}}_\theta = -\hat{\mathbf{u}}_z \sin\theta + \cos\theta[\hat{\mathbf{u}}_x \cos\phi + \hat{\mathbf{u}}_y \sin\phi]$$
$$= \hat{\mathbf{s}}_\phi \times \hat{\mathbf{s}} = (\partial\hat{\mathbf{s}}/\partial\theta)$$
$$\hat{\mathbf{s}}_\phi = -\hat{\mathbf{u}}_x \sin\phi + \hat{\mathbf{u}}_y \cos\phi \tag{A.42}$$

S^{12} is related to the orientation dependent second virial coefficient, $B_2(1,2)$ by

$$B_2(1,2) = (2\pi/3)h_{12}S^{12} \tag{A.43}$$

where h_{12} is the support functions for two bodies in contact. For two chiral bodies

$$h_{12} = h_1^e + h_2^e + \delta h_1 - \delta h_2 \tag{A.44}$$

whereas for a chiral body and a sphere of radius a,

$$h_{12} = a + h_1^e + \delta h_1 \tag{A.45}$$

Each of the four support functions directly above is expressed with respect to a common surface normal, $\hat{\mathbf{s}}$, emanating outward from the contact point on body 1.

B. APPENDICES TO PART II

A. Free Energy of Molecular Crystals

In this appendix, we discuss how the free energy of the solid phase of a given hard-core model system is related to the free energy of a known reference state. We use an artificial thermodynamic integration procedure to relate the free energy per particle, f_0, of the solid at a particular density ρ_0 to the free energy of an Einstein crystal of the same structure, at the same density. [229] The potential energy of this reference Einstein crystal is given by

$$U_E(\lambda_1, \lambda_2) = \lambda_1 \sum_{i=1}^{N} \Delta\mathbf{r}_i^2 + \lambda_2 \sum_{i=1}^{N} \sin^2\theta_i \tag{B.1}$$

where $\Delta\mathbf{r}_i \equiv \mathbf{r}_i - \mathbf{r}_i^0$ and θ_i denote, respectively, the translational and angular displacement of particle i from its equilibrium position and orientation. λ_1 and λ_2 are the coupling parameters which determine the strength of the harmonic force. The Einstein crystal is a convenient reference system, because its free energy per particle is known in closed form. For large values of the coupling constants, the configurational part of this free energy is approximately given by

$$f_E \approx -\frac{k_B T}{N} \ln \left[N^{-3/2} \left(\frac{\pi}{\beta \lambda_1} \right)^{3/2(N-1)} \left(\frac{2\pi}{\beta \lambda_2} \right)^N \right] \quad (\lambda_{1,2} \to \infty) \quad \text{(B.2)}$$

where we have imposed the additional constraint that the center of mass of the system is kept fixed. In order to construct a reversible path from the Einstein crystal to a hard-core crystal at the same density, we introduce a generalized potential energy function U_{λ_1,λ_2}

$$U_{\lambda_1,\lambda_2} = U_0 + U_E(\lambda_1, \lambda_2) \quad \text{(B.3)}$$

where U_0 is the potential energy function of the hard-particle system in the absence of any harmonic springs. For sufficiently large values of λ_1 and λ_2, the free energy of this interacting Einstein crystal reduces to the free energy of the ideal Einstein crystal. This equality only holds if all configurations of the ideal Einstein crystal are also acceptable configurations of the interacting system. In practice, a small fraction, P_0, of the configurations of the ideal Einstein crystal would result in hard-core overlaps. However, it is easy correct for this effect. [229] In order to compute the free energy f_0 of the hard-particle crystal at the reference state ρ_0, we perform thermodynamic integration to compute the change in free energy as we slowly switch off the Einstein crystal coupling constants λ_1 and λ_2 from their maximum value $\lambda_{1,2}^{max}$. The final expression that relates f_0 to the free energy of the ideal Einstein crystal is [229]

$$f_0 = f_E(\lambda_1^{max}, \lambda_2^{max}) + \frac{k_B T}{N} P_0 - k_B T \frac{\ln V}{N}$$
$$- \int_0^{\lambda_1^{max}} d\lambda_1 \langle \Delta r^2 \rangle_{\lambda_1,\lambda_2} - \int_0^{\lambda_2^{max}} d\lambda_2 \langle \sin^2 \theta \rangle_{\lambda_1,\lambda_2} \quad \text{(B.4)}$$

where $\langle \Delta r^2 \rangle_{\lambda_1,\lambda_2}$ is the mean-square displacement of a particle from its lattice site, at a given value of λ_1 and λ_2 and at fixed center of mass. The term $(-k_B T \ln V)/N$ corrects the free energy for this fixed center of mass constraint. Of course, the thermodynamic integration to the Einstein crystal must be performed at a density where the crystalline phase is at least mechanically stable. Once the absolute free energy of a phase is known at one density, we can use Eq. (4.69) to compute it at any other density that can be reached by reversible expansion or compression from the reference state.

B. System-Size Dependence of Nematic Order Parameter

When we determine the nematic order parameter in a numerical simulation, we find that the value of this quantity depends on the system size. It is, however, important to note that the different expressions for the the nematic order parameter that are used in the literature do not all have the same system-size dependence. In this section, we describe an approximate method to estimate the N-dependence of the different eigenvalues of the $\mathbf{Q}$ tensor

$$\mathbf{Q} = \frac{1}{N} \sum_i \left(\frac{3}{2} \hat{\mathbf{u}}_i \hat{\mathbf{u}}_i - \frac{\mathbf{I}}{2} \right) \tag{B.5}$$

(where $\hat{\mathbf{u}}_i$ is a unit vector specifying the orientation of molecule i; we assume cylindrically symmetric molecules). We consider an idealized model of an orientationally ordered fluid, namely one in which the $\mathbf{Q}$ tensors of different particles are uncorrelated, that is,

$$< \mathbf{Q}_i \mathbf{Q}_j > = < \mathbf{Q}_i > < \mathbf{Q}_j > \tag{B.6}$$

This situation will occur, for instance, in noninteracting gas of molecules in a magnetic field. Although in a real liquid crystal (or isotropic fluid) there are short-range orientational correlations the present analysis is still qualitatively valid in such a system if one reinterprets the $\mathbf{Q}_i$ not as molecular $\mathbf{Q}$ tensors but as the average $\mathbf{Q}$ tensor of a domain of size ξ^3, where ξ is the correlation length of order parameter fluctuations. The eigenvalue equation to be solved is

$$\mathbf{Q} \cdot \mathbf{v}_n = \lambda_n \mathbf{v}_n \tag{B.7}$$

where λ_n is the nth eigenvalue and $\mathbf{v}_n$ the nth eigenvector. It is convenient to study the equivalent problem of finding the eigenvalues of the tensor $\mathbf{M}$:

$$\mathbf{M} \equiv \frac{1}{N} \sum_i \hat{\mathbf{u}}_i \hat{\mathbf{u}}_i \tag{B.8}$$

The eigenvectors of $\mathbf{M}$ are the eigenvectors of $\mathbf{Q}$ and the eigenvalues μ of $\mathbf{M}$ are related to those of $\mathbf{Q}$ by $\mu_n = 2/3\lambda_n + 1/3$. The eigenvalue equation for $\mathbf{M}$ then becomes

$$\det|\mathbf{M} - \mu\mathbf{I}| = 0 \tag{B.9}$$

or

$$\begin{vmatrix} \frac{1}{N}\sum_i x_i x_i & \frac{1}{N}\sum_i x_i y_i & \frac{1}{N}\sum_i x_i z_i \\ \frac{1}{N}\sum_i y_i x_i & \frac{1}{N}\sum_i y_i y_i & \frac{1}{N}\sum_i y_i z_i \\ \frac{1}{N}\sum_i z_i x_i & \frac{1}{N}\sum_i z_i y_i & \frac{1}{N}\sum_i z_i z_i \end{vmatrix} = 0 \qquad (B.10)$$

or

$$-\mu^3 + \mu^2 + c_1\mu + c_0 = 0 \qquad (B.11)$$

where

$$c_1 =$$
$$-\frac{1}{N^2}\sum_{i,j}\left[(x_i x_i y_j y_j - x_i y_i x_j y_j)\right. \qquad (B.12)$$
$$\left. +(x_i x_i z_j z_j - x_i z_i x_j z_j) + (y_i y_i z_j z_j - y_i z_i y_j z_j)\right]$$
$$c_0 =$$
$$\frac{1}{N^3}\sum_{i,j,k}\left[(x_i x_i y_j y_j z_k z_k - x_i x_i y_j z_j y_k z_k\right. \qquad (B.13)$$
$$\left. -y_i y_i x_j z_j x_k z_k - z_i z_i x_j y_j x_k y_k) + 2x_i y_i y_k z_k z_j x_j\right]$$

Solving this equation yields μ_n as a function of all orientations:

$$\mu_n = f_n(x_1, y_1, z_1, \ldots, x_N, y_N, z_N) \qquad (B.14)$$

where f_n is a *nonlinear* function of all orientations. To obtain $< \mu_n >$, one should average the roots of this nonlinear equation. This is in general not possible. We simplify this problem by solving the equation with the average coefficients $< c_1 >$ and $< c_0 >$. Of course, for a nonlinear problem, this is *not* equivalent to a procedure where the μ's are averaged after the equation has been solved. However, as discussed below, we have tested the quality of this approximation in a few specific cases and find that the pre-averaging has no noticeable effects on our qualitative or even, for that matter, quantitative conclusions.

The averaged equation for μ now reads

$$-\mu^3 + \mu^2 - \mu\frac{N-1}{N}\left(< x^2 >(< x^2 > +2 < z^2 >)\right)$$
$$+\frac{(N-1)(N-2)}{N^2} < x^2 >< y^2 >< z^2 > = 0 \qquad (B.15)$$

where we have used a coordinate system such that the z-axis coincides with the axis of cylindrical symmetry. Moreover, we have used the fact that there is no short-range orientational correlation. Using the relation

$$< x^2 >=< y^2 >= (1 - S)/3$$
$$\text{and}$$
$$< z^2 >= (2S + 1)/3$$

(B.16)

where S is the nematic order parameter, we obtain the following equation for $\mu = 2/3\lambda + 1/3$:

$$\lambda^3 - \frac{3}{4}\lambda\frac{1 + S^2(N - 1)}{N} - \left[S^3/4 + \frac{3(S^2 - S^3)}{4N} + \frac{1 - 3S^2 + 2S^3}{4N^2}\right] = 0$$

(B.17)

This cubic equation can be solved in a closed form.

$$\lambda_n = r \cos(\phi_n + (n - 1)\frac{2\pi}{3}), \quad n = -1, 0, 1$$

(B.18)

with

$$r = \sqrt{\frac{1 + S^2(N - 1)}{N}}$$

(B.19)

and

$$\phi = \frac{1}{3} \arccos\left(\frac{S^3 + 3(S^2 - S^3)/N + (1 - 3S^2 + 2S^3)/N^2}{r^3}\right)$$

(B.20)

Let us consider the isotropic case ($S=0$) first. In that case,

$$r = 1/\sqrt{N}$$
$$\text{and}$$
$$\phi = \tfrac{1}{3} \arccos\left(\frac{1}{\sqrt{N}}\right)$$

(B.21)

for N not too small (say $N > 20$),

$$\arccos\left(\frac{1}{\sqrt{N}}\right) \approx \frac{\pi}{2} - \frac{1}{\sqrt{N}}$$

(B.22)

and hence

$$\phi = \frac{\pi}{6} - \frac{1}{3\sqrt{N}}$$

(B.23)

The expressions for the eigenvalues λ_n then become

$$\lambda_0 = \tfrac{1}{\sqrt{N}} \cos\left(-\frac{\pi}{2} - \frac{1}{3\sqrt{N}}\right) \approx -\frac{1}{3N}$$
$$\lambda_+ = \tfrac{1}{\sqrt{N}} \cos\left(-\frac{\pi}{6} - \frac{1}{3\sqrt{N}}\right) \approx -\frac{1}{\sqrt{N}}\left(\frac{\sqrt{3}}{2} + \frac{1}{6\sqrt{N}}\right)$$
$$\lambda_- = \tfrac{1}{\sqrt{N}} \cos\left(-\frac{5\pi}{6} - \frac{1}{3\sqrt{N}}\right) \approx -\frac{1}{\sqrt{N}}\left(\frac{-\sqrt{3}}{2} + \frac{1}{6\sqrt{N}}\right)$$

(B.24)

We therefore arrive at the important conlusion that, although all eigenvalues vanish in the thermodynamic limit $(N \to \infty)$, the largest (in absolute value) eigenvalues $(\lambda_\pm)$ vanish as $1/\sqrt{N}$, whereas the middle eigenvalue (λ_0), vanishes as $1/N$. An example of the N-dependence of λ_-, λ_0 and λ_+ is shown in Fig. B.1. Next we consider the case $S \neq 0$. Then

$$\cos 3\phi = \left(\frac{S^3 + 3(S^2 - S^3)/N + (1 - 3S^2 + 2S^3)/N^2}{r^3} \right) \tag{B.25}$$

which, to leading order in $1/N$, is equal to

$$\cos 3\phi \approx 1 - \frac{3}{2N} + 3\frac{2S - 1}{2NS^2} \tag{B.26}$$

Clearly, for N and S not too small, we can expand the cosine to obtain the following expression for ϕ:

$$\phi^2 = \frac{1}{3N} + \frac{2S - 1}{3NS^2} \tag{B.27}$$

from which it follows that $\phi = \mathcal{O}(1/\sqrt{N})$. For $S \neq 0$, r is given by

$$r = \sqrt{S^2 + \frac{1 - S^2}{N}} = S + \frac{1 - S^2}{2NS} + \mathcal{O}\left(\frac{1}{N^2}\right) \tag{B.28}$$

Hence $r = S + \mathcal{O}(1/N)$. If we now look at λ_-, λ_0 and λ_+, we find that for $S \neq 0$,

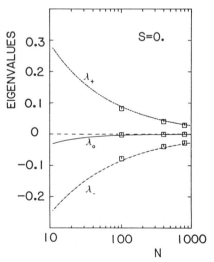

Figure B.1.   System-size dependence of the three roots of the average **Q** tensor, as given by Eq. (B.16) for an isotropic fluid ($S = 0$) with no short-range correlations. In the same figure we have also indicated numerical results for an ideal isotropic gas for N= 100, 400 and 800 (OPEN SQUARES). Note that the numerical results agree quite well with the (approximate) analytical expressions.

$$\lambda_+ = S + \mathcal{O}(\tfrac{1}{N})$$
$$\lambda_0 = -\tfrac{S}{2} + \mathcal{O}(\tfrac{1}{\sqrt{N}}) \qquad\qquad (B.29)$$
$$\lambda_- = -\tfrac{S}{2} + \mathcal{O}(\tfrac{1}{\sqrt{N}})$$

Hence, in the nematic phase, λ_+ yields a better estimate of S than $-2\lambda_0$. Note that the fact that we find three different eigenvalues does not indicate a biaxial phase, because this result was derived assuming a uniaxial phase. The apparent biaxiality is a system size effect. The dependence of λ_-, λ_0 and λ_+ on S is shown in Fig. B.2.

Using the above analysis, we can give a simple "geometrical" interpretation to the fluctuations in the nematic order parameter. A graphical representation of the eigenvalues λ_-, λ_0 and λ_+ is shown in Fig. B.3. The three eigenvalues are equal to the projections of a triad of vectors of equal

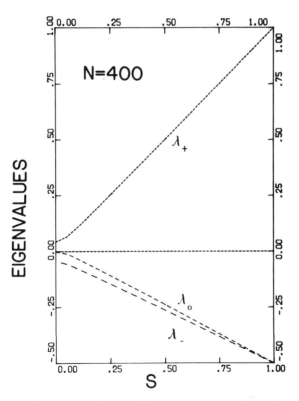

Figure B.2. Relation between the "real" (infinite system order parameter S, and the three roots of the average **Q** tensor, as computed using Eq. (B.16), for $N=400$. Note that at low values of S, λ_0 approaches zero more rapidly than either λ_+ or λ_-. At larger values of S, λ_+ approaches S.

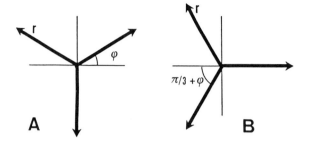

Figure B.3. Graphic representation of the computation of the eigenvalues of the **Q** tensor (Eq. B.16). The three eigenvalues are given by the projection of the three vectors on the horizontal axis. In the isotropic phase (a), $r=\mathcal{O}(1/\sqrt{N})$ while $\phi = \pi/6 - \mathcal{O}(1/\sqrt{N})$ In the nematic phase (b), $r=S + \mathcal{O}(1/N)$, while $\phi = \mathcal{O}(1/\sqrt{N})$. Fluctuations in the order parameter are caused by fluctuations in both r and ϕ.

length which make angles of 120° with one another. Fluctuations in the order parameters are driven by fluctuations in both the length of the vectors (r) and the angle (ϕ). It is important to note that the fluctuations in λ_0 contain both effects, that is, a contribution due to biaxial fluctuations (ϕ) and fluctuations in the size of the order parameter (r). As one approaches the isotropic–nematic transition, all fluctuations increase in magnitude and ϕ starts to rotate from $\pi/6$ to 0, while r grows from $l/\sqrt{N}$ to S. In the immediate vicinity of the I–N transition, all three eigenvalues of **Q** depend strongly on the size of the system.

Finally, we should return to the pre-averaging approximation that we made in order to solve Eq. (B.10). There we computed the eigenvalues of the averaged tensor order parameter, rather than the average of eigenvalues of the fluctuating tensor order parameter. It is of course interesting to know how seriously this averaging of a nonlinear equation affects the predicted average values of λ_n. To this end, we have carried out the following numerical test. We have generated the **Q** tensor for 10^4 configurations of a system of N ideal gas molecules ($N = 100, 400, 800$). In Fig. B.1 we compare the predicted eigenvalues computed from Eq. (B.16) with the average eigenvalues obtained by direct diagonalization of **Q** for all independent ideal gas configurations. Clearly, the agreement is quite good.

REFERENCES

[1] J.-P. Hansen and I. R. McDonald, *Theory of Simple Liquids,* 2nd edition, Academic Press, New York, (1986).
[2] P. G. de Gennes, *The Physics of Liquid Crystals,* Oxford University Press, (1974).
[3] L. Onsager, *Ann. N. Y. Acad. Sci.* **51**, 627 (1949).

[4] G. Ciccotti, D. Frenkel and I. R. McDonald, Eds., *Simulations of Liquids and Solids,* North-Holland, Amsterdam, (1987).

[5] J. Vieillard-Baron, *J. Chem. Phys.* **56**, 4729 (1972).

[6] J. Vieillard-Baron, *Mol. Phys.* **28**, 809 (1974).

[7] K. Mortensen, W. Brown and B. Nordén, *Phys. Rev. Lett.* **68**, 2340 (1992).

[8] J. W. Perram, M. S. Wertheim, J. L. Lebowitz and G. O. Williams, *Chem. Phys. Lett.* **105**, 277 (1984).

[9] J. W. Perram and M. S. Wertheim, *J. Comp. Phys.* **58**, 409 (1985).

[10] Ph. De Smedt, J. Talbot and J. L. Lebowitz, *Mol. Phys.* **59**, 625 (1986).

[11] M. P. Allen and D. J. Tildesley, *Computer Simulation of Liquids,* Clarendon Press, Oxford, (1987).

[12] T. Boublik and I. Nezbeda, *Collect. Czech. Chem. Comm.* **51**, 2301 (1986).

[13] D. A. McQuarrie, *Statistical Mechanics,* Harper and Row, New York, (1976).

[14] T. Boublik, *Mol. Phys.* **27**, 1415 (1974).

[15] T. Kihara, *Adv. Chem. Phys.* **5**, 147 (1963).

[16] B. Widom, *J. Chem. Phys.* **39**, 2808 (1963).

[17] R. S. C. She and G. T. Evans, *J. Chem. Phys.* **85**, 1513 (1986).

[18] W. A. Steele, *J. Chem. Phys.* **39**, 3197 (1963).

[19] L. Blum and A. J. Torruella, *J. Chem. Phys.* **56**, 303 (1972).

[20] L. Blum, *J. Chem. Phys.* **57**, 1862 (1972).

[21] L. Blum, *J. Chem. Phys.* **58**, 3295 (1973).

[22] Y. D. Chen and W. A. Steele, *J. Chem. Phys.* **50**, 1428 (1969).

[23] A. Perera, P. G. Kusalik and G. N. Patey, *J. Chem. Phys.* **87**, 1295 (1987).

[24] J. Talbot, A. Perera and G. N. Patey, *Mol. Phys.* **70**, 285 (1990).

[25] A. Perera, G. N. Patey, and J. J. Weis, *J. Chem. Phys.* **89**, 6941 (1987).

[26] W. B. Streett and D. J. Tildesley, *Proc. R. Soc. A* **348**, 485 (1976).

[27] P. A. Monson and M. Rigby, *Chem. Phys. Lett.* **58**, 1699 (1978).

[28] P. A. Monson and M. Rigby, *Mol. Phys.* **38**, 1699 (1979).

[29] V. N. Kabadi and W. A. Steele, *Ber. Bunsenges Phys. Chem.* **89**, 9 (1985).

[30] V. N. Kabadi, *Ber. Bunsenges Phys. Chem.* **90**, 332 (1985).

[31] V. N. Kabadi and W. A. Steele, *Ber. Bunsenges Phys. Chem.* **89**, 2 (1985).

[32] F. Gazi and M. Rigby, *Mol. Phys.* **62**, 1103 (1987).

[33] B. Kumar, C. James and G. T. Evans, *J. Chem. Phys.* **88**, 7071 (1988).

[34] J. Talbot, D. Kivelson, M. P. Allen, G. T. Evans and D. Frenkel, *J. Chem. Phys.* **92**, 3048 (1990).

[35] R. G. Cole and G. T. Evans, *Annu. Rev. Phys. Chem.* **37**, 106 (1986).

[36] D. Kivelson and P. A. Madden, *Annu. Rev. Phys. Chem.* **31**, 523 (1980).

[37] D. Frenkel, *J. Phys. Chem.* **91**, 4912 (1987).

[38] M. Rigby, *Mol. Phys.* **66**, 1261 (1989).

[39] M. Rigby, *Mol. Phys.* **68**, 686 (1989).

[40] P. A. Monson and M. Rigby, *Mol. Phys.* **35**, 1337 (1978).

[41] B. Tjipto-Margo and G. T. Evans, *J. Chem. Phys.* **93**, 4254 (1990).

[42] H. Reiss, H. L. Frisch, and J. L. Lebowitz, *J. Phys. Chem.* **31**, 369 (1959).

[43] E. Helfand, H. Reiss, H. L. Frisch, and J. L. Lebowitz, *J. Chem. Phys.* **33**, 1379 (1960).

[44] H. Reiss, *Statistical Mechanics and Statistical Methods in Theory and Applications: A Tribute To Elliott W. Montroll* Plenum Press, New York, (1976).

[45] H. Reiss, *J. Phys. Chem.* **96**, 4736 (1992).

[46] M. Wojcik and K. E. Gubbins, *Mol. Phys.* **53**, 397 (1984).

[47] K. H. Naumann and Y. P. Chen and T. W. Leland, *Ber. Bunsenges. Phys. Chem.* **85**, 1029 (1981).

[48] Y. Song and E. A. Mason, *Phys. Rev.* **41**, 3121 (1990).

[49] B. Barboy and W. M. Gelbart, *J. Chem. Phys.* **71**, 3053 (1979).

[50] B. Barboy and W. M. Gelbart, *J. Stat. Phys.* **22**, 685 (1980).

[51] B. Barboy and W. M. Gelbart, *J. Stat. Phys.* **22**, 709 (1980).

[52] J. K. Percus and G. J. Yevick, *Phys. Rev.* **110**, 1 (1958).

[53] N. F. Carnahan and K. E. Starling, *J. Chem. Phys.* **51**, 635 (1969).

[54] A. Isihara, *J. Chem. Phys.* **19**, 1142 (1951).

[55] B. M. Mulder, *Liq. Crys.* **1**, 539 (1986).

[56] B. Tjipto-Margo and G. T. Evans, *J. Chem. Phys.* **94**, 4546 (1991).

[57] R. S. C. She, C. James and G. T. Evans, *J. Chem. Phys.* **85**, 1525 (1986).

[58] D. M. Tully-Smith and H. Reiss, *J. Chem. Phys.* **53**, 4015 (1970).

[59] H. Reiss and R. V. Casberg, *J. Chem. Phys.* **61**, 1107 (1974).

[60] C. Gray and K. E. Gubbins, *Theory of Molecular Fluids,* Clarendon Press, Oxford, (1984).

[61] I. Nezbeda and S. Labik and A. Malijevsky, *Collect. Czech. Comm.* **54**, 1137 (1989).

[62] S. Labik and A. Malijevsky and W. R. Smith, *Mol. Phys.* **73**, 87 (1991).

[63] B. Tjipto-Margo, G. T. Evans, M. P. Allen and D. Frenkel, *J. Phys. Chem.* **96**, 3942 (1992).

[64] V. R. Bhethanabotla and W. Steele, *Mol. Phys.* **60**, 249 (1987).

[65] M. Whittle and A. J. Masters, *Mol. Phys.* **72**, 247 (1991).

[66] L. Verlet, *Phys. Rev.* **159**, 98 (1967).

[67] R. Eppenga and D. Frenkel, *Mol. Phys.* **52**, 1303 (1984).

[68] M. C. Duro and J. A. Martin-Pereda and L.M. Sesé, *Phys. Rev.* **A37**, 284 (1988).

[69] J. Vieillard-Baron, *These de Doctorat A637,* PhD thesis, Orsay, (1970).

[70] M. Weber, *Lehrbuch der Algebra I,* New York, (1962).

[71] W. H. Press and B. P. Flannery and S. A. Teukolsky and W. T. Vetterling, *Numerical Recipes: The Art of Scientific Computing,* Cambridge University Press, (1986).

[72] L. Qin, J. Wesemann and P. Siders, *Langmuir* **5**, 1358 (1989).

[73] M. He and P. Siders, *J. Phys. Chem.* **94**, 7280 (1990).

[74] J. W. Perram, We thank John Perram for sending us this material prior to publication.

[75] D. Frenkel and B. M. Mulder, *Mol. Phys.* **55**, 1171 (1985).

[76] F. H. Ree and W. G. Hoover, *J. Chem. Phys.* **40**, 939 (1964).

[77] M. P. Allen and D. Frenkel and J. Talbot, *Comp. Phys. Rep.* **9**, 301 (1989).

[78] W. W. Wood, *J. Chem. Phys.* **48**, 415 (1968).

[79] I. R. McDonald, *Mol. Phys.* **23**, 41 (1972).

[80] W. R. Cooney, S. M. Thompson and K. E. Gubbins, *Mol. Phys.* **66**, 1269 (1989).

[81] J. A. C. Veerman and D. Frenkel, *Phys. Rev.* **A45**, 5633 (1992).

[82] D. Frenkel, *J. Phys. Chem.* **91**, 4912 (1987).

[83] D. Frenkel, *J. Phys. Chem.* **92**, 5314 (1988); erratum to [82].

[84] D. Frenkel, *Mol. Phys.* **60**, 1 (1987).

[85] D. Frenkel, *Mol. Phys.* **65**, 493 (1988); erratum to [84].

[86] M. Wertheim and J. Talbot, We would like to thank Mike Wertheim and Julian Talbot for letting us use this unpublished material.

[87] C. F. Curtiss, *J. Chem. Phys.* **24**, 225 (1956).

[88] C. F. Curtiss and C. Muckenfuss, *J. Chem. Phys.* **26**, 1619 (1957).

[89] C. Muckenfuss and C. F. Curtiss, *J. Chem. Phys.* **29**, 1257 (1958).

[90] C. F. Curtiss and J. S. Dahler, *J. Chem. Phys.* **38**, 2352 (1963).

[91] N. F. Sather and J. S. Dahler, *J. Chem. Phys.* **35**, 2029 (1961).

[92] J. S. Dahler and N. F. Sather, *J. Chem. Phys.* **38**, 2363 (1963).

[93] S. I. Sandler and J. S. Dahler, *J. Chem. Phys.* **43**, 1750 (1965).

[94] S. I. Sandler and J. S. Dahler, *J. Chem. Phys.* **47**, 2621 (1967).

[95] D. K. Hoffman, *J. Chem. Phys.* **50**, 4823 (1969).

[96] D. K. Hoffman and J. S. Dahler, *J. Stat. Phys.* **1**, 521 (1969).

[97] S. Chapman and T. G. Cowling, *The Mathematical Theory of Non-Uniform Gases, 3rd edition,* Cambridge University Press, (1970).

[98] J. H. Ferziger and H. G. Kaper, *Mathematical Theory of Transport Processes of Gases,* North-Holland, Amsterdam, (1972).

[99] R. G. Cole and D. R. Evans and D. K. Hoffman, *J. Chem. Phys.* **82**, 2061 (1985).

[100] H. Mori, *Prog. Theor. Phys.* **33**, 423 (1965).

[101] G. T. Evans, *Mol. Phys.* **74**, 775 (1991).

[102] M. H. Ernst and J. R. Dorfman and E. G. D. Cohen, *Physica* **31**, 493 (1965).

[103] M. H. Ernst, J. R. Dorfman, W. R. Hoegy and J. M. J. van Leeuwen, *Physica* **45**, 127 (1969).

[104] M. H. Ernst, *Physica* **32**, 209 (1966).

[105] M. H. Ernst, *Physica* **32**, 273 (1966).

[106] J. R. Dorfman and H. van Beijeren, *Statistical Mechanics, Part B: Time Dependent Processes,* Plenum, New York, (1977).

[107] D. Forster, *Phys. Rev.* **A9**, 943 (1974).

[108] G. F. Mazenko, *Phys. Rev.* **A9**, 360 (1974).

[109] G. F. Mazenko and S. Yip, *Statistical Mechanics, Part B: Time Dependent Processes,* Plenum, New York, (1977).

[110] W. Sung and J. S. Dahler, *J. Chem. Phys.* **78**, 6264 (1983).

[111] W. Sung and J. S. Dahler, *J. Chem. Phys.* **78**, 6280 (1983).

[112] W. Sung and J. S. Dahler, *J. Chem. Phys.* **80**, 3025 (1984).

[113] G. T. Evans, *J. Chem. Phys.* **88**, 5035 (1988).

[114] G. T. Evans, *J. Chem. Phys.* **91**, 1252 (1989).

[115] T. Kagen and L. Maksimov, *Zh. Eksp. Teor. Fiz.* **51**, 5035 (1966).

[116] T. Kagen and L. Maksimov, *Sov. Phys. JETP* **24**, 1272 (1967).

[117] D. R. Evans, G. T. Evans and D. K. Hoffman, *J. Chem. Phys.* **93**, 8816 (1990).

[118] J. V. Sengers, *Phys. Fluids* **9**, 1333 (1966).

[119] J. V. Sengers, M. H. Ernst and D. T. Gillespie, *J. Chem. Phys.* **56**, 5583 (1972).

[120] J. R. Mehaffey and R. C. Desai and R. Kapral, *J. Chem. Phys.* **66**, 1665 (1977).

[121] S. R. deGroot and P. Mazur, *Non-Equilibrium Thermodynamics,* North-Holland, Amsterdam, (1962).

[122] M. Theodosopulu and J. S. Dahler, *J. Chem. Phys.* **60**, 3567 (1974).

[123] M. Theodosopulu and J. S. Dahler, *J. Chem. Phys.* **60**, 4048 (1974).

[124] J. S. Dahler and M. Theodosopulu, *Adv. Chem. Phys.* **31**, 155 (1975).

[125] S. Jagannathan and J. S. Dahler and W. Sung, *J. Chem. Phys.* **83**, 1808 (1985).

[126] R. G. Cole and D. K. Hoffman and G. T. Evans, *J. Chem. Phys.* **80**, 5365 (1984).

[127] H. Grad, *Pure Appl. Math.* **2**, 331 (1949).

[128] B. J. Berne and G. D. Harp, *Adv. Chem. Phys.* **17**, 63 (1970).

[129] P. Madden and G. T. Evans, *J. Chem. Phys.* **89**, 685 (1988).

[130] G. T. Evans and P. A. Madden, *Mol. Phys.* **74**, 1171 (1991).

[131] D. Kivelson and T. Keyes, *J. Chem. Phys.* **57**, 4599 (1972).

[132] G. T. Evans and R. G. Cole and D. K. Hoffman, *J. Chem. Phys.* **77**, 3209 (1982).

[133] G. Harrison, *The Dynamical Properties of Supercooled Liquids,* Academic Press, New York, (1976).

[134] R. A. MacPhail and D. Kivelson, *J. Chem. Phys.* **90**, 6555 (1989).

[135] S. Murad, D. P. Singh, H. J. M. Hanley and D. J. Evans, *Mol. Phys.* **72**, 487 (1991).

[136] G. T. Evans and D. Kivelson, *J. Chem. Phys.* **84**, 385 (1986).

[137] G. T. Evans and D. R. Evans, *J. Chem. Phys.* **81**, 6039 (1984).

[138] J. Talbot, M. P. Allen, G. T. Evans, D. Frenkel and D. Kivelson, *Phys. Rev.* **A39**, 4330 (1989).

[139] B. J. Alder, D. M. Gass and T. E. Wainwright, *J. Chem. Phys.* **53**, 3813 (1970).

[140] B. J. Alder and T. E. Wainwright, *J. Chem. Phys.* **31**, 459 (1959).

[141] B. J. Alder and T. E. Wainwright, *Phys. Rev.* **A1**, 18 (1970).

[142] D. Frenkel and J. F. Maguire, *Mol. Phys.* **49**, 503 (1983).

[143] J. J. Magda, H. T. Davis, and M. Tirrell, *J. Chem. Phys.* **85**, 6674 (1986).

[144] J. J. Magda, H. T. Davis, and M. Tirrell, *J. Chem. Phys.* **88**, 1207 (1988).

[145] A. Alavi and D. Frenkel, *Phys. Rev.* **A45**, 5355 (1992).

[146] M. P. Allen and D. Frenkel, *Phys. Rev. Lett.* **58**, 1748 (1987).

[147] T. Keyes and D. Kivelson, *J. Chem. Phys.* **56**, 1057 (1974).

[148] T. D. Gierke and W. H. Flygare, *J. Chem. Phys.* **61**, 2231 (1974).

[149] G. R. Alms, T. D. Gierke and W. H. Flygare, *J. Chem. Phys.* **61**, 4083 (1974).

[150] G. J. Vroege and H. N. W. Lekkerkerker, *Phase Transitions in Lyotropic Colloidal and Polymer Liquid Crystals, Rep. Prog. Phys.* **55**, 1241 (1992).

[151] H. Workman and M. Fixman, *J. Chem. Phys.* **58**, 5024 (1973).

[152] R. Evans, *Adv. Phys.* **28**, 143 (1979).

[153] T. Morita and K. Hiroike, *Prog. Theor. Phys.* **25**, 537 (1961).

[154] W. F. Saam and C. Ebner, *Phys. Rev.* **A15**, 2566 (1977).

[155] H. L. Frisch, N. River and D. Wyler, *Phys. Rev. Lett.* **54**, 2061 (1985).

[156] R. M. Gibbons, *Mol. Phys.* **17**, 81 (1969).

[157] G. Lasher, *J. Chem. Phys.* **53**, 4141 (1970).

[158] M. A. Cotter, *Phys. Rev.* **A10**, 625 (1974).

[159] K. M. Timling, *J. Chem. Phys.* **61**, 465 (1974).

[160] M. A. Cotter, *J. Chem. Phys.* **66**, 1098 (1977).

[161] B. M. Mulder and D. Frenkel, *Mol.Phys.* **55**, 1193 (1985).

[162] P. H. Fries and G. N. Patey, *J. Chem. Phys.* **82**, 429 (1985).

[163] R. Pynn, *Solid. State Comm.* **14**, 29 (1974).

[164] R. Pynn, *J. Chem. Phys.* **60**, 4579 (1974).

[165] F. Lado, *Mol. Phys.* **54**, 407 (1985).

[166] M. Baus, J.-L. Colot, X.-G. Wu and H. Xu, *Phys. Rev. Lett.* **59**, 2148 (1987).

[167] J.-L. Colot, X.-G. Wu, H. Xu and M. Baus, *Phys. Rev.* **A38**, 2022 (1988).

[168] P. Tarazona, *Mol. Phys.* **52**, 81 (1984).

[169] P. Tarazona and R. Evans, *Mol. Phys.* **52**, 847 (1985).

[170] W. A. Curtin and N. W. Ashcroft, *Phys. Rev.* **A32**, 2909 (1985).

[171] W. A. Curtin and N. W. Ashcroft, *Phys. Rev. Lett.* **56**, 2775 (1986).

[172] A. Poniewierski and R. Hołyst, *Phys. Rev. Lett.* **61**, 2461 (1988).

[173] R. Hołyst and A. Poniewierski, *Phys. Rev.* **A39**, 2742 (1989).

[174] A. M. Somoza and P. Tarazona, *J. Chem. Phys.* **91**, 517 (1989).

[175] A. M. Somoza and P. Tarazona, *Phys. Rev.* **A41**, 965 (1990).

[176] J. L. Lebowitz and J. W. Perram, *Mol. Phys.* **50**, 1207 (1983).

[177] L. Mederos and D. E. Sullivan, *Phys. Rev.* **A39**, 834 (1989).

[178] S. D. P. Flapper and G. Vertogen, *Phys Lett* **79A**, 87 (1980).

[179] S. D. P. Flapper and G. Vertogen, *Phys. Rev.* **A24**, 2089 (1981).

[180] S. D. Lee, *J. Chem. Phys.* **87**, 4972 (1987).

[181] S. D. Lee, *J. Chem. Phys.* **89**, 7036 (1988).

[182] J. D. Parsons, *Phys. Rev.* **A19**, 1225 (1979).

[183] D. Frenkel and B. M. Mulder, *Mol. Phys.* **55**, 1171 (1985).

[184] J. F. Marko, *Phys. Rev.* **A39**, 2050 (1989).

[185] U. P. Singh and Y. Singh, *Phys. Rev.* **A33**, 2725 (1986).

[186] U. P. Singh and U. Mohanty and Y. Singh, *Physica* **158A**, 817 (1989).

[187] G. J. Zarragoicoechea, D. Levesque and J. -J. Weis, *Mol. Phys.* **75**, 989 (1992).

[188] D. Frenkel, H. N. W. Lekkerkerker and A. Stroobants, *Nature* **332**, 822 (1988).

[189] M. Freiser, *Mol. Cryst. Liq. Cryst.* **14**, 165 (1971).

[190] R. Alben, J. R. McColl and C. S. Shih, *Solid State Comm.* **11**, 1081 (1972).

[191] C. S. Shih and R. Alben, *J. Chem. Phys.* **57**, 3055 (1972).

[192] R. Alben, *Phys. Rev. Lett.* **30**, 778 (1973).

[193] W. M. Gelbart and B. Barboy, *Acc. Chem. Res.* **13**, 290 (1980).

[194] B. M. Mulder, *Phys. Rev.* **A39**, 360 (1989).

[195] R. Hołyst and A. Poniewierski, *Mol. Phys.* **69**, 193 (1990).

[196] B. M. Mulder, *Liq. Cryst.* **8**, 527 (1990).

[197] M. P. Allen, *Liq. Cryst.* **8**, 499 (1990).

[198] G. T. Evans, *Mol. Phys.* (1992); in press.

[199] M. Hosino, H. Nakano and H. Kimura, *J. Phys. Soc. Jpn.* **46**, 1709 (1979).

[200] M. Hosino and H. Nakano and H. Kimura, *J. Phys. Soc. Jpn.* **47**, 740 (1979).

[201] A. Stroobants, H. N. W. Lekkerkerker and D. Frenkel, *Phys. Rev.* **A36**, 2929 (1987).

[202] B. M. Mulder, *Phys. Rev.* **A35**, 3095 (1987).

[203] X. Wen and R. B. Meyer, *Phys. Rev. Lett.* **59**, 1325 (1987).

[204] R. Hołyst and A. Poniewierski, *Mol. Phys.* **71**, 561 (1990).

[205] M. P. Taylor, R. Hentschke and J. Herzfeld, *Phys. Rev. Lett.* **62**, 800 (1989).

[206] J. M. Caillol and J. J. Weis, *J. Chem. Phys.* **90**, 7403 (1989).

[207] H. Xu, *Mol. Phys.* **77**, 311 (1992).

[208] A. M. Somoza and P. Tarazona, *Phys. Rev.* **A40**, 6069 (1989). At the time of publication the erratum of [265] had not appeared. This changes the comparison of simulation and theory in this paper.

[209] A. M. Somoza and P. Tarazona, *Phys. Rev.* **A40**, 4161 (1991).

[210] J. A. C. Veerman and D. Frenkel, *Phys. Rev.* **A43**, 4334 (1991).

[211] R. Hołyst, *Phys. Rev.* **A42**, 3438 (1990).

[212] A. Poniewierski and R. Hołyst, *Phys. Rev.* **A41**, 6871 (1990).

[213] A. Poniewierski and T. J. Sluckin, *Phys. Rev.* **A43**, 6837 (1991).

[214] A. M. Somoza and P. Tarazona, *Phys. Rev. Lett.* **61**, 2566 (1988).

[215] A. Poniewierski, *Phys. Rev.* **A45**, 5605 (1992).

[216] M. P. Taylor and J. Herzfeld, *Phys. Rev.* **A40**, 1678 (1989).

[217] G. T. Evans, *Mol. Phys.* **76**, 1359 (1992).

[218] R. Deblieck and H. N. W. Lekkerkerker, *J. Phys. Lett.* **41**, L-351 (1980).

[219] A. Stroobants and H. N. W. Lekkerkerker, *J. Phys. Chem.* **88**, 3699 (1984).

[220] R. Alben, *J. Chem. Phys.* **59**, 4299 (1973).

[221] Y. Rabin, W. E. McMullen and W. M. Gelbart, *Mol. Cryst. Liq. Cryst.* **89**, 67 (1982).

[222] P. Palffy-Muhoray, J. R. De Bruyn and D. A. Dunmur, *J. Chem. Phys.* **82**, 5294 (1985).

[223] J. A. Cuesta and D. Frenkel, *Phys. Rev.* **A42**, 2126 (1990).

[224] J. M. Kosterlitz and D. Thouless, *J. Phys. C* **6**, 1181 (1973).

[225] M. Romeiro, *J. Math. Phys.* **19**, 802 (1978).

[226] A. Z. Panagiotopoulos, *Mol. Phys.* **61**, 813 (1987).

[227] D. Frenkel, In G.Ciccotti and W.G.Hoover, eds., *Proc. 97th Int. School of Physics Enrico Fermi,* North-Holland, Amsterdam, 1986, pp. 151–188.

[228] W. G. Hoover and F. H. Ree, *J. Chem. Phys.* **47**, 4873 (1967).

[229] D. Frenkel and A. J. C. Ladd, *J. Chem. Phys.* **81**, 3188 (1984).

[230] E. J. Meijer, D. Frenkel, R. A. LeSar and A. J. C. Ladd, *J. Chem. Phys.* **92**, 7570 (1990).

[231] E. J. Meijer and D. Frenkel, *J. Chem. Phys.* **94**, 2269 (1991).

[232] J. S. van Duijneveldt and D. Frenkel, *J. Chem. Phys.* **96**, 4655 (1992).

[233] A. Stroobants, H. N. W. Lekkerkerker and D. Frenkel, *Phys. Rev. Lett.* **57**, 1452 (1986).

[234] J. A. C. Veerman and D. Frenkel, *Phys. Rev.* **A41**, 3237 (1990).

[235] W. G. T. Kranendonk and D. Frenkel, *Mol. Phys.* **72**, 699 (1991).

[236] W. G. Hoover and F. H. Ree, *J. Chem. Phys.* **49**, 3609 (1968).

[237] M. P. Allen and M. R. Wilson, *J. Computer Aided Molecular Design* **3**, 335 (1989).

[238] D. Frenkel, *Mol. Phys.* **54**, 145 (1985).

[239] H. Nakano, M. Hosino and H. Kimura, *J.Phys. Soc. Jpn* **46**, 1709 (1979).

[240] F. Dowell and D. E. Martire, *J. Chem. Phys.* **68**, 1088 (1978).

[241] F. Dowell and D. E. Martire, *J. Chem. Phys.* **68**, 1094 (1978).

[242] F. Dowell and D. E. Martire, *J. Chem. Phys.* **69**, 2322 (1978).

[243] F. Dowell, *Phys. Rev.* **A28**, 3526 (1983).

[244] A. Stroobants, *Phys. Rev. Lett.* **69**, 2388 (1992).

[245] T. E. Strzelecka, M. W. Davidson and R. L. Rill, *Nature* **331**, 457 (1988).

[246] F. Livolant, A. M. Levelut, J. Doucet and J. P. Benoit, *Nature* **339**, 724 (1989).

[247] D. Frenkel, *J. Phys. Chem.* **92**, 3280 (1988).

[248] D. Frenkel, *Liq. Cryst.* **5**, 929 (1989).

[249] H. Azzouz, J. M. Caillol, D. Levesque and J. J. Weis, *J. Chem. Phys.* **96**, 4551 (1992).

[250] D. Frenkel and R. Eppenga, *Phys. Rev.* **A31**, 1776 (1985).

[251] R. F. Kayser and H. J. Raveche, *Phys. Rev.* **A17**, 2067 (1978).

[252] Wu-Ki Tung, *Group Theory in Physics,* World Scientific, Singapore, (1985).

[253] J. C. Tolédano and P. Tolédano, *The Landau theory of Phase Transitions,* Lecture Notes in Physics, **Vol. 3**, World Scientific, Singapore, (1985).

[254] W. Maier and A. Saupe, *Z. Naturforsch.* **A13**, 564 (1958).

[255] M. Hammermesh, *Group Theory and its Applications to Physics,* Addison-Wesley, Reading, MA, (1962).

[256] R. G. Priest, *Phys. Rev.* **A7**, 720 (1973).

[257] J. P. Straley, *Phys. Rev.* **A8**, 2181 (1973).

[258] A. Poniewierski and J. Stecki, *Mol. Phys.* **38**, 1931 (1979).

[259] J. Stecki and A. Poniewierski, *Mol. Phys.* **41**, 1451 (1980).

[260] A. Poniewierski and J. Stecki, *Phys. Rev.* **A25**, 2368 (1982).

[261] D. Frenkel, In D. Levesque, J. P. Hansen, and J. Zinn-Justin, Eds., *Liquids, Freezing and the Glass Transition,* North-Holland, Amsterdam, 1991, Chapter 9,

[262] D. Forster, *Hydrodynamic Fluctuations, Broken Symmetry and Correlation Functions,* Frontiers in Physics, **Vol. 47**, Benjamin, Reading, MA, (1975).

[263] A. M. Somoza and P. Tarazona, *Mol. Phys.* **72**, 911 (1991).

[264] M. P. Allen and D. Frenkel, *Phys. Rev.* **A37**, 1813 (1988).

[265] M. P. Allen and D. Frenkel, *Phys. Rev.* **A42**, 3641 (1990); (Erratum).

[266] M. P. Allen and A. J. Masters, *Mol. Phys.* (1992). in press.

[267] M. P. Allen, *Phys. Rev. Lett.* **65**, 2881 (1990).

[268] S. Hess, D. Frenkel and M. P. Allen, *Mol. Phys.* **74**, 765 (1991).

[269] K.-S. Chu and D. S. Moroi, *J. Phys. Coll. C1* **36**, 99 (1975).

[270] M. Doi, *J. Phys.* **36**, 607 (1975).

[271] M. Doi and S. F. Edwards, *J. Chem. Soc., Faraday. Trans. 2* **74**, 568 (1978).

[272] M. Doi and S. F. Edwards, *J. Chem. Soc., Faraday. Trans. 2* **74**, 918 (1978).

[273] M. Doi and S. F. Edwards, *The Theory of Polymer Dynamics,* Clarendon Press, Oxford, (1986).

[274] D. Levesque and J. J. Weis, in K. Binder, ed., *The Monte Carlo Method in Condensed Matter Physics,* Springer Verlag, Berlin, 1992 Chapter 6

[275] T. Biben and J. P. Hansen, *Phys. Rev. Lett.* **66**, 2215 (1991).

[276] T. R. Kirkpatrick, *J. Chem. Phys.* **89**, 5020 (1988).

[277] M. D. Amos and G. Jackson, *Mol. Phys.* **74**, 191 (1991).

[278] A. L. Archer and G. Jackson, *Mol. Phys.* **73**, 881 (1991).

[279] G. Jackson, W. G. Chapman and K. E. Gubbins, *Mol. Phys.* **65**, 1 (1988).

[280] H. Goldstein, *Classical Mechanics, 2nd edition,* Addison-Wesley, Reading, MA, (1980).

TRIPLET EXCITONS IN WEAK ORGANIC CHARGE-TRANSFER CRYSTALS

J. KRZYSTEK*

Institute of Physics, Polish Academy of Sciences, 02-668 Warsaw, Poland

J.U. VON SCHÜTZ

3. Physikalisches Institut, Universität Stuttgart, 70550 Stuttgart, Germany

CONTENTS

* *Present address:* Department of Chemistry, University of Washington, Seattle WA 98105 USA.

Advances in Chemical Physics, Volume LXXXVI, Edited by I. Prigogine and Stuart A. Rice.
ISBN 0-471-59845-3 © 1993 John Wiley & Sons, Inc.

GLOSSARY OF MOST OFTEN USED ABBREVIATIONS

CT	charge transfer
CT character	charge transfer character
Cw	continuous wave
DF	delayed fluorescence
DF-ODMR	delayed fluorescence-detected ODMR
DSC	differential scanning calorimetry
ENDOR	electron nuclear double resonance
ESE	electron spin echo
ESR	electron spin resonance
FID	free induction decay
Fs	fine structure = zfs
Hfs	hyperfine structure
Homo	highest occupied molecular orbital
Hf	high (magnetic) field (e.g. hf-ODMR)
ISC	intersystem crossing
LAC	level anticrossing
LCAO	linear combination of atomic orbitals
Lumo	lowest unoccupied molecular orbital
Mw	microwave
NMR	nuclear magnetic resonance
OEP	optical electron polarization
OSP	optical spin polarization = OEP
ONP	optical nuclear polarization
ODMR	optically detected magnetic resonance
P	phosphorescence
PMDR	phosphorescence microwave double resonance
QNS	quasielastic neutron scattering
Rf	radiofrequency
Rf-ONP	radiofrequency-induced ONP
RYDMR	reaction yield-detected magnetic resonance
SLR	spin lattice relaxation
SOC	spin orbit coupling
S_0 state	ground singlet state
S_1 state	lowest excited singlet state
T_1 state	lowest excited triplet state
T_c	critical temperature (of a phase transition)
TDMR	triplet doublet magnetic resonance
Zf, zfs	zero (magnetic) field (e.g., zf-ODMR), zf-splitting

DONORS

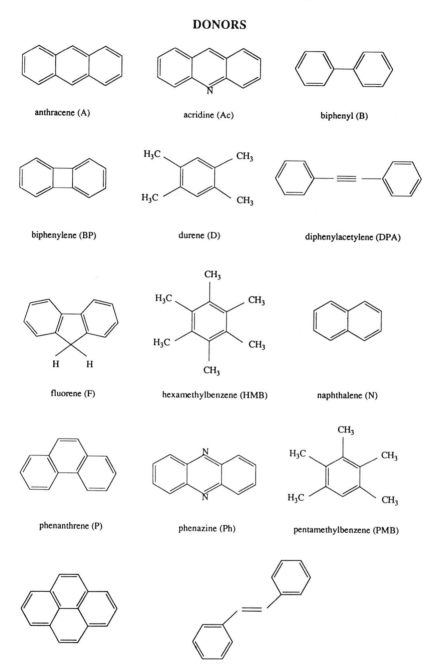

anthracene (A)

acridine (Ac)

biphenyl (B)

biphenylene (BP)

durene (D)

diphenylacetylene (DPA)

fluorene (F)

hexamethylbenzene (HMB)

naphthalene (N)

phenanthrene (P)

phenazine (Ph)

pentamethylbenzene (PMB)

pyrene (Py)

trans-stilbene (tS)

ACCEPTORS

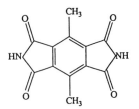

dimethylpyromellitimide (DMPM)

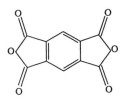

pyromellitic dianhydride (PMDA)

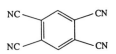

1,2,4,5-tetracyanobenzene (TCNB)

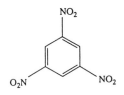

tetracyanoquinodimethan (TCNQ)

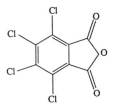

tetrachlorophthalic anhydride (TCPA)

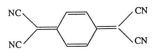

1,3,5-trinitrobenzene (TNB)

I. INTRODUCTION

A. General

The general purpose of this review is to discuss the properties of triplet excitons in organic "weak" charge transfer (CT) crystals. Its basis is a collection of data contained in a number of papers published in recent years, partly by the authors of this work. The CT crystals concerned are built of pairs of organic (mostly planar aromatic) molecules, forming so-called charge transfer complexes. One molecule in a complex acts as an acceptor, the other may be an electron donor. The reason for calling some of these complexes "weak" depends on the degree of charge transfer and is explained in Section I.B. At this point it is important to stress that in discussing the available data, we concentrate on the properties of the **mobile** triplet species in weak CT crystals rather than those of the localized species abundant in them, often observable at low temperatures when thermal detrapping is not possible. These localized states however, being distributed over isolated monomers, pairs or triads of molecules encompassed in the crystal lattice, can be very useful in understanding certain properties of the mobile excitons, particularly their electronic structure.

Isolation of a complex can also be achieved by dissolving it in a low temperature glass. Historically, this was the earliest attempt to understand the properties of the CT states. Let us thus first concentrate on an isolated CT complex as found in a rigid solution.

B. Electronic States of an Isolated Charge Transfer Complex

In terms of simple molecular orbital theory, the transfer of an electron in a complex takes place from the highest occupied orbital (homo) of one (donor) molecule to the lowest unoccupied orbital (lumo) of the other (acceptor) (Fig. 1.1). Both molecules acquire opposite electric charges

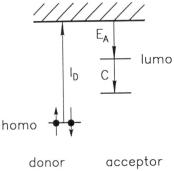

Figure 1.1. Schematic representation of a charge transfer from the highest occupied molecular orbital of the donor (homo) to the lowest unoccupied orbital of the acceptor (lumo). I_D is the ionization potential of the donor, E_A is the electron affinity of the acceptor and C is the Coulomb energy gained by the electron transfer.

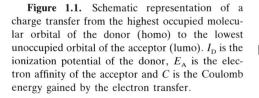

and attract each other additionally by Coulomb forces. If the sum of the electron affinity of the acceptor and the Coulombic energy gained by the charge transfer is greater than the ionization potential of the donor, the ground state of the complex will be ionic and such a complex can be described as "strong". In the other case, the energy gain is not enough to override the ionization potential, and the electronic ground state of the complex remains neutral. Molecules in this state interact with the usual forces in molecular systems (e.g., van der Waals interactions), the charge transfer taking place only in the excited electronic state. Such complexes are usually called "weak" and crystals formed by them are the main topic of this chapter.

It was early observed that a solution of two aromatic colorless compounds of distinctly different ionization potentials and/or electron affinities acquires color as a consequence of a new optical absorption band appearing in the visible region of the electromagnetic spectrum. The first satisfactory explanation of this phenomenon was delivered by Mulliken [1]. His theory is still useful in describing the low-lying electronic states of CT complexes and therefore a brief outline is presented below in order to make the following discussion more easily comprehensible.

According to Mulliken, the ground state of a CT complex is a result of the resonance interaction between the two zero order states (D, A) and D^+A^-). The first one corresponds to a pair of molecules held together by weak van der Waals forces only and is called a "no-bond" zero order state. The second one involves a full electron transfer from the donor to the acceptor and is therefore called an "ionic" or "CT" zero order state. The wave functions of the resulting (physical) ground (S_0) and first excited singlet (S_1) states are given by the following linear combinations:

$$\Psi_0(DA) = a\Psi_0(D, A) + b\Psi_0(D^+A^-) \tag{1.1}$$

$$\Psi_1(DA) = a'\Psi_1(D, A) + b'\Psi_1(D^+A^-) \tag{1.2}$$

The interaction between the no-bond and CT zero order states lowers the energy of the ground state relative to the vacuum level by a small factor ΔS (Fig. 1.2) and increases the energy of the first excited state by the same factor, given by the following equation:

$$\Delta S = \frac{V_1^2}{{}^1E(D^+A^-) - {}^1E(D, A)} \tag{1.3}$$

where ${}^1E(D^+A^-)$ and ${}^1E(D, A)$ are the energies of the CT and no-bond zero order states and V_1 denotes the resonance interaction energy between them.

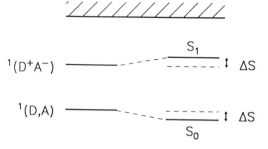

Figure 1.2. Resonance interaction between the zero order singlet states of a CT complex. Explanations in text.

The same resonance interaction which shifts the energies of the electronic states of a complex apart also mixes them, that is, gives the ground S_0 state some CT character, and the S_1 state some local character, although generally the relations $a \gg b$ and $a' \ll b'$ hold, that is, the ground state remains local in character and the excited state remains largely ionic. The CT character of the S_0 state can be estimated if the element V_1 is known, according to the formula

$$c_1^2 = \left\{ \frac{V_1}{{}^1E(\mathrm{D^+A^-}) - {}^1E(\mathrm{D, A})} \right\}^2 \tag{1.4}$$

This problem is treated further in Section II.B.1.

The first excited (S_1) singlet state of a weak CT complex is its more important characteristic, being of predominantly ionic nature. The absorption of light within a complex leads to excitation into this state giving the characteristic color. This involves an electron transfer from the donor to the acceptor molecule, hence the term "charge transfer complex".

We should not forget that besides the excited state of the complex appearing as a direct result of a intermolecular electron transfer, there remain further excited states characteristic for each of molecules forming the complex. In the first approximation, these states are mixed with neither the ground state nor the CT state of the complex and remain neutral in character. We have thus a manifold of donor singlet states $S_{1D}, S_{2D}, \ldots$ and acceptor singlet states $S_{1A}, S_{2A}, \ldots$. Their energy is usually higher than the CT state since the first singlet–singlet transition in uncomplexed aromatic compounds generally lies in the UV region. Since in most spectroscopic experiments we are interested only in the lowest lying states of the CT complex, the energy order of these states is usually quite easy to predict:

$$S_0 \text{ (neutral)} \ll S_1(\text{CT}) < S_2 \text{ (neutral: (D*A) or (DA*))}$$

The Mulliken theory provides an explanation of the nature of the two lowest singlet states of a CT complex. The situation is not quite so simple for the **triplet** states which exist in parallel with the singlets (with the exception of the ground state). In particular, the properties of the lowest excited triplet state T_1 of a CT complex may differ strongly from those of the S_1 state for the reasons discussed below.

It is well known that the difference between a singlet state and the corresponding triplet state is equal to $2K$, K denoting the exchange integral:

$$K = \left\langle \phi_n^*(1)\phi_m(2) \left| \frac{e^2}{r_{12}} \right| \phi_n(2)\phi_m^*(1) \right\rangle \tag{1.5}$$

(ϕ_i, $i = m, n$ are the orbitals occupied by the unpaired electrons 1 and 2 forming the triplet state, r_{12} is the average distance between them). In the ionic zero order state (D^+A^-), the exchange integral is very small due to a delocalization of the electron density over two molecules which enlarges r_{12}. Thus the triplet energy of this zero order state is almost equal to its singlet energy (in fact in the CT zero order state K may be negative, which means that its triplet energy can be slightly higher than the singlet energy [2]). Conversely, the exchange integral is generally much larger for the local zero order states, that is, (D^*A) or (DA^*), hence the locally excited triplet zero order state may lie either higher or close to, or even lower than the triplet CT zero order state. This situation is illustrated by Fig. 1.3 and may be summarized as follows [3]:

Case I

a. $^3E(D^*A) > {}^3E(D^+A^-)$ $(\simeq {}^1E(D^+A^-))$ for the systems in which the first excited donor triplet state lies below the corresponding acceptor state, that is, $^3E(D^*) < {}^3E(A^*)$, or

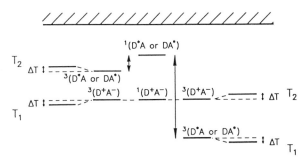

Figure 1.3. Mixing of the triplet zero order states in the case I (left) and III (right) after [3]). The singlet and triplet energies of the ionic zero order state are assumed equal. The arrows represent twice the exchange integral values for cases I and III. Other explanations in text.

b. $^3E(DA^*) > {}^3E(D^+A^-)$ for the systems in which the first excited acceptor state lies below the corresponding donor state, that is, $^3E(A^*) < {}^3E(D^*)$.

Case II

a. $^3E(D^*A) \simeq {}^3E(D^+A^-)$ (for $^3E(D^*) < {}^3E(A^*)$) or

b. $^3E(DA^*) \simeq {}^3E(D^+A^-)$ (for $^3E(A^*) < {}^3E(D^*)$)

Case III

a. $^3E(D^*A) < {}^3E(D^+A^-)$ (for $^3E(D^*) < {}^3E(A^*)$) or

b. $^3E(DA^*) < {}^3E(D^+A^-)$ (for $^3E(A^*) < {}^3E(D^*)$).

In Fig. 1.3, the energies of the excited states are related to the energy of the ground state of the complex which can be different from the energies of the ground states of the uncomplexed donor or acceptor. The effect of the changes of the ground state energy is discussed later.

The wave function of the lowest (T_1) triplet state can be represented analogously to the S_1 state by the linear combination of the zero order triplet states:

$$^3\Psi_1(DA) = c_1 \, {}^3\Psi(D^+A^-) + c_2 \, {}^3\Psi(D^*A) + \cdots \qquad (1.6)$$

or

$$^3\Psi_1(DA) = c_1 \, {}^3\Psi(D^+A^-) + c_2 \, {}^3\Psi(DA^*) + \cdots \qquad (1.7)$$

depending on which of the two possible local zero order states has a lower energy and interacts with the CT zero order state, neglecting in the first approximation the higher lying states. The mixing coefficients c_i strongly depend on the relative energy of the interacting zero order states. We may accordingly expect $c_1 > c_2$ for Case I, $c_1 \simeq c_2$ for Case II, and $c_1 < c_2$ for Case III. The lowest triplet state of a CT complex is thus expected to possess a charge transfer character (defined usually as c_1^2) varying in a broad range, from $c_1^2 \simeq 0$ to $c_1^2 \simeq 1$. As will be shown in this review, this corresponds to the physical situation encountered in the investigations of triplet states of numerous complexes, first isolated in low temperature glasses and later in the form of excitons in the crystalline phase.

The interaction in the triplet manifold between the locally excited state (let us assume it is the D*A state) and the ionic (D^+A^-) zero order state shifts the perturbed (T_1 and T_2) states apart by $2\Delta T$, analogously to the interaction between the two lowest singlet zero order states. ΔT is given by an expression analogous to Eq. (1.3):

$$\Delta T = \frac{V_3^2}{{}^3E(D^+A^-) - {}^3E(D^*A)} \qquad (1.8)$$

where $^3E(D^+A^-)$ and $^3E(D^*A)$ are the energies of the CT and locally excited zero order triplet states, and V_3 is the resonance interaction energy between them.

C. Optical Spectra of Isolated CT Complexes

The early optical experiments on isolated complexes involved mostly measurements of CT absorption and fluorescence, thus concentrating principally on the properties of the ionic S_1 state.

The first information about the T_1 state was obtained from phosphorescence experiments on complexes dissolved in low temperature glasses. These early investigations of Briegleb and co-workers [4] involved complexes characterized by the T_1 state predominantly localized on the donor molecules, for example, phenanthrene (P) or naphthalene (N) with the acceptor being tetrachlorophthalic anhydride (TCPA), that is, the situation corresponding to case III of Iwata et al. [3]. A review of these experiments is given in [5].

The local character of the T_1 states of the complexes investigated by Briegleb et al. is demonstrated by:

i. the spectral position of the phosphorescence from the complex which almost coincides with that of the uncomplexed donor; this proves that the energy of the (D^*A) zero order state is almost unperturbed by any interaction with the higher lying (D^+A^-) zero order state, and the CT character of the T_1 state is negligible (the triplet energy of the acceptor TCPA is much higher than those of all the donors used and thus the acceptor triplet state does not perturb the triplet energy of the complex);

ii. the resolved vibronic structure which is very similar to that of the donor phosphorescence.

The first observation of a different kind of phosphorescence was reported in [3] for a series of complexes formed in glasses by tetracyanobenzene (TCNB) as an acceptor and benzene and its methylated derivatives used as donors. The triplet energy of TCNB ($22,650\,\text{cm}^{-1}$) is lower than those of the donors, so the local zero order state of a CT complex interacting with the CT state is of the type (DA^*). It was found that for the complexes with multi-methylated donors like tetramethylbenzene (durene, D), pentamethylbenzene (PMB) and hexamethylbenzene (HMB) the phosphorescence is broad and structureless, very much like the fluorescence, and positioned at lower energies than the phosphorescence of TCNB (maximum at $18,300\,\text{cm}^{-1}$ in HMB/TCNB) (Fig. 1.4). Moreover, it does not differ much on the energy scale from the fluorescence (maximum at $20,000\,\text{cm}^{-1}$ in the same case). The authors inter-

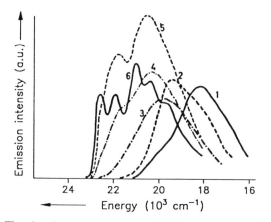

Figure 1.4. The phosphorescence spectra of TCNB complexes in EP solution at 77 K; (1) HMB/TCNB, (2) D/TCNB, (3) mesithylene/TCNB, (4) toluene/TCNB, (5) benzene/TCNB, (6) TCNB only (from [3]).

preted this emission as a CT phosphorescence, that is, originating from a T_1 state of CT character which increases gradually for the given series of donors from benzene through mesitylene to multi-methylated compounds. This interpretation was confirmed by electron spin resonance (ESR) experiments on the same complexes. Both kinds of experiments were reviewed in the monograph of Nagakura [6]. Consequently, only those that are relevant to the subject of this chapter are mentioned here.

D. Charge Transfer Crystals

It was found early that charge transfer complexes readily crystallize in regular structures of well-defined molecular stoichiometries (usually but not always 1:1). The common characteristic of the weak CT crystals is their structure (for an early review, see [7]). The vast majority of them are built of infinite stacks of alternating donor and acceptor molecules, placed with their planes approximately parallel to each other at a typical distance of 3.3–3.6 Å (Fig. 1.5). Since the overlap of molecular orbitals between the aromatic molecules considered here is maximal in the direction perpendicular to their planes, interactions between the molecules forming such a crystal are usually stronger in the direction defined by the stacking axis. This makes these crystals examples of low-dimensional solids as far as their structure is concerned. It is shown later that other properties are a consequence of this particular crystal structure, being also predominantly low-dimensional. The two most prominent properties are the excitation energy transfer and electrical conductivity. It should be mentioned that only a few one-component organic crystals have

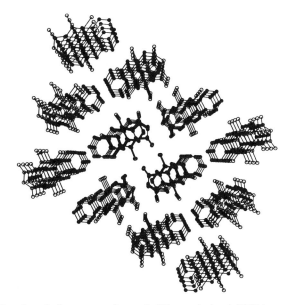

Figure 1.5. A typical structure of a weak CT crystal: the A/TCPA crystal viewed along the stack axis. TCPA molecules are oriented in one of the two possible positions (from [11]).

structures similar to the CT crystals, the well-known examples being 1,2,4,5-tetrachlorobenzene [8] and 1,4-dibromonaphthalene [9].

The structure of weak CT crystals is also interesting from another point of view: there is usually a large degree of motional freedom of the component molecules, particularly in the molecular planes, which is due to the different sites of the molecules. Consequently, in many cases molecules undergo librations around their out-of-plane axes and also to a smaller extent around other axes, which means that the crystals exhibit dynamic disorder. Lowering the temperature often leads to freezing in these movements, which in turn can lead to phase transitions. The best known example of a phase transition in a CT crystal is probably the one at about 206 K in A/TCNB (anthracene/TCNB) [10]. Sometimes more than one phase transition takes place in the commonly monitored temperature range (1.2–300 K), as is the case in A/TCPA [11]. Another kind of orientational disorder, namely static disorder, is encountered if there are at least two translationally inequivalent positions of the molecules in the lattice. This kind of disorder takes place if the molecules are distributed randomly over both orientations. This phenomenon is often met if the two orientations differ by a 180° rotation which is the case for molecules of symmetry lower than D_{2h} (e.g. fluorene (F) or TCPA). Both dynamic and static disorder may have far-reaching consequences for the molecular dynamics of the crystal; this is discussed in Chapter 5.

As mentioned above, the stacks forming a weak CT crystal are of mixed nature, containing both kinds of molecules in an alternating manner. This is in striking contrast to the "strong", or ionic CT crystals, also called CT salts to underscore the ionic nature of their ground state. In CT salts there exist separate, uniform stacks consisting of either donor or acceptor molecules. This class of crystals is of considerable interest due to their electrical and magnetic properties. It has been subject of several reviews in recent years (see e.g., [12]) and so remains out of the scope of this work.

Yet another group is formed by ionic CT crystals built of mixed stacks (notable examples are tetrathiofulvalene/chloranil [13] and tetrathioful-valene/2-bromo-5-methyldicyanoquinonediimine [14]). Their most interesting property is that under external pressure some of them undergo a phase transition into the neutral phase, that is, their ground state becomes neutral as in standard mixed-stack systems. The reader is referred to the original works for more details.

E. Excitons in CT Crystals

The change of environment of a CT complex from glass to a CT crystal has immediate consequences for the spectroscopic properties. The most obvious difference is the radical change in the lineshape and position of lines in the optical emission and excitation spectra. The reason for this is that the molecules are now in an almost uniform environment, which drastically reduces the linewidth. On the other hand, other mechanisms, notably electron-phonon coupling, cause a secondary broadening of the lines. A new kind of optical emission, P-type delayed fluorescence appears under certain conditions. This proves that the excitation becomes mobile in the crystalline phase because such emission is a result of an annihilation process $T_1 + T_1 \rightarrow S_1$. Quite often this form of depopulation of the triplet state can compete with those prevailing in isolated complexes, like radiative or nonradiative decay or E-type delayed fluorescence (originating from the thermal activation $T_1 \rightarrow S_1$), thus having a strong influence on the triplet kinetics. Another important consequence of the emerging mobility of the triplet excitation is a change of shape and position of ESR signals (motional narrowing and other averaging effects).

There is a significant number of articles devoted to the problem of excitation mobility in molecular crystals. It is therefore not the aim of this chapter to repeat the already known facts unless they directly concern triplet excitons in CT crystals. A short introduction in the subject is nevertheless necessary in anticipation of the further discussion of this topic.

The excitation mobility that gives rise to the numerous phenomena described above has an origin in the periodic structure of a· crystal. Namely, a configuration corresponding to a single molecule n in an excited electronic state while all the others are in the ground state, represented by the following wave function [15]:

$$\Psi_n^* = |\ldots \phi_{n-1} \phi_n^* \phi_{n+1} \ldots\rangle = \ldots |\phi_{n-1}\rangle |\phi_n^*\rangle |\phi_{n+1}\rangle \ldots \quad (1.9)$$

is not stationary. From the well-known Bloch's theorem, it follows that such a system containing N molecule has N eigenstates of the form of a linear combination,

$$\Psi^*(k) = 1/\sqrt{N} \sum_n \Psi_n^* e^{i\overline{k}an} \quad (1.10)$$

where $\overline{k}$ is a wave vector, $\overline{a}$ is a lattice constant. There is always a finite probability of the excitation energy transferring to a neighboring molecule and this transfer can take place many times before the excitation becomes relaxed in any way. The probability of an energy transfer from the molecule n to the molecule m, let us say in a linear molecular chain, is given by the following transfer integral:

$$\beta_{nm} = \langle \phi_n^* \phi_m | V_{nm} | \phi_n \phi_m^* \rangle \quad (1.11)$$

The intermolecular interaction term V_{nm} consists of Coulomb and exchange terms of which only the latter are significant in the triplet state. The total electronic Hamiltonian of the crystal can be written for the simplest case of one molecule in the elementary cell as follows:

$$\mathbf{H} = \mathbf{H}_n + 1/2 \sum_{n,m} V_{nm} \quad (1.12)$$

where $\mathbf{H}_n$ denotes the Hamiltonian of the single molecule n. The effect of this Hamiltonian acting on the wave functions of the form (1.10) is to give N eigenvalues differing by the wave vector $\overline{k}$. The energy of the excited state of the crystal is then given by

$$E(\overline{k}) = E_0 + 2\beta_{nm} \cos \overline{k}a \quad (1.13)$$

where E_0 is the energy of a molecule in the absence of energy transfer. Since $\cos \overline{k}a$ can assume values from -1 to $+1$, an excitonic band is created of width $4\beta_{nm}$. The exciton propagates in the crystal as a wave packet which it is sometimes convenient to describe as a quasi-particle, analogously to the photon.

The model of a coherent movement of an exciton in a simple linear chain of molecules is valid only if:

i. the element β_{nm} is large in comparison with the element $\beta_{nm'}$, describing the interaction of molecule n with molecule m' belonging to a neighboring chain (otherwise the model has to be extended into more dimensions); and

ii. β_{nm} is large in comparison with the corresponding exciton-phonon and exciton-intramolecular vibration matrix elements, which is usually the case only at very low temperatures when phonons and vibrations are not active.

An alternative idea of excitonic movement describes it as an incoherent hopping between different sites, with the phase memory lost between the jumps. This concept is particularly useful when the exciton–phonon interaction becomes comparable with the excitonic bandwidth [16], that is, at high temperatures. Probably the most useful formalism describing this kind of exciton motion is the Haken–Strobl model [17], allowing an exact mathematical solution of the problem.

There are two principal kinds of excitons known in solid-state physics, namely Frenkel and Wannier excitons. It is the first kind that appears in most molecular crystals. Its most important characteristic is that the electron and the hole reside on the same molecule (the electron transferred to the lowest unoccupied orbital of the molecule, the hole residing on its vacant highest occupied orbital) [18,19]. The Wannier type, in which the electron is separated from the hole by several lattice constants, is met in certain inorganic crystals of high dielectric constant [20].

Charge transfer excitons are particularly interesting in view of the above. They represent an intermediate step between the Frenkel and Wannier cases, with the electron and hole separated on average by one intermolecular distance, and are expected to possess certain distinct properties. It should be noted that the existence of CT excitons has long been postulated also in one-component molecular crystals such as anthracene [21,22] but they are usually difficult to observe by direct methods due to the prevalence of Frenkel excitons. CT crystals offer an opportunity to study their properties experimentally.

One peculiarity of the CT excitons is the intrinsic sensitivity to lattice phonons for reasons easy to understand qualitatively: as the wave function of a CT state of a complex is spread over (at least) two molecules, it is very sensitive to any changes of intermolecular distance and configuration. Therefore, it may be expected that the propagation properties of CT excitons will be different from those of Frenkel excitons.

Another feature of CT excitons is a possibility of "self-trapping". Upon excitation from the neutral ground state, the molecules forming the complex acquire opposite electric charges and attract each other much more strongly than in the ground state. This leads to a change of intermolecular equilibrium distance, clearly detectable by the significant Stokes shift of the CT emission relative to absorption. It has been proven theoretically [23] that such an ionic pair may have a lower energy than the surrounding predominantly nonionic pairs and thus may be trapped in the lattice. Despite the theoretical proof, the experimental evidence for this effect is still rather scarce and this very interesting phenomenon remains to be explored further.

When discussing triplet excitations in weak CT crystals, a certain paradox must be pointed out: not all of them are indeed CT excitons. Unlike singlet excitons, which closely resemble CT excitons having a CT character approaching 100%, triplet excitons in CT crystals represent a more complex situation. The most important property is their CT character, which can vary over a very broad range. Accordingly, we may expect that in certain aspects, they will approach the Frenkel limit ($c_1^2 = 0$), and in others, the CT limit ($c_1^2 = 1$). The properties of triplet excitons in CT crystals have not been extensively discussed to our knowledge, in contrast to the singlet excitons, which have been the object of at least one review [24]. It is our opinion that they deserve more attention. Thus, the principal aim of this work has been to collect all the available knowledge about properties of triplet excitons in weak CT crystals and to systematize them. The data available are a result of three general kinds of experiments, of which the first two play the more important role:

  i. optical experiments depending on detection of $S_0 \leftrightarrow T_1$ transitions;
 ii. magnetic resonance, depending on observation of transitions between the triplet sublevels;
iii. photoconductivity measurements, depending on the interaction between triplet excitons with doublet states (charge carriers).

In all cases, much information on mobile triplet excitons has become available only in recent years, principally due to improvements in experimental techniques; mobile triplet excitons are generally more difficult to detect and investigate than isolated triplet states, localized either on defects in crystals or as complexes in glasses.

II.   CHARGE TRANSFER CHARACTER OF TRIPLET EXCITONS

A.   General

It was generally pointed out in Section I that the CT character of triplet states and hence also of mobile triplet excitons may vary over a broad range. In Section II we show the experimental methods of determining the CT character of the triplet state. These methods are in principle the same for the localized and mobile species and can be divided into optical spectroscopy and magnetic resonance. We concentrate first on optical experiments (Section II.B) and later on magnetic resonance (Section II.C). In doing so, we will usually discuss first the properties of the triplet state in general (i.e., an isolated state) only to indicate later the profound changes appearing when the excitation becomes mobile in a crystal.

B.   CT Character and the Optical Spectra of the Triplet State

1.   Singlet-Triplet Energy Separation

As was pointed out in Section I.B, the resonance interaction between the ionic and "no-bond" zero order triplet states of a complex determines the actual energy of the lowest T_1 state. The reverse is also true to a certain extent, that is, the experimentally determined energy of the T_1 state points at its character. The energy of the T_1 state (or in a crystal of the T_1 band) is therefore a very important observable. In principle, it can be determined from two types of optical experiments: observation of the spectral position of the T_1 state emission (i.e., phosphorescence), or the $T_1 \leftarrow S_0$ absorption spectrum. In the case of a crystal, there usually appear several triplet states, most of them isolated and located on structural defects, their energy determined by the nature of the deformation. Much more important, however, is the energy of the triplet band, as it characterizes the bulk of the crystal. The precise position of the singlet and triplet bands is therefore a very important parameter.

The easiest way to determine the position of the triplet band, the observation of a direct band phosphorescence, is difficult if not impossible in most CT crystals, as the population of the triplet band is efficiently decreased by numerous processes discussed in Section IV. At low temperatures, it is commonly dominated by the trap phosphorescence which, as said before, is only an approximate indication of the triplet band energy since it originates from defects situated below the band.

As an alternative to emission experiments a singlet-triplet absorption measurement is in principle always possible, but this absorption is usually very weak (the $T_1 \leftarrow S_0$ transition is spin-forbidden) and not always technically feasible. For this reason it is usually recorded indirectly by monitoring the emission intensity while scanning the excitation wavelength. This method is known as excitation spectroscopy, and is also analogously used to determine the energy of the S_1 band by monitoring the fluorescence. If mobile excitons are to be characterized, the emission monitored should be the delayed fluorescence as it originates from bimolecular processes involving mobile species (Section IV). If, as is often the case, delayed fluorescence cannot be used for monitoring absorption, trap phosphorescence can be used as a measure of $T_1 \leftarrow S_0$ absorption. Care must be applied, however, in order to make sure that the defects responsible for this emission are populated via the triplet excitonic band, and not directly from the S_0 or S_1 state.

The energies of the S_1 and T_1 bands of those CT crystals for which information is available in the literature are collected in Table I. Their precise values were determined if possible from the so-called zero-phonon lines (see Section II.B.2) in the absorption. In singlet states, such lines

TABLE I

Excited Singlet (E_S) and Triplet (E_T) Band Energies, Singlet-Triplet Gap $(E_S - E_T)$, Shift of Triplet Emission versus Local (Donor or Acceptor) T_1 Energy, and CT Character of Triplet Excitons

Crystal	E_S (cm^{-1})	E_T (cm^{-1})	$E_S - E_T$ (cm^{-1})	T_1 shift (cm^{-1})	CT character (%)
A/PMDA	18320 [33]	15543 ± 3 [31]	2777	+790	7 [31]
A/TCNB	19000 ± 100 [34]	15380 ± 3 [31]	≈3600	+627	5 [35] 9 [31]
A/TCPA	20680 ± 50 [36]	15289 ± 3 [30]	≈5400	+536	3 [37] 8 [31]
		15329 ± 3 [30]		+576	
Ph/TCNQ	16600 [38]	8865 [39]	7735		<10 [38]
F/TCNQ	16500 [38]	9863 [39]	6637		<10 [38]
N/TCNB	23510 [26]	21352 ± 5 [27]	2158	+144	30 [40]
Nd$_8$/TCNB	23500 ± 100 [32]	21448 ± 15 [32]	≈2000		27 [41]
		21670 ± 100 [32]	≈1800		high CT?
N/TCPA	≈25200 [42]	21470 ± 5 [27]	3757	+262	6 [27] <10 [43]
		21617 ± 5 [27]	3610	+409	?
N/PMDA		21720 ± 50 [32]		+512	CT?
P/TCPA	22800 ± 100 [44]	21162 ± 2 [28]	≈1600	−440	30 [44]
P/PMDA	20770 ± 40 [45]	19940 ± 40 [45]	≈830	−1660	76 [46]
B/TCNB	22500 ± 50 [47]	≈20800 [47]	≈1700	−1550	54 [48]
DPA/TCNB	22200 ± 200 [49]	≈20900 [49]	≈1300	−1450	54 [50]
F/TCNB	21660 ± 100 [51]	≥20500 [51]	≤1100	−1850	70 [51]
HMB/PMDA	20370 ± 100 [52]	20000 ± 100 [52]	≈ 370	−2300	≈100 [52]
HMB/TCNB	21325 ± 100 [52]	≥20250 [53]	≤1100	−2400	90 [54]
HMB/TCPA	22300 ± 200 [55]	22300 ± 200 [55]	≈0	−1000	≈100 [54]
Pyr/PMDA	18100 ± 100 [32]	16510 ± 5 [32]	≈1600	−260	30–35 [56]
		16740 ± 30 [32]		≈0	high CT?
Pyr/TNB	19000 ± 200 [57]	≥16900 [57]	≤2100	≈0	10 [57]
Ph/PMDA	20300 ± 100 [32]	15480 [32]	≈4800	+30	loc. Ph
Ac/PMDA		16080 ± 10 [32]		+330	loc. Ac

have been detected in the $S_1 \leftarrow S_0$ absorption of two CT crystals only: A/PMDA (A/pyromellitic dianhydride) [25] and possibly N/TCNB [26]. The energies of the S_1 band of the other crystals shown in Table I are only approximate, taken from the positions of the onset of the wide singlet-singlet absorption bands at low temperatures. In the case of $T_1 \leftarrow S_0$ absorption, sharp zero-phonon lines could be detected in some (but by no means all) crystals. In cases where these lines could not be detected, the position of the T_1 band is estimated from the onset of the broad $T_1 \leftarrow S_0$ absorption, analogously to the S_1 band. If no triplet excitation data are available, the very approximate energy of the T_1 band is taken from the onset of a shallow trap phosphorescence spectrum at the lowest temperature possible. The lack of accuracy is then indicated by large estimated errors.

In some crystals listed in Table I more than one $T_1 \leftarrow S_0$ transition has been discovered (A/TCPA, perdeuterated naphthalene/TCNB (Nd_8/TCNB), N/TCPA, Pyr/PMDA). Additional absorption bands have also been found in N/TCNB [27] and P/TCPA [28], but are not presented in Table I because their precise spectral positions are difficult to determine due to the broadness of the lines. In A/TCPA, the additional zero-phonon line is found at an energy 40 cm^{-1} [29–31], and in N/TCPA 147 cm^{-1} [27] higher than the fundamental one, and both are accompanied by their own vibronic progressions and phonon sidebands. In the other crystals, no zero-phonon lines can be detected in the additional absorptions. In the cases of N/TCNB, N/TCPA, P/TCPA and Pyr/PMDA [32], this extra absorption has been interpreted as an indication of a second, usually higher lying triplet band of a more pronounced CT character (with the exception of P/TCPA) than the lower lying one. The strongest argument that this additional absorption leads to a creation of mobile triplet excitons is the appearance of the P-type delayed fluorescence at an energy higher than that of the exciting photons. This indicates bimolecular processes and hence exciton mobility. In the case of A/TCPA, the origin of the second triplet band may be looked for in the particular crystal structure consisting at low temperature of two symmetry-independent lattices [11]. This is discussed in detail in Section III along with the triplet exciton dynamics in this crystal. Due to the distinct characteristics of the triplet excitation spectra in these systems (sharp zero-phonon lines, each accompanied by its vibronic progression), the presence of a second, higher lying triplet band in both A/TCPA and N/TCPA seems to be established without much doubt. In the other crystals mentioned, this interpretation of the excitation experiments needs to be confirmed by some other independent method.

The difference between the energies of the S_1 and T_1 states is often called the singlet-triplet energy gap. It follows from Table I that it can vary in different CT crystals from a few thousand cm^{-1} (5400 cm^{-1} in the case of the A/TCPA crystal [11]) to almost zero in HMB/TCPA [53]. The CT character of the triplet state in these two crystals has been determined independently by optical and ESR methods, and found to vary from close to zero (3–8% in A/TCPA [37,31]) to almost 100% in HMB/TCPA [54]. Thus, the singlet-triplet gap gives a certain indication of the CT character of the T_1 state in CT crystals, which may be intuitively expected from the definition for the exchange integral (Eq. 1.5). This is further confirmed by a plot of the S_1-T_1 energy gap versus CT character for a number of crystals for which the corresponding data are available (Fig. 2.1). In this figure, the energy differences between the S_1 and the fundamental T_1 bands in several CT crystals are plotted. The value of the singlet-triplet gap has certain far-reaching consequences in triplet exciton kinetics (e.g., influencing the population of the triplet band and the excitonic lifetime by determining the efficiency of the thermal activation of triplet excitons back to the S_1 band), which is discussed in Section IV.

Figure 2.1 shows that the S_1-T_1 gap gives only qualitative information on the CT character; although the general trend of decreasing energy gap with increasing CT character is clearly discernible for crystals of different CT character, in some crystals of a very similar CT character, the S_1-T_1 gap can vary significantly. This is particularly the case for crystals with a small CT character of triplet excitons and hence with a large singlet-triplet gap. For example, although in the crystals formed by anthracene with three different acceptors (PMDA, TCNB, TCPA) the CT character

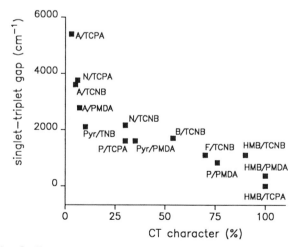

Figure 2.1. S_1-T_1 energy gap values in different CT crystals versus CT character of the triplet excitons.

of triplet excitons is similar and very small (3–8%), the singlet-triplet energy gap differs from about 2770 cm^{-1} in A/PMDA to 3600 cm^{-1} in A/TCNB to 5400 cm^{-1} in A/TCPA. As the triplet band energy in these crystals (related as usual to the energy of the ground S_0 state) differs by a much smaller factor (15,543, 15,380, 15,289 cm^{-1}, respectively [31]), the difference between these systems must lie in energy changes in the singlet manifold relative to the vacuum level, that is, involving either the S_0 or the S_1 state, or both. This is further confirmed by a blue shift of the phosphorescence in all the CT crystals containing anthracene as a donor in comparison with anthracene itself (Table I). The blue shift, calculated from the energies of excitonic triplet bands in the CT crystals and those of the corresponding donors or acceptors (whichever has a lower energy) also appears in crystals having naphthalene as a donor. This means that it is the energy of the S_0 state which becomes lowered in comparison to energy of the local (D*A) zero order state, due to the resonance interaction with the singlet CT zero order state. The mixing of the corresponding triplet zero order local state with the triplet zero order CT state is minimal in naphthalene systems, as shown by the negligible CT character of triplet excitons, so the triplet band energy remains approximately constant, the small differences being attributed to the solvent shift in different crystals.

A detailed explanation of the blue shift of phosphorescence in the A/PMDA crystal was given by Haarer et al. [58]. In this crystal, the shift is about 800 cm^{-1}. Of this value, 300 cm^{-1} can be attributed to the solvent shift. In order to account for the remaining 500 cm^{-1}, the energies of unperturbed (or zero order) states have to be considered on the basis of Eqs. (1.3) and (1.8) (Fig. 2.2). The shift of phosphorescence is blue only

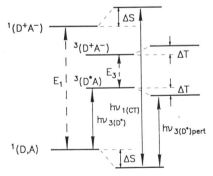

Figure 2.2. Diagram of the unperturbed (zero order: D, A, D$^+$A$^-$, D*A) and perturbed energy levels of a CT complex. E_1 and E_3 are the energy denominators in Eqs. (1.3) and (1.8). $h\nu_{1(CT)}$ is the measured $S_0 \leftrightarrow S_1$ transition energy in a CT complex, $h\nu_{3(D*)}$ is the measured $S_0 \leftrightarrow T_1$ transition energy in the donor, $h\nu_{3(D*)pert}$ is the measured $S_0 \leftrightarrow T_1$ transition energy in a CT complex, ΔS and ΔT are explained in Section I.B (after [58]).

if $\Delta S > \Delta T$, that is, if the S_0 state is stabilized stronger than the T_1 state. The denominator in Eqs. (1.3) and (1.8) in A/PMDA was estimated as $\approx 18,000 \text{ cm}^{-1}$ for the singlet, and $\approx 2500 \text{ cm}^{-1}$ for the triplet manifold. In order to have $\Delta S > \Delta T$, $(V_1)^2/(V_3)^2$ must be larger than 7, that is, V_1/V_3 must be larger than 2.5. In other words, the resonance interaction between the singlet zero order states must be at least 2.5 times stronger than that between the triplet zero order states.

Haarer et al. [58] showed that this is indeed the case by Hückel calculations of the overlap of the orbitals of anthracene and PMDA involved in the creation of the electronic states of the complex, based on the mutual molecular configuration derived from the A/PMDA crystal structure. Their result is $V_1/V_3 \simeq 12$, which easily explains the lowering of the S_0 state and thus the observed blue shift of phosphorescence in the crystal.

As mentioned in Section I.B, the mixing of the neutral (no-bond) zero order singlet state with the higher lying ionic zero order state not only lowers the energy of the ground state, but also gives it an admixture of CT character. This admixture can be also estimated from the energy lowering of the ground state (Eq. 1.4). In A/PMDA, c_1^2 was found by Haarer et al. to be $\approx 4\%$. A similar estimation made for the A/TCNB crystal yielded a similarly negligible CT character of the ground state of $\approx 2.5\%$ [34].

In CT crystals with anthracene and naphthalene serving as donors, the T_1 emission shift is towards higher energies, that is, "blue". In most crystals with higher lying donor triplet bands, however, the shift is toward lower energies, that is, "red". This red shift is easily understood on the basis of the Mulliken's theory. As the triplet energy of the donor (or in some cases acceptor) increases, it approaches the energy of the CT zero order state and the resonance interaction pushes the energy of the perturbed (i.e., physical) T_1 state down, sometimes to a much lower energy than that of the corresponding donor or acceptor. The T_1 state acquires at the same time a significant CT character (Table I).

2. Exciton-Phonon Coupling

A deeper insight into the properties of the triplet state in CT crystals is offered by an analysis of the shape, width and intensity of their optical spectra. It was observed early on that the $S_1 \leftarrow S_0$ absorption and $S_0 \leftarrow S_1$ emission of CT complexes in solution is broad and almost structureless [59]. As the S_1 state in weak CT complexes is practically a CT state, these attributes are characteristic for any CT absorption or emission, and accordingly are found also in the triplet emission of those complexes in which the T_1 state is to a large degree a CT state [3]. The high sensitivity of the CT wave function (and hence the CT state energy) to the intermolecular distance and configuration was suggested in [60] as a way to explain this phenomenon. In solution, these parameters may fluctuate around the equilibrium values which results in the inhomogenuous broadening of the spectra.

This interpretation, however true for the isolated complexes in solution, fails in the case of crystals. The mutual configuration of the component molecules in a crystal is rather well defined and linewidths on the order of a few cm^{-1} are to be expected. Contrary to this, linewidths of a few thousand cm^{-1} are quite common in the observed spectra of CT crystals. Another mechanism must therefore be responsible for this broadening in crystals. This mechanism is the exciton-phonon coupling.

As was mentioned briefly in Section I, the transfer of an electron in a CT exciton implies a strong coupling to the lattice phonons. This coupling is quantitatively expressed by the Huang–Rhys factor S [61], which represents the average number of phonons coupled to an exciton. A system with $S < 1$ is usually described as characterized by a weak coupling (e.g., excitons in one-component molecular crystals), for $1 \leq S \leq 10$ the coupling is intermediate, and for $S > 10$ it is classified as strong. The exciton-phonon coupling causes an appearance of additional bands in the optical transition. In absorption, they appear at the energy corresponding to the sum of the purely electronic transition (called a zero-phonon line) and the energy of the coupled phonon. In emission, they show up at the difference of these energies. These lines, known as phonon sidebands, form a progression. In the strong coupling case, the large number of phonons coupled to the exciton makes the phonon sidebands merge into one broad line, called a phonon wing. The relation between the intensity of the zero-phonon line and the phonon sidebands or wing is then given by [62]

$$S = -\ln\left(\frac{I_{ZPL}}{I_{ZPL} + I_{PH}}\right) \qquad (2.1)$$

where I_{ZPL} is the intensity of the zero-phonon line, and I_{PH} is the intensity of the phonon sidebands and wing. It follows from this that the intensity of the zero-phonon line decreases exponentially with the increase of S, with the corresponding increase of the phonon wing. The Huang–Rhys factor S is temperature-dependent according to this formula (valid for low temperatures) [62],

$$S(T) = S(0)\left[1 + \frac{2\pi^2 T^2}{3T_D^2}\right] \qquad (2.2)$$

where $S(0)$ is the Huang–Rhys factor extrapolated to $T = 0$ K, and T_D is the Debye temperature.

Of all the weak CT crystals known to us so far, in only two cases have zero-phonon lines been discovered in the $S_0 \leftrightarrow S_1$ transitions. The best known case is the A/PMDA crystal, which has since been extensively examined by a variety of methods [25,33,63–67]. The detection of the zero-phonon line in the $S_1 \leftarrow S_0$ absorption and reflectivity of this crystal can be attributed to the moderate value of $S \approx 6$ for the S_1 excitons, which makes them an example of intermediate exciton-phonon coupling. Also, the exciton seems to be coupled to one predominant phonon frequency of 26 cm^{-1}, unlike some other cases (e.g., P/PMDA, [33]). The good crystal quality achieved in A/PMDA may also play a role, considering the small width of this line (2.2 cm^{-1} [25]). In explaining the moderate value of the S factor in A/PMDA, Haarer [33] developed a theoretical Coulombic model, which led to good agreement between the theory and experiment. It also yielded the relation between the Huang–Rhys factor and the CT character of the exciton, which is given by the following formula:

$$S = \left(\frac{7}{32\pi}\right)^2 \frac{c_1^4}{\epsilon^2 \omega_n^3 d^4 \hbar m} \tag{2.3}$$

where c_1 is the CT admixture in the excited state wave function, ϵ is the dielectric constant of the medium, ω_n is the frequency of the coupled phonon, d is the average distance between the molecules, and m is the mass of the molecule (in this particular case PMDA).

Because the Huang–Rhys factor is dependent on temperature (Eq. 2.2), it should be extrapolated to $T = 0$ K before calculating the CT character. From Eq. (2.3) it follows that the S factor is proportional to the square of the CT character of the excited state (defined itself as c_1^2). Equation (2.3) was obtained after multiple approximations, so the estimation of CT character from the S factor is a semi-quantitative one only. However, it leads to good results when applied to triplet excitons, as will be shown below.

In contrast to singlet excitons in CT crystals, usually characterized by a large CT character and consequently a large S value (which makes the zero-phonon lines disappear in most cases under the phonon wings), triplet excitons may have a varying degree of charge-transfer character, hence some of them may fall into the intermediate exciton-coupling category. In fact, zero-phonon lines with the accompanying phonon wings have been observed in the $S_0 \leftrightarrow T_1$ transition spectra in several cases. Chronologically, perhaps the first observation of this kind was the report by Beckmann and Small [68] of the phosphorescence of the Pyr/PMDA complex embedded in the N/PMDA crystal. This was followed by the phosphorescence studies of Haarer [69] of the N/TCPA crystal. In both cases, trap phosphorescence was observed and thus the results are relevant for isolated trap states but not necessarily for mobile excitons. Later, sharp zero-phonon transitions were directly observed in the $T_1 \leftarrow S_0$ absorption of thick N/TCPA and N/tetrabromophthalic anhydride crystals [42].

The first observation of a zero-phonon singlet–triplet band transition and the accompanying phonon structure in an excitation spectrum was reported by Kaesdorf [28] for the CT crystal P/TCPA. The zero-phonon line at 21,162 cm^{-1}, giving the energy of the T_1 excitonic band, is rather weak in this system in comparison with the accompanying phonon wing. However, the progression of very weak phonon sidebands can also be observed (Fig. 2.3). The Huang–Rhys factor extrapolated to 0 K was found to be 6.0 ± 0.2 (see Table II); the exciton-phonon coupling in P/TCPA belongs therefore to the intermediate category. Since the exciton appears to be coupled to more than one phonon mode, the model of Haarer [33] is not applicable for determination of the CT character of the triplet state in this case.

A more detailed analysis of the exciton-phonon coupling was reported in the case of a series of crystals formed by anthracene as a donor with various acceptors [31]. In all cases the zero-phonon lines were detected in the delayed fluorescence excitation spectra. In A/TCPA, two zero-phonon transitions were found, each of them accompanied by its own phonon structure (Fig. 2.4). The Huang–Rhys factors determined from the intensities of the zero-phonon lines and phonon wings were found to

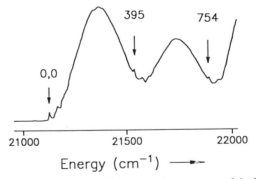

Figure 2.3. Phonon structure in the triplet excitation spectrum of the P/TCPA crystal at 5.3 K. The arrows point at the zero-phonon line in the 0–0 transition and its vibronic progression (from [28]).

TABLE II
Huang–Rhys Factors (Extrapolated to 0 K) for Triplet Excitons and CT Character of the T_1 State Derived From Them

Crystal	$S(0)$	c_1^2 (%)	Ref.
P/TCPA	6.0 ± 0.2		[28]
A/PMDA	0.9	6.7	[31]
A/TCNB	2.1	9.6	[31]
A/TCPA	1.4	8.0	[31]
N/TCPA	0.9 ± 0.3	≈ 6	[27]
Ph/PMDA	0.65		[32]

lie in the 0.9–2.1 range (Table II). The exciton-phonon coupling in these systems can be thus classified as weak-intermediate. Similar observation was made in the crystal N/TCPA [27], where the exciton-phonon coupling is weak ($S = 0.9 \pm 0.3$). In other systems investigated (N/PMDA [32], F/TCNB [51], P/PMDA [33] or HMB with various acceptors [52–55]), the zero-phonon lines could not be found in the triplet excitation spectra, which consist exclusively of very broad phonon wings. Not surprisingly, the magnetic resonance studies on these crystals (except N/PMDA) reported in Section II.C.3 found a large CT character of the triplet state.

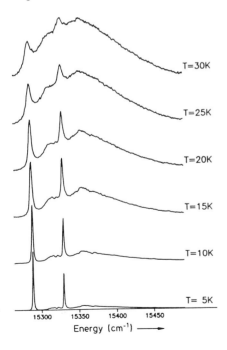

Figure 2.4. Temperature dependence of the zero-phonon lines in the triplet excitation spectra of the A/TCPA crystal. The doublets originate from the specific crystal structure of A/TCPA (from [27]).

The broad absorption bands also appear independent from the sharp zero-phonon transitions in the crystals N/TCNB, Nd_8/TCNB, P/TCPA and Pyr/PMDA. Should they indeed correspond with the higher triplet bands, as interpreted by the authors, excitons in these bands would be characterized by a higher CT character than in the fundamental bands. However, the presence of triplet excitons in these bands has not been confirmed so far by other methods.

C. CT Character and Paramagnetic Properties of the Triplet State

1. Zero Field and High Field Splitting of Triplet Sublevels

The two unpaired electrons in the triplet state of an isolated molecule (or complex) are coupled by magnetic dipolar interaction, whose Hamiltonian is given by

$$\mathbf{H} = \frac{1}{2}(g_e\beta)^2\left\{\frac{\bar{\mathbf{S}}_1 \cdot \bar{\mathbf{S}}_2}{r^3} - 3\frac{(\bar{\mathbf{r}} \cdot \bar{\mathbf{S}}_1)(\bar{\mathbf{r}} \cdot \bar{\mathbf{S}}_2)}{r^5}\right\} \tag{2.4}$$

where g_e is the free electron g factor, β is the Bohr magneton, $\bar{\mathbf{S}}_{1,2}$ are the spin operators of the two electrons, and $\bar{\mathbf{r}}$ is the distance vector between the electrons.

This Hamiltonian acts on the spin part of the two-electron triplet state wave function $\Psi(1,2)$. As a result, the degeneracy of the triplet function is lifted and a splitting into three sublevels takes place (Fig. 2.5). This effect is called zero field splitting (zfs), zero field meaning that the splitting appears in the absence of any external (magnetic or electric) field. Alternatively the term fine structure (fs) is often used.

The zfs splitting is represented by a 3×3 tensor $\mathbf{F}$ whose elements F_{ij} in a laboratory frame are given by [46]

$$F_{ij} = \langle\Psi(1,2)|\mathbf{H}_{ij}|\Psi(1,2)\rangle \tag{2.5}$$

where

$$\mathbf{H}_{ij} = \frac{1}{2}(g_e\beta)^2\frac{3x_{12}^i x_{12}^j - r_{12}^2\delta_{ij}}{r_{ij}^5}$$

x_{12}^i is the ith component of r_{12}, and δ_{ij} is Kronecker's delta.

The tensor $\mathbf{F}$ can be diagonalized by a transformation to a new set of coordinates corresponding to the principal axes. In this axis system, the tensor is represented by three diagonal principal values X, Y and Z, and is traceless, that is, only two of the principal values are independent ($X + Y + Z = 0$). As an alternative to the principal values X, Y and Z, the trace of the tensor $\mathbf{F}$ is often represented by zfs parameters D and E, defined as follows:

$$D = \frac{3}{4}(g_e\beta)^2\langle\Psi(1,2)|\frac{r_{12}^2 - 3z^2}{r_{12}^5}|\Psi(1,2)\rangle \qquad (2.6)$$

$$E = \frac{3}{4}(g_e\beta)^2\langle\Psi(1,2)|\frac{y^2 - x^2}{r_{12}^5}|\Psi(1,2)\rangle \qquad (2.7)$$

The zfs parameters are related to the principal values

$$D = \frac{Y + X}{2} - Z = -\frac{3}{2}Z \qquad (2.8)$$

$$E = \frac{Y - X}{2} \qquad (2.9)$$

Another zfs parameter D^m, "mean D parameter", was used in early articles on the paramagnetic properties of the triplet state for reasons explained in Section II.C.3. It is defined as $D^m = \sqrt{D^2 + 3E^2}$.

In an external magnetic field $\bar{\mathbf{B}}$, the spin Hamiltonian for a molecule in the triplet state is the sum of the zfs and Zeeman term:

$$\mathbf{H} = g_e\beta\bar{\mathbf{B}}\cdot\bar{\mathbf{S}} + D(S_z^2 - \tfrac{1}{3}\bar{S}^2) + E(S_x^2 - S_y^2) + \cdots \qquad (2.10)$$

$$\text{(Zeeman term)} \qquad \text{(zero-field splitting)}$$

(neglecting at this point the hyperfine and nuclear Zeeman terms).

The result of the zero field and high field Hamiltonian acting on the spin part of the two-electron triplet wave function $\Psi(1,2)$ is shown in Fig. 2.5. There is a pronounced difference in the zero field and high field

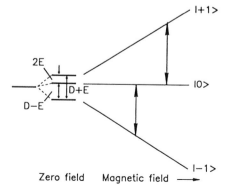

Figure 2.5. The effect of the zero field (Eq. 2.4) and high field (Eq. 2.10) Hamiltonian acting on a triplet wave function of a molecule. In high field the molecule is oriented so that $B \parallel x$. The solid arrows represent the three allowed transitions in zero field and the two allowed transitions in high field.

picture. While in zero field there are three transitions possible and equally allowed as a function of electromagnetic radiation frequency, in high magnetic field only two transitions are allowed, corresponding to the change of magnetic spin quantum number m_s of $\Delta m_s = \pm 1$. The third transition corresponding to $\Delta m_s = \pm 2$ is forbidden although it is observable in certain conditions (Section II.C.3).

2. *Experimental Approaches*

The lowest excited triplet states of most CT complexes are typically $\pi\pi^*$ states. The zero field splitting of such a triplet state is on the order of 0.1 cm^{-1} which corresponds to a frequency of 3 GHz. Thus in principle three transitions between the triplet sublevels should be observed by measuring the resonance absorption of microwaves in the GHz range. From the transition frequencies, the fine structure parameters can be determined directly. In practice, it is technically difficult to measure

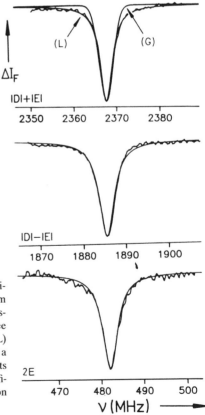

Figure 2.6. An example of ODMR experiment: fluorescence-detected ODMR spectrum of mobile triplet excitons in the A/TCPA crystal at 1.2 K. The spectrum consists of three transitions. In the top transition $(|D| + |E|)$ (L) means a fit to a Lorentz curve, (G) is a fit to a Gauss curve. For the other two signals the fits were made to Lorentz curves only. The significance of the lineshape is discussed in Section III.B (from [37]).

microwave absorption while changing the frequency over a sufficiently broad range, so a technique was developed allowing detection of these transitions indirectly, called optical detection of magnetic resonance (ODMR). Generally speaking, ODMR depends on the fact that the triplet sublevels are characterized by different radiative optical decay constants. In a typical ODMR experiment, triplet concentration is monitored while the microwave frequency is swept continuously. The resonance absorption of the microwaves changes the populations of the two coupled triplet sublevels, which in turns leads to a change of the triplet or singlet emission intensity [70] (Fig. 2.6). ODMR techniques have advanced considerably in recent years: the original method used the direct triplet phosphorescence for monitoring purposes; later developments monitored triplet-triplet absoprtion [71], delayed fluorescence [72], $S_0 \leftrightarrow S_1$ transitions via prompt fluorescence [73] and singlet-singlet and singlet-triplet absorption [74]. ODMR has been extensively reviewed elsewhere (see e.g., [75]). Its great merit lies in the precision in determining the zfs parameters of the triplet state. Also, it makes it possible to examine optical properties of the individual triplet sublevels rather than the triplet state as a whole. A drawback of this technique for fine structure determination is that while giving precise values of zfs parameters, it makes it very difficult to obtain information on the orientation of the zfs tensor **F** relative to the laboratory frame.

A schematic drawing of an ODMR experiment is shown in Fig. 2.7(a)

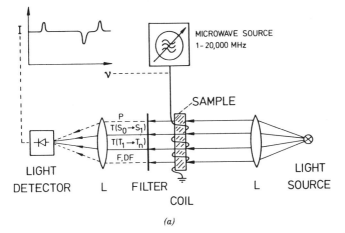

(a)

Figure 2.7. (a) A typical setup of a ODMR experiment. L, lens. Between the filter and one lens, five of the potential detection methods are presented schematically: phosphorescence, singlet-singlet absorption, triplet-triplet absorption, fluorescence, and delayed fluorescence. (b) A block diagram of an ODMR setup. Explanations in the text.

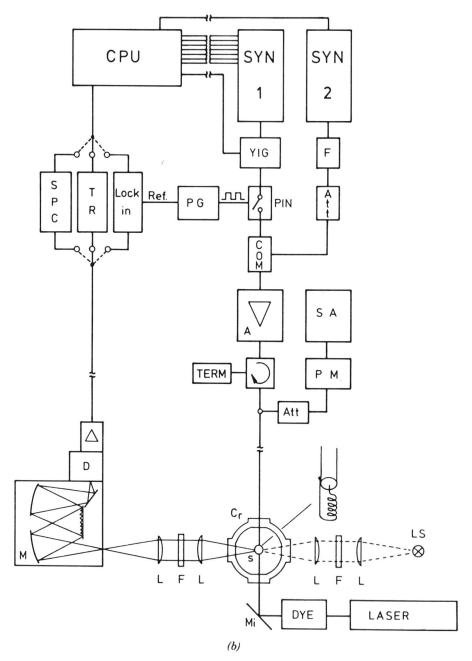

(b)

Figure 2.7. (*Continued*).

while figure 2.7(b) depicts the block diagram of an advanced ODMR setup (Stuttgart University). The optical paths are shown in the lower part. Light excitation is achieved either by an argon-ion laser with a tunable dye laser via the mirror (Mi) or by an arc or halogen lamp (LS), selected in wavelength by the filter (F) and focused on the sample (S) via the lenses (L). The sample is positioned in a coil placed inside a variable-temperature helium cryostat (Cr). The emitted or transmitted light is focused on the entrance slit of a monochromator (M), equipped with a light detector (D) in form of a diode or a photomultiplier. The electric output signal passes a preamplifier to a single photon counter (SPC), to a transient recorder (TR) in the case of fast-varying amplitudes, or a lock-in amplifier, respectively, to be averaged by the central processing unit (CPU). The CPU controls also the microwave part of the setup. Synthesizer 1 (SYN1) provides the frequency-variable microwaves. This frequency source is a "direct" synthesizer, allowing phase-coherent frequency changes within the microwave time scale in contrast to "phase-locked-loop" generators with locking times on the order of 10^{-2} sec. Fast frequency changes demand a prebuffer, driving the synthesizer (Ailtech 380) parallel in BCD code. Unwanted sidebands and harmonics of SYN1 are supressed by a tunable YIG filter (YIG), synchronized to the center output frequency. Amplitude modulation is done by a PIN-diode switch (PIN), driven by a pulse generator (PG) synchronizing the lock-in. For EEDOR and hole-burning experiments the power of a second frequency source (SYN2) is added via the filter (F), the attenuator (Att) and the combiner (COM). Power amplification is achieved by the amplifier (A), protected against mismatch by a circulator with a 50 Ohm load (TERM). Via the attenuator (Att) power spectral purity and frequencies are monitored by a power meter (PM) and a spectrum analyzer (SA). The microwaves are finally fed to the coil surrounding the sample by a semirigid 50 Ohm transmission line.

An alternative to zero field experiments is offered by high field techniques, among which ESR plays the most prominent role. Chronologically it was the first magnetic resonance method applied to investigate the paramagnetic properties of aromatic molecules [76] and later also CT complexes [77]. The general principle of this experiment is to apply microwaves of a constant frequency to the sample and measure their absorption directly as a function of the external magnetic field. One expects to observe a doublet of signals equally spaced about approximately the g-value of the free electron ($g_e \simeq 2$), as follows from Fig. 2.5. This is exactly what is encountered in the ESR spectra of single crystals where the molecules are oriented in a uniform pattern. An example of such a spectrum is shown in Fig. 2.8.a.

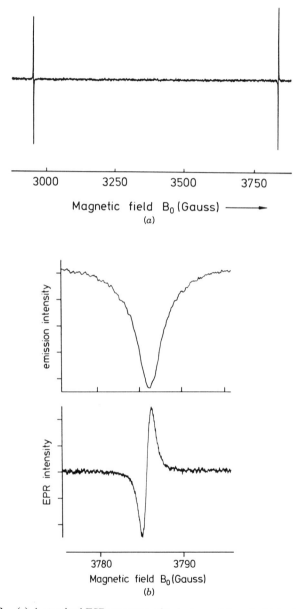

Figure 2.8. (a) A standard ESR spectrum of a single crystal excited into the T_1 state. In this case, mobile triplet excitons at room temperature in the Ph/TCNQ crystal are observed. The magnetic field is perpendicular to the molecular planes of TCNQ and phenazine. (b) bottom: one of the ESR signals in a narrow field range. (b) top: a high field fluorescence-detected ODMR signal at the same crystal orientation and experimental conditions as the ESR signal (from [38,78]).

ESR spectra of randomly oriented molecules (either in low temperature glasses or in powdered samples) besides showing a different shape of
the allowed $\Delta m_s \pm 1$ signals also exhibit forbidden ($\Delta m_s \pm 2$) transitions.
In fact, the latter are usually stronger by at least an order of magnitude
than the allowed ones, but yield only the "mean D" parameter D^m (the
high intensity of the forbidden transitions is a result of their isotropy).

A version of ESR which is a combination of standard ESR and ODMR
is known as high field ODMR (hf-ODMR). This technique depends on
monitoring the optical emission intensity as a function of the external
magnetic field. Similarly to zf-ODMR, different kinds of optical emission
may be used for monitoring purposes. The hf-ODMR transitions appear
exactly at the same resonance fields where the ESR signals occur, as
shown in Fig. 2.8.b. The advantage of this method over ESR lies in the
usually higher sensitivity compared to the direct microwave absorption
measurements.

The two allowed high field (ESR or hf-ODMR) transitions are highly
anisotropic, that is, their resonance fields depend strongly on the orientation of a molecule (or complex) relative to the magnetic field, as follows
from Eq. (2.10). In experiments performed on single crystals, characteristic rotational patterns of resonance fields appear with a 180° period (Fig.
2.9). From these patterns the zfs parameters (or principal values) can be
extracted, usually by numerical fit procedures. Because of crystal orientation uncertainties, the zfs values obtained from high field experiments are
usually less accurate than those derived from a straightforward zero field
measurement. However, high field methods have an advantage of yielding the orientation of the zfs tensor relative to the laboratory coordinate
system, which is usually taken as either the crystal axes system, or the

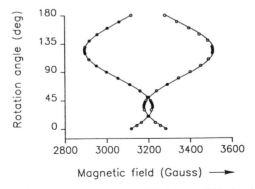

Figure 2.9. A typical angular dependence of triplet state ESR signals in a single crystal.
In this case mobile excitons in the F/TCNB at room temperature are followed with the
crystal rotated about its needle axis (from [51]).

molecular frame (in cases where the detailed structure of the crystal is known). The most accurate method of evaluating all the elements of the zfs tensor appears to be the following: measure the zfs parameters directly by a zero field experiment (ODMR) in order to obtain accurate values of the zfs parameters D and E, then use them as constants in the fitting procedure to obtain the angles between the zfs principal and the laboratory axes, as described in [79]. Of course care should be applied to assure that the same triplet species is observed in both the zero and high field experiments.

3. Zero Field Splitting and the CT Character of the Triplet State

Iwata et al. [3] originally proposed that an electron transfer taking place in the triplet state of a complex should reduce its zfs parameters. Shortly afterwards, an ESR experiment on the D/TCNB complex in a low temperature glass was performed [77]. In the ESR spectrum, the forbidden ($\Delta m_s = \pm 2$) transition of the complex appears at a distinctly higher magnetic field than that of TCNB itself, which means that the "mean D" parameter D^m of the complex (the only zfs parameter possible to extract from the $\Delta m_s = \pm 2$ signal) is much lower than that of TCNB.

In the subsequent paper of Hayashi et al. [80], a more detailed analysis of the dependence of the D^m parameter on the CT character of the T_1 state was presented. According to [80], D^m is the expectation value of a dipolar Hamiltonian $\mathbf{F}$ acting on a two-electron function $\Psi(1,2)$:

$$D^m = \langle \Psi(1,2) | \mathbf{F} | \Psi(1,2) \rangle \qquad (2.11)$$

It was shown in Section I.B that the triplet wave function of a CT complex can be written in the first approximation as a linear combination:

$$\Psi = c_1 \Psi(D^+A^-) + c_2 \Psi(D^*A \text{ or } DA^*) + \cdots \qquad (2.12)$$

Neglecting cross terms of the type $\langle \Psi(D^+A^-) | \mathbf{F} | \Psi(D^*A \text{ or } DA^*) \rangle$ and overlap integrals $\langle \Psi(D^+A^-) | \Psi(D^*A \text{ or } DA^*) \rangle$ which have been shown to be very small [58], the parameter D^m can be written as

$$D^m = c_1^2 D^m(D^+A^-) + (1 - c_1^2) D^m(DA^*) \qquad (2.13)$$

(the triplet energy of the acceptor TCNB is much lower than those of all the donors used by Hayashi et al., hence only the local zero order state (DA*) is considered to interact with the ionic zero order state).

The CT character c_1^2 can be therefore determined from the D^m value obtained in the ESR experiment, according to the formula

$$c_1^2 = \frac{D_{loc}^m - D_{exp}^m}{D_{loc}^m - D_{CT}^m} \qquad (2.14)$$

where D_{exp}^m is the experimentally found D^m value of the complex, D_{loc}^m is the experimentally found D^m value of the acceptor, and D_{CT}^m is the D^m value of the CT zero order state.

As the CT zero order state is only a hypothesis appearing in Mulliken's theory, its zfs parameters are not available experimentally. The D^m value of the zero order CT state of the HMB/TCNB complex was theoretically calculated in [80] by a point-dipole approximation to be $\simeq 0.022\,\mathrm{cm}^{-1}$. This value when applied to the series of TCNB complexes with benzene and its methylated homologues gives a CT character of the T_1 states which varies from about 5% (benzene/TCNB) to as much as 95% (HMB/TCNB).

A very similar approach to the same problem was reported independently by Beens et al. [81] for the complexes of triazine and N,N-diethylaniline (donors) with benzonitrile (acceptor). In these complexes the allowed ($\Delta m_s = \pm 1$) transitions were detected in the ESR spectra of rigid solutions along with the forbidden ones, therefore both D and E parameters could be determined separately. $|D_{CT}|$ and $|E_{CT}|$ values of the complex N,N-diethylaniline/benzonitrile were calculated by the Hückel method to be 0.024 and 0.002 cm^{-1}, respectively, a result very similar to that of Hayashi et al. [80].

The problem of estimating the CT character of the triplet state from zfs parameters was tackled once more by Möhwald and Sackmann [48]. The authors pointed out two inconsistencies of the method of Hayashi et al. [80]:

• the parameter D^m can be used in estimating the CT character only if $D \gg E$;

• D^m is not sensitive to a change of sign of D and E, thus if D_{loc} and D_{CT} have different signs, Eq. (2.14) may lead to a significant error.

Möhwald and Sackmann [48] proposed therefore neglecting D_{CT} and E_{CT} in the estimations altogether, arguing that since all the previous calculations of these parameters have yielded very small values of $|D_{CT}| \simeq 0.02\,\mathrm{cm}^{-1}$, this approximation is justified. Using simplified formulas similar to those of Hayashi et al.,

$$c_1^2 = \frac{D_{loc} - D_{exp}}{D_{loc}} \qquad (2.15a)$$

and

$$c_1^2 = \frac{E_{loc} - E_{exp}}{E_{loc}} \qquad (2.15b)$$

they estimated the CT character of triplet states in four CT crystals: A/TCNB (3%), Pyr/TCNB (11%), B/TCNB (biphenyl/TCNB) (54%) and D/TCNB (73%).

To overcome the simplifications contained in Eqs. (2.15), Möhwald and Böhm [82] and Dalal et al. [83] tried subsequently to solve the problem of the sign of the zfs parameters. Their conclusions were that the sign of the D parameter of the ionic zero order state in CT complexes is negative while one of the T_1 states of the series of CT complexes of TCNB with different donors is positive. Substituting the negative value of D_{ion} in the corresponding formula leads to a good agreement between the CT character obtained from the fine structure and from the hyperfine structure (Section VI). However, determination of the absolute sign of the zfs parameters is not possible in a simple ESR or ODMR experiment. It can be done by evaluating the forbidden hyperfine transitions in the triplet ESR spectra [84] or by an ENDOR experiment. Also, the literature data on the subject are scarce and confusing; while Möhwald and Böhm [82] reported a positive sign for a number of TCNB complexes with several donors, more recent ENDOR data for A/TCNB yielded a negative value for D_{exp}, which is also very close to D_{loc} [85].

Another problem with estimating the CT character from the fine structure by the methods discussed above is that it is derived from the D and E parameters only, that is, only two independent parameters of the zfs tensor. As was mentioned earlier, the full description of a zfs tensor requires five independent parameters to account not only for its magnitude which is given by its trace, but also for its orientation relative to the crystal axis system. It was pointed out by Keijzers and Haarer [46] that the symmetries and orientations of the local and ionic zfs tensors of the same CT complex may in principle differ. In the lowest triplet state of a planar aromatic molecule, the symmetry of the zfs tensor generally corresponds to the symmetry of the molecule, that is, the principal axes of the tensor correspond to the molecular symmetry axes. In particular, for molecules such as anthracene and naphthalene, the z principal axis is generally parallel to the z (out-of-plane) molecular axis while the x and y axes correspond to the long and short molecular axes, respectively. There are quite a few exceptions to this rule, the most notable being perhaps pyrene, in which the z principal axis points in the direction of the long in-plane molecular axis, and phenanthrene in which it also lies in the molecular plane. The symmetry and orientation of the CT tensor is a more complex problem; it depends on the CT character of the triplet state as well as on the mutual orientation of the participating molecules. Thus, the symmetry of the corresponding local zfs tensor is generally preserved for those triplet states of CT crystals which have a small CT character, for example, P/TCPA [44], Pyr/trinitrobenzene (Pyr/TNB) [57]. Similarly, the zfs axes of the T_1 state in F/TCNQ (F/tetra-cyanoquinodimethan) and Ph/TCNQ (phenazine/TCNQ) crystals point in the direction of the TCNQ symmetry axes just as they do in phenanthrene and pyrene [40]: the z principal axis lies in the molecular plane.

Investigations of the T_1 state of the Nd_8/TCNB crystal, however, detected small deviations of the zfs principal axes from the naphthalene molecular axes of up to 3°, which can be the result of a nonnegligible CT character of about 30% [41]. In systems characterized by a large CT character of the triplet excitons, this deviation may be expected to cause a grave error in estimating the CT character from the D and E parameters alone. In such a case, one needs to evaluate all five independent elements of the complete zfs tensor in the crystal axis or molecular axis frame, that is, two diagonal and three nondiagonal elements or, alternatively, after transformation of the tensor to the principal axis frame, the two principal values + three Euler angles necessary for the transformation. This was noticed by Keijzers and Haarer [46] in their investigations of the triplet state of the P/PMDA crystal.

4. Fine Structure of Mobile Excitons in CT Crystals

The experimental methods described in previous chapters for determining the CT character of an isolated triplet state are also valid for mobile excitation migrating through a crystal. However, the mobility of the excitons may have a pronounced effect on the fine structure tensor. It is known in the literature [86] that the ESR spectra of thermally activated mobile triplet excitons in certain TCNQ CT salts exhibit very strong spin exchange effects at high concentrations. As a result of this exchange, the fine structure becomes averaged as the temperature is raised. At sufficiently high temperatures, a single line in the $g = 2$ region appears instead of the expected doublet of signals for an arbitrary orientation of the crystal. This effect, however, has not been observed in weak CT crystals because of the much lower concentration of mobile triplets under usual conditions (the singlet–triplet energy gap in weak CT crystals is much larger than in CT salts, and the excitons are not populated thermally).

Another example of averaging effects in optically excited triplet excitons in one-component molecular crystals occurs in anthracene. The elementary cell of anthracene contains two magnetically inequivalent molecules [87], hence one expects two pairs of signals to appear in the ESR spectrum. This is exactly what happens, but only at low temperatures for isolated (trapped) triplet states. At higher temperatures, when the excitons become mobile, only one pair of ESR signals is observed, appearing exactly at the middle position between the trap lines [88]. This effect of averaging two molecular zfs tensors in anthracene differing in their orientation was also observed at 1.2 K by delayed fluorescence-detected ODMR. In this experiment [89], a completely new set of zero field transitions between the triplet sublevels of the excitonic state was found upon raising the microwave power.

It is therefore apparent that special care must be applied when determining the properties of mobile excitons (particularly their CT character) from their zfs tensors. In particular, a comparison of the zfs tensors with those of the shallow defects always present in the crystal, which represent the isolated triplet states, can be very useful in determining the effects linked to the mobility of excitons. Also, the most meaningful information on excitons is usually obtained at very low temperatures when many of the averaging effects, for example, thermally activated librations of the molecules, are absent.

The presence of the temperature-dependent processes affecting the zfs tensors is illustrated in Table III. In this table, zfs parameters of mobile excitons are presented in the low and high temperature limits for CT crystals for which respective data are available. Also, two one-component molecular crystals are included in order to show that the temperature effects on the zfs parameters are not specific to CT crystals alone.

It follows from Table III that in most crystals, the zfs parameters decrease with raising the temperature, the notable exception being N/TCNB. Besides those collected in Table III, other CT crystals show temperature changes of the zfs parameters: in the crystals Pyr/TNB [57]

TABLE III
High-(D^H, E^H) and Low Temperature (D^L, E^L) zfs Parameters of Mobile Triplet Excitons in CT Crystals (in cm^{-1})

Crystal	$|D^L|$	$|E^L|$	T^L (K)	$|D^H|$	$|E^H|$	T^H (K)	$\Delta D\%$	$\Delta E\%$	Ref.
A/TCNB	0.06763	0.00772	1.2	0.0667	0.00775	300	−1.4	+0.4	[72,91]
A/PMDA	0.0695	0.0079	1.6	0.0665	0.00765	300	−4.3	−3.2	[92,93]
A/TCPA	0.07093	0.00804	1.2	0.0690	0.00799	300	−2.7	−0.6	[37]
Nd$_8$/TCNB	0.0639^a	0.0063^a	130	0.0684	0.00745	300	+6.6	+15.4	[41]
N/TCPA	0.0962	0.0144	77	0.0922^a	0.0142^a	200	−4.2	−1.4	[43]
F/TCNQ	0.06144	0.00740	120	0.05982	0.00700	280	−2.6	−5.4	[38,94]
Ph/TCNQ	0.06149	0.00802	1.2	0.05973	0.00752	300	−2.9	−6.2	[38,78]
A	0.0518	0.00144	1.2	0.0524	0.00136	300	+1.2	−5.6	[89,88]
β-DCA	0.06997	0.00626	1.2	0.06688	0.00676	300	−4.4	−8.2	[95,96]

^{a}Calculated from data in Refs. 41 and 43, respectively. β-DCA, β-9,10-dichloroanthracene.

and F/TCNB [51], the D and E values decrease with elevating temperature while in B/TCNB the trend is reversed; the zfs parameters actually increase with elevating temperature [90], similar to N/TCNB. This observation is very interesting, as the B/TCNB crystal is almost isomorphic in structure with F/TCNB, which illustrates the complexity of the processes leading to the temperature changes of the zfs parameters.

These averaging processes can be separated into the following classes, although in most cases several processes may simultaneously contribute to the experimentally observed changes of zfs parameters:

a. Thermal expansion of the lattice. This process changes the intermolecular distance and configuration and thus leads to a modification of the electron density distribution in the excited state. Thermal expansion is of general importance, not just for CT crystals. In CT crystals, the average distance between the two electrons forming the triplet state is enhanced, consequently the zfs parameters are reduced with increasing temperature. This phenomenon is not linked to the mobility of excitons.

b. Dynamic orientational disorder. Molecular librations prominent in many molecular crystals at high temperatures may lead in principle to two different processes depending on the nature of librations. In A/TCNB, the information on librations available from temperature-dependent X-ray experiments suggests a model of averaging tensors corresponding to the extreme complex configurations, analogous to the "jumping spin" formalism [97]. In another CT crystal, N/TCNB, the librations of the naphthalene molecule lead to a decrease of the overlap between the donor and acceptor thus reducing the CT character of the T_1 state and consequently increasing the parameter D [41]. Both models are discussed in Section V which covers the molecular dynamics in CT crystals.

c. Exciton dynamics. This process was mentioned above for the case of anthracene crystals. In CT crystals due to the low dimensionality of the structure, the excitons move preferentially along one kind of stack which is usually built of uniformly oriented complexes. Therefore in most cases there is no averaging of the zfs tensors belonging to two differently oriented stacks. In many cases, however, a temperature-dependent orientational disorder takes place within a stack [7]. An exciton moving in the stack encounters an increasing number of misoriented molecules and the averaging process becomes increasingly effective. The averaged value of the D parameter is always smaller hence this process can account for the changes of the zfs parameters in most CT crystals.

In the anthracene crystal, both orientations of the molecules are energetically equivalent. There are other crystals known in which this is not the case. Each of the inequivalent sublattices then gives rise to a separate triplet energy band. An exciton may then move between the bands as well as within them. If each of the bands has a different zfs tensor, such a process will definitely influence the zfs parameters, and may be analyzed from the point of view of a "jumping spin" formalism. In particular, the observed resonance fields will become the mean values of both states weighted by their populations (depending on the temperature). This model was first proposed for the crystal A/TCNB [98]. The upper triplet state was estimated from a fit procedure to lie $589 \, cm^{-1}$ higher than the lowest (T_1) one, and its CT character was estimated at 30%. The same interpretation was subsequently used to account for the temperature changes of the zfs parameters in the N/TCPA crystal [43]. At the time as the ESR experiments on A/TCNB and N/TCPA crystals were performed, the existence of the higher lying triplet bands was merely a hypothesis. In the case of A/TCNB, there is no reason for the existence of a higher band from the point of view of its structure, and it has not been found since by any experimental method other than ESR, so the parameters characterizing the upper triplet state (energy and zfs parameters) remain the result of quite an arbitrary multi-parameter fit. However, in the case of N/TCPA, a higher lying triplet band was recently detected by triplet excitation spectroscopy [27] and its energy ($147 \, cm^{-1}$ higher an origin than T_1) agrees remarkably well with the result of the numerical fit of the temperature dependence of ESR resonances ($150 \, cm^{-1}$ [43]).

The presence of a second, higher lying triplet band was also established by excitation spectroscopy in the A/TCPA crystal [29,30]. In this crystal, the zfs parameters have the usual tendency to increase with decreasing temperature and this trend continues down to very low temperature (4.2 K) [37]. A model based on the "jumping spin" formalism explains the observed changes in this case and yields a reasonable value for the activation energy, which is close to the energy difference of the two bands ($40 \, cm^{-1}$). This is discussed further in Section III, which covers triplet exciton dynamics.

Taking into consideration the above aspects of the dependence of the zfs parameters on external factors, we have collected the available data on mobile triplet excitons in CT crystals which we present in Tables IV and V.

In Table IV we gathered the zfs parameters of the excitons obtained at the lowest possible temperature at which the excitation is indeed mobile. The zfs parameters obtained by zero field experiments are given preference. In certain cases no information on mobile species was available, therefore the data concerning shallow traps in these systems are presented.

TABLE IV
Zfs Parameters of Mobile Triplet Excitons in CT Crystals

Crystal	D or $|D|$ (cm^{-1})	E or $|E|$ (cm^{-1})	c_1^2 (%)	Method	T (K)	Ref.
A/TCNB	0.06763	−0.00772		ODMR	1.2	[72]
	0.06730	−0.00782				
	±0.0678(3)	∓0.0077(3)	4	ESR	30	[91]
A/PMDA	+0.0695(3)	−0.0079(3)	≤5	ESR	1.6	[92]
A/TCPA	±0.07093	∓0.00804	3	ODMR, ESR	1.2	[37]
A/DMPM	0.0701	0.0081		RYDMR	300	[99]
Nd$_8$/TCNB	±0.0639	∓0.0063	32	ESR	130	[41]
N/TCPA	±0.0962(2)	∓0.0144(1)	≤10	ESR	77	[43]
P/PMDA	+0.0228(1)	−0.0015(1)	76 ± 5	ESR	130	[46]
P/TCPA	±0.090(10)	∓0.017(5)	30	ESR	100	[44]
F/TCNB	±0.0298(1)	∓0.0097(1)	70	ESR, ODMR	300	[51]
F/TCNQ	±0.06144	∓0.00740	≈0	ODMR, ESR	120	[38]
B/TCNB	±0.0527	±0.0168	54	ESR	300	[100]
Ph/TCNQ	0.06149(5)	0.00802(5)	≈0	ODMR	1.2	[38]
	±0.06137(6)	∓0.00796(8)		ESR	4.2	[78]
Pyr/TCNB	$D^m = 0.081$		11[a]	ESR	10	[48]
Pyr/TNB	±0.0596(3)	∓0.0310(3)	10	ESR	100	[57]
Pyr/PMDA	0.05550(15)	0.00883(6)	30[a]	ODMR	1.2	[56]
D/TCNB	±0.034	∓0.006	73[a]	ESR	10	[48]
BP/TCNB	±0.1197	∓0.0037	≈0[a]	ESR	250	[101]
DPA/TCNB	±0.0448	±0.0125	54	ESR	300?	[50]
HMB/PMDA	0.0162	0.0029	≃100[a]	ODMR	1.2	[52]
HMB/TCNB	0.0285(1)	0.0066(1)	90[a]	ODMR	1.2	[54]
HMB/TCPA	0.01525(5)	0.00342(5)	≃100[a]	ODMR	1.2	[54]

[a]Trapped exciton.

TABLE V
Comparison of the Room Temperature zfs Parameters of Mobile Triplet Excitons in Certain CT Crystals Obtained by High Field Methods (In Some Cases the Data Pertain to Lower than Room Temperature)

Crystal	D (cm^{-1})	E (cm^{-1})	Method	Ref.
A/TCNB	±0.0669	∓0.0076	ESR	[102]
	±0.0670(5)	∓0.0079(5)	ESR	[35]
	±0.06668(10)	∓0.00775(10)	hf-ODMR	[91]
	0.0670(5)	−0.0076(5)	ESR, hf-ODMR (260 K)	[103,104]
N/TCNB	±0.0684	±0.00745	ESR	[40]
	0.0697	0.0082	ESR	[41]
	±0.0681(2)	∓0.0077(2)	ESR (280 K)	[105]
B/TCNB	±0.0527	±0.0168	ESR	[100]
	+0.0551(5)	−0.0145(5)	ESR (288 K)	[90]
	+0.0565(5)	−0.0151(5)	ESR, hf-ODMR	[104]
	±0.0534(1)	±0.0152(1)	ESR	[56]

In Table V we collected additional information on the zfs parameters of mobile excitons in the three most-studied CT systems, A/TCNB, N/TCNB and B/TCNB. These data have been obtained by high field methods, and the often significant discrepancies between them point out the imperfection of these methods for determining zfs parameters in comparison with zero field techniques.

In the following, we discuss the individual crystals from the point of view of their excitonic zfs parameters and CT character.

Triplet excitons in CT crystals can be arbitrarily divided into three classes depending on their CT character. The first class has a negligible CT character and thus is very similar to typical Frenkel excitons. It is shown, however, in Section III that due to the very special crystal structure of CT complexes, they have specific properties different from those of the excitons in one-component crystals. This neutral-character class includes excitons in CT crystals with anthracene, pyrene and naphthalene serving as donors, with various acceptors. It also includes excitons in crystals consisting of TCNQ serving as an acceptor, with various donors.

The three CT crystals for which the largest amount of information on mobile triplet excitons is available belong to the neutral-character class, A/TCNB, A/PMDA and A/TCPA. In all these cases, the existence of mobile excitons has been demonstrated by magnetic resonance over a very wide temperature range, and the particular importance of the zero field experiment on A/TCNB and A/TCPA at liquid helium temperatures should be stressed at this point. In A/TCPA, mobile excitons can be followed continuously from 1.2 to 300 K by both ODMR and ESR techniques, allowing the zfs parameters to be determined very accurately. In all these crystals, the excitonic zfs parameters are very similar to the molecular parameters of anthracene itself ($D = \pm 0.0694$, $E = \mp 0.0084 \, \text{cm}^{-1}$) [88]. This means that within an exciton, the excitation resides almost exclusively on the anthracene moiety and the exciton migrates in the crystal within a sublattice formed by the anthracene molecules. However, there is no averaging of the molecular zfs tensors as there is in the anthracene crystal (shallow traps in these CT systems have very similar zfs parameters to the excitons). The small differences between the CT character values obtained by optical methods (6–8%, Table II) and by magnetic resonance (3–5%, Table IV) may be attributed to the rather crude approximations made in deriving the corresponding formulas, which make both methods only semi-quantitative. Besides the approximations discussed earlier, in the case of anthracene-like CT crystals there emerges the problem of finding the correct D_{loc} and E_{loc} parameters to use in Eqs. (2.15). The review of literature data on anthracene yields a very wide range of zfs molecular parameters, depending on the environment of the molecule [36]. It appears that the most logical molecular zfs values are those of isolated anthracene in a low temperature glass. This is a general problem in all the crystals where triplet excitons are of almost neutral character (anthracene, naphthalene, pyrene, with various acceptors).

Triplet excitons of practically neutral character have also been detected over a wide temperature range by both zero and high field methods in two other weak CT crystals, both containing TCNQ as an acceptor, Ph/ and F/TCNQ [38]. In these two cases, triplet excitation is localized on the acceptor in contrast to the anthracene-like crystals. The reason for this is most probably the very low energy of the T_1 state of TCNQ, estimated theoretically to be on the order of $7600-10,300 \text{ cm}^{-1}$ [106]. Recent experiments of Frankevich et al. [39] who measured the magnetic field effects on the intensity of prompt fluorescence as a function of excitation power and temperature, confirm these estimates, yielding the energy of the T_1 state at 9865 cm^{-1} in F/TCNQ and 8865 cm^{-1} in Ph/TCNQ. As a result of this low energy, the local triplet zero-order state of TCNQ (DA*) has practically no interaction with the ionic (D^+A^-) state, situated much higher (about $16,500 \text{ cm}^{-1}$), and the CT character of the triplet excitons is almost zero. This result cannot be verified by the method of Hayashi et al. or Möhwald and Sackmann because the zfs parameters of TCNQ itself are not known, but the similarity of the zfs parameters in both Ph/ and F/TCNQ crystals suggests that the CT character is negligible and the experimentally determined zfs parameters of the two complexes characterize the hitherto unknown triplet state of TCNQ [38].

An almost neutral CT character is also typical for triplet excitons in crystals formed from pyrene as a donor with different acceptors (TCNB, TNB, PMDA). The available data, however, are scarce. The Pyr/TCNB crystal undergoes a violent phase transition at 171 K and shatters upon cooling (the ESR experiment reported in [48] was conducted apparently on a powdered sample, hence only the D^m parameter was quoted). The mobile excitons in Pyr/TNB were detected in the 90–250 K temperature range and have zfs parameters reduced by about 10% in relation to pyrene [57]. The triplet energy of the seldom-used acceptor TNB is, however, unknown, so the estimation is tentative only. In the last crystal referred to, Pyr/PMDA, magnetic resonance experiments did not reveal mobile excitons, but their presence was indicated by the appearance of delayed fluorescence in the emission spectra [32]. The CT character of 30% given in Table IV was estimated for one of two shallow traps observed by ODMR at very low temperature (the other one has a 35%

CT character [56]). As the optical data suggest a sizable CT character for the mobile excitons in this crystal, the value of 30–35% may be thought of as being representative.

The local (donor-like) character of triplet excitons is also characteristic for the two crystals with naphthalene, N/TCPA and N/TCNB. In the first case, the CT character is small (<10%) [43], which was confirmed by optical measurements [27], while for N/TCNB a controversy has dominated the literature for a long time [40, 41, 107–109]. It is now clear that the mobile excitons in this crystal have a CT character of about 30% [40, 41], while various defects present in the crystal may have a CT character varying from 0 to as much as 50% [110].

A very similar CT character is attributed to the excitons in the P/TCPA crystal [44]. The zfs parameters shown in Table IV were not very accurately determined because the excitonic ESR signals in this crystal were too weak to be followed in the wide enough angular range, but they point to a local phenanthrene-like character of the excitons with a 30% admixture of the CT zero order state.

A class of CT crystals with triplet excitons possessing a moderate-to-high CT character is represented by B/TCNB, F/TCNB, P/PMDA, and possibly D/TCNB and DPA/TCNB (diphenylacetylene/TCNB). The CT character of triplet excitons in the two structurally similar crystals B/TCNB and F/TCNB has been found to be 54 and $\simeq 70\%$, respectively [48, 51], by comparison with the zfs parameters of TCNB (the T_1 states of both biphenyl and fluorene lie much higher than that of TCNB). As for the P/PMDA crystal, the zfs tensor of the isolated triplet state was an object of the excellent work of Keijzers and Haarer [46], which is described in more detail below. Its CT character was estimated as $76 \pm 5\%$. For the other two crystals, that is, D/TCNB and DPA/TCNB, there is so far no information available on the optical properties, hence the relatively high CT character of the triplet excitons in these two crystals remains to be confirmed.

Keijzers and Haarer [46] very precisely determined the complete zfs tensor of a shallow triplet trap in P/PMDA by ESR after having made certain that it does not differ significantly from that of mobile excitons. They also calculated the complete zfs tensor of the ionic zero order state in the molecular frame with the help of a point-dipole model. Finally, they compared both tensors with the molecular tensor of phenanthrene. All the tensors are shown in Table VI in terms of principal values and orientation angles.

It follows from Table VI that although in the experimentally determined zfs tensor, the largest value of $0.0154\,\mathrm{cm}^{-1}$ corresponds to the x molecular axis of phenanthrene, exactly as is the case in phenanthrene

TABLE VI

Zfs tensors of an Isolated Triplet State in P/PMDA at 20 K ($\mathbf{F}^{exp}$), Ionic Zero-Order State of the Same Crystal ($\mathbf{F}^{ion}$) and Phenanthrene ($\mathbf{F}^{loc}$) (after [46])

| Tensor | Principal values $(cm^{-1} \times 10^{-4})$ | Orientation relative to phenanthrene (°) | | |
		x	y	z
$\mathbf{F}^{exp}$	97.2	90	27	64
	57.2	91	63	153
	−154.4	0	90	90
$\mathbf{F}^{ion}$	57.0	73	163	91
	53.3	19	73	97
	−110.3	83	89	7
$\mathbf{F}^{loc}$	669.9	90	90	0
	131.2	90	0	90
	−800.1	0	90	90

itself, the other two axes deviate strongly from the molecular frame, the effect attributed by the authors to the CT interaction in the triplet state. This is further confirmed by the calculated ionic tensor: the CT interaction makes the largest principal value point in a direction almost parallel to the z (out-of-plane) axis of phenanthrene, while the other axes do not correspond to the molecular frame. The ionic tensor therefore has a different symmetry from the local one and the sign of the parameter D_{ion} is also opposite to D_{loc}. Application of Eqs. (2.15) may lead to an error in this case, so Keijzers and Haarer estimated the CT character of the T_1 state according to the formula

$$\mathbf{F}_{ij}^{exp} = c_1^2 \mathbf{F}_{ij}^{ion} + (1 - c_1^2) \mathbf{F}_{ij}^{loc} \qquad (2.16)$$

where $\mathbf{F}_{ij}$ are the five independent elements of the complete zfs tensor.

The fitting procedure yields the CT character of $76 \pm 5\%$. The authors draw attention to the symmetry properties of the ionic zfs tensor which they describe as "cigar-shaped"; an almost axial symmetry (principal values $X \simeq Y$, which means that the asymmetry parameter E is close to zero) with the z principal axis pointing in the direction approximately perpendicular to the molecular plane. In contrast, the local phenanthrene tensor is described as "pancake-like". We return to this point at the end of this section.

It should be added at this point that the specific ("zigzag") structure of

P/PMDA [111] allowed Keijzers and Haarer [112] to solve the problem of the spacial distribution of the triplet excitation. The question can be formulated as follows: is the triplet exciton limited to the pair of molecules forming the complex, as is the case in the complexes isolated in solutions and glasses, or does it spread onto more than two molecules? The answer is that at least in P/PMDA it is spread over three molecules in the following manner: $(PMDA^{-0.38}P^{+0.76}PMDA^{-0.38})$. Since this result is inferred from the zfs tensor of the excitation localized on a defect, further discussion is deferred until Section VI which covers defects in CT crystals.

A special class of CT crystals is made up by three crystals with HMB as a donor and TCNB, PMDA, and TCPA as acceptors. The question is whether the CT character is very high as suggested by the very broad and structureless triplet excitation and phosphorescence spectra, and particularly the negligible singlet-triplet energy gap [55]. Unfortunately, it is this small energy difference which also makes a magnetic resonance detection of mobile triplet excitons in these systems very difficult because of the efficient thermal deactivation from the triplet band back to the singlet manifold which lowers the steady-state concentration of the excitons, and causes a consequent lifetime shortening (see Section IV). At very low temperatures, where thermally activated processes are no longer active, the excitons become trapped and the traps dominate in the ESR and ODMR spectra. Consequently, only the zfs parameters of these traps can be presented in Table IV. What makes the HMB crystals unique is the smallness of their zfs parameters: in the case of HMB/TCNB $|D| = 0.0285$, $|E| = 0.0066 \, cm^{-1}$. This puts them in the range of zfs parameters which had been calculated for the purely ionic zero order triplet state. Calculating the mean D^m parameter from the D and E values and comparing it with the theoretically calculated D^m parameter of the ionic $HMB^+/TCNB^-$ zero order state $(0.022 \, cm^{-1}$, [80]) yields a CT character of about 90%, essentially in agreement with the results of Hayashi et al. on the isolated complex. In the case of HMB/TCPA and HMB/PMDA crystals, no theoretically calculated zfs parameters of their respective ionic zero order states are available, and their zfs parameters are in fact smaller than those of the ionic zero order state of $HMB^+/TCNB^-$. Since the implementation of the simplified formulae of Möhwald and Sackmann (Eqs. (2.15)) lead to a large error in the case of small zfs parameters, the CT character of excitons in these two crystals can only be roughly estimated as very close to 100%. The triplet excitons in the HMB CT crystals are therefore true CT excitons.

In comparing the values of the zfs parameters of HMB/TCPA excitons to those of the ionic zero order state, an assumption is made that both have the same sign. As it was pointed out in Section II.C.3 some authors argue that this is not the case. The controversy has not yet been solved, although we believe the signs need not be different. If, as suggested by Möhwald and Böhm [82], the signs of the D_{ion} and D_{exp} are opposite, one should expect in principle to find crystals in which triplet excitons possess a CT character larger than that in the HMB-like crystals and having D and E close to, or equal to zero, that is, with no zero field splitting observed. In ESR, this effect would appear as a single line at $g \simeq 2$ instead of a doublet placed symmetrically around this point. To our best knowledge, such excitons have not been found yet except in some CT salts, where the disappearance of zfs is a result of averaging effects due to excitonic exchange but not of a large electron delocalization effect [86].

In the HMB/TCNB crystal, the complete zfs tensor of a shallow defect was determined from ESR and ODMR experiments [54] analogous to those of Keijzers and Haarer. The ODMR data allowed increased precision of the measurements. A deviation of the principal axes from the molecular frame was established, but the tensor is much less symmetrical than in the case of P/PMDA: $|D| = 0.0285$, $|E| = 0.0066\,\mathrm{cm}^{-1}$, $|E| \simeq 1/4$ $|D|$, despite the undoubtedly large CT character of about 90% (estimated by the method of Hayashi due to the lack of the complete data regarding the local TCNB tensor). An even more surprising symmetry of the zfs tensor of triplet excitons was found in the F/TCNB crystal, in which the E parameter reaches its maximally allowed value of $1/3D$; $D = \pm 0.0298$, $E = \mp 0.0097\,\mathrm{cm}^{-1}$ (the estimated CT character 70%). Despite the large CT character, this tensor is certainly not "cigar-shaped", but rather "disk-shaped", with the disk perpendicular to the molecular plane. It seems therefore that the conclusions of Keijzers and Haarer on the symmetry of the ionic zfs tensor, while certainly valid for the specific case of the P/PMDA crystal, do not necessarily hold for other crystals which have different structures and hence intermolecular configurations.

III.   TRIPLET EXCITON DYNAMICS

A.   General

As stated in Section I, the mobility of triplet excitation, which is a direct result of the periodicity of the crystal structure, has a pronounced influence on the characteristics of both singlet-triplet optical spectra and magnetic transitions between the triplet sublevels.

The creation of triplet energy bands in crystals, which changes the pattern of singlet-triplet absorption and emission spectra in comparison with isolated complexes, was discussed in Section II.B.2. The excitonic mobility causes a totally new phenomenon called P-type delayed fluorescence (DF) to appear in crystals. P-type DF is a result of an annihilation of two triplet species and appears at the spectral position of the prompt fluorescence, but has a lifetime closely related to that of the mobile triplet species. Another consequence of the mobility of triplet excitation is the possibility of an exciton getting trapped on some defect. This process often competes with the triplet-triplet annihilation and with direct decay of the mobile exciton. Both triplet-triplet annihilation and trapping significantly influence the lifetime and population of the mobile excitons, in other words the exciton kinetics. These processes are discussed extensively in Section IV which covers kinetic effects in CT crystals.

In Section III we deal in detail with the effects of triplet exciton mobility on the magnetic resonance spectra. In Section III.B we discuss the averaging of the hyperfine structure of ESR signals as a result of the triplet energy mobility in the CT crystals. Section III.C treats the averaging of fine structure (zero field splitting) tensors as a consequence of the excitonic motion. In Section III.D, we embark on a discussion of the coherent effects exhibited (or not exhibited) by triplet excitons in CT crystals. Section III.E is devoted to the interaction of mobile triplet excitons with nuclear spins causing the relaxation of the latter and creating optical nuclear polarization (ONP). Finally, Section III.F describes the effects of mobile triplet excitons interacting with doublet ($S = 1/2$) species, that is, charge carriers.

B. Averaging of Hyperfine Structure in ESR Spectra

The first observations of the mobility of triplet excitons in the CT crystals was made by Möhwald and Sackmann on B/TCNB [100] and A/TCNB [102]. In both cases, the mobility was seen as the principal reason for the very small linewidths of the room temperature ESR signals: 0.1 G in B/TCNB and 0.4 G in A/TCNB. In comparison, triplet species in these crystals immobilized at low temperatures show the ESR linewidths of 2.2–16 G (B/TCNB) and 10 G (A/TCNB). The significant narrowing of the excitonic ESR signals was attributed to the effect of motional narrowing first discussed in [113]. In the case of triplet excitons, this phenomenon may be interpreted as being caused by the time averaging of the hyperfine interactions between the electronic spin $S = 1$ and the nuclear spins of the various nuclei contained in the complexes visited by the exciton during its lifetime. In other words, an exciton propagating in a crystal encounters many complexes characterized by different nuclear spin configurations. As a result, the excitonic ESR linewidth and lineshape are identical to those resulting from a rapid spin exchange between the two (or more) nuclear spin configurations. The theory of

ESR linewidth in the presence of such an exchange is given in [114]. According to [114] in the fast exchange regime (exciton hopping frequency larger than the hyperfine constant) and in the presence of multiple spin states (equivalent to multi-site exchange), the initial inhomogeneously broadened Gaussian envelope characteristic for an immobilized triplet state signal changes its shape into Lorentzian and becomes narrowed as the result of the excitonic motion. The hopping frequency can be estimated from the inhomogeneously broadened linewidth of the immobile species ΔB_0 and the motionally narrowed excitonic linewidth ΔB according to

$$\nu_{hop} \geq \frac{g\beta}{2\sqrt{3}\hbar} \frac{(\Delta B_0)^2}{\Delta B} \qquad (3.1)$$

In fact, the shape of the room-temperature excitonic signals in A/TCNB is Lorentzian [35,102] which confirms the concept of motional narrowing in this crystal.

The ESR linewidth and lineshape of triplet excitons are thus measures of the efficiency of the hyperfine averaging process, and this in turn is a function of their mobility. Möhwald and Sackmann [48] noticed a pronounced difference in the exciton dynamics in the two systems studied (B/ and A/TCNB). The ESR linewidth in B/TCNB is thermally activated with $\Delta E_a = 700 \, cm^{-1}$ and decreases from 2.2–16 G at low temperature (the actual linewidth depends on the crystal orientation and/or sample quality) to 0.1 G at room temperature (Fig. 3.1). In contrast in A/TCNB, in which the situation is somewhat complex because of the phase transition at $T_c = 206 \, K$, the linewidth remains constant above T_c (0.4 G) and is only weakly temperature-dependent below T_c (minimum 1.5 G at 100 K, Fig. 3.2). No thermal activation of the ESR linewidth exists in A/TCNB ($\Delta E_a < 10 \, cm^{-1}$) except in the vicinity of T_c at certain crystal orientations relative to the magnetic field. Möhwald and Sackmann estimate the excitonic hopping rates in B/TCNB as $5 \times 10^6 \, s^{-1}$ at 150 K and $5 \times 10^7 \, s^{-1}$ at 300 K [100] and in A/TCNB as $9 \times 10^4 \, s^{-1}$ at 100 K (below T_c) and $1.3 \times 10^6 \, s^{-1}$ at 300 K (above T_c) [102]. The same authors formulate interesting ideas about the properties of triplet energy transport in the two crystals studied [48]. Firstly, they attribute the phenomenon of thermal activation of excitonic hopping rate in B/TCNB to self-trapping due to a significant CT character of triplet excitons in this crystal (~50%). The idea of self-trapping was briefly discussed in Section I.

Self-trapping has long been considered as an inherent property of CT excitons from a theoretical point of view (see e.g., [23]) although the

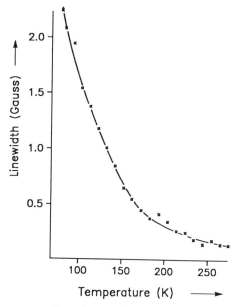

Figure 3.1. Temperature dependence of the ESR triplet exciton linewidth in the B/TCNB crystal (from [100]).

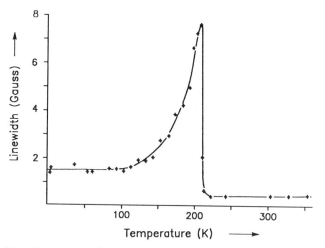

Figure 3.2. Temperature dependence of the triplet exciton ESR linewidth in the A/TCNB crystal (from [102]).

experimental evidence for it has been rather elusive [26,115]. Möhwald and Sackmann [48] estimate the change of intermolecular distance in B/TCNB necessary to distort the crystal lattice in a sufficient degree to account for the activation energy of 700 cm^{-1} to be 0.13 Å. The plausibility of the self-trapping observation in CT crystals by ESR is further discussed in this section, but it should be noticed that in A/TCNB, in which triplet excitons have almost zero CT character, no activation of excitonic mobility has been observed. Secondly, Möhwald and Sackmann [48] discuss the problem of the dimensionality of excitonic mobility in CT crystals. They argue that since in B/TCNB the triplet excitation is to a large extent spread over two molecules, it should propagate principally along the stacks of which the crystal is built. In contrast, in A/TCNB, in which the excitation is localized exclusively on anthracene, the triplet state of TCNB positioned about 8000 cm^{-1} higher than that of anthracene acts as a barrier for excitonic propagation, which is thus two-dimensional in the directions perpendicular to the stacks. This conclusion, although intuitively plausible, turns out to be somewhat oversimplified in view of the subsequent research on the same crystal.

Ponte Goncalves verified the results of Möhwald and Sackmann on A/TCNB in [35]. He found the phase transition at 202 K and noticed that the number of room temperature excitonic ESR signals (1 pair) doubles below T_c (Fig. 3.3). The linewidth is 0.8 G above T_c, independent of the crystal orientation, and ~1.5 G below T_c. However, at the orientations for which the two pairs of low temperature signals overlap, the merged signals are narrowed to the room temperature value of 0.8 G (Fig. 3.4) at any temperature.

The conclusions of Ponte Goncalves can be summarized as follows. Below T_c the crystal consists of two symmetry-related but magnetically nonequivalent sublattices D_1 and D_2 (this was later confirmed by X-ray diffraction by Stezowski [10]). The excitons move primarily within each of the sublattices with the frequency $\nu_{11} = \nu_{22}$ which can be estimated from the Eq. (3.1). Using this formula, Ponte Goncalves estimated the exciton hopping rate within each of the sublattices as $\geq 6.4 \times 10^8$ s^{-1}, independent of temperature. The temperature and angular effects shown in Figs. 3.3 and 3.4 point to the presence of an additional hopping taking place between the sublattices, $D_1 \leftrightarrow D_2$, with the frequency $\nu_{12} = \nu_{21}$. In the case of slow exchange ($\nu_{12} = \nu_{21} < B_{12}^0$, where B_{12}^0 is the splitting between the signals originating from each of the sublattices in the absence of exchange) the following formulas apply:

$$\nu_{12} = \nu_{21} = \frac{g\beta}{2\sqrt{2}\hbar} [(B_{12}^0)^2 - (B_{12})^2]^{1/2} \tag{3.2}$$

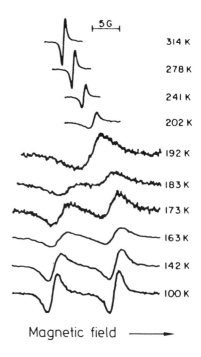

Figure 3.3. Variation of the upper half of the excitonic ESR spectrum of A/TCNB with temperature. Magnetic field $\parallel x_0$ (from [35]).

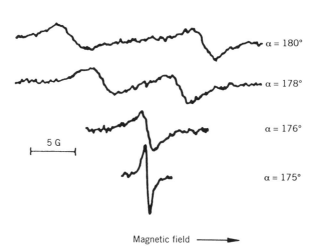

Figure 3.4. Angular dependence of the upper half of the excitonic ESR spectrum in A/TCNB at 163 K with the magnetic field rotated near the $a'b$ plane (from [35]).

B_{12} is the actual splitting between the signals at a given orientation, and

$$\nu_{12} = \nu_{21} = \frac{\sqrt{3}g\beta}{2\hbar} (\Delta B - \Delta B_0) \qquad (3.3)$$

ΔB and ΔB_0 are the linewidths in the presence and absence, respectively, of the exchange. Thus the effect of the $D_1 \leftrightarrow D_2$ hopping is to broaden the individual signals by up to 1.5 G, which was also observed by Möhwald and Sackmann, and to shift them toward each other. This hopping is obviously thermally activated, but a temperature-independent residue is still present at the lowest temperature. Ponte Goncalves estimates this contribution as $\nu_{12}^0 \sim 10^7 \, s^{-1}$ and the total hopping frequency just below T_c as $\nu_{12} \sim 10^8 \, s^{-1}$. The activation energy of this process is $\Delta E_{12} \sim 700 \, cm^{-1}$.

It follows from the above results that the excitons in A/TCNB move preferentially within each of the low temperature sublattices D_1 and D_2 with a temperature-independent frequency on the order of $10^9 \, s^{-1}$, and between the sublattices with a much smaller, thermally activated frequency of $\sim 10^7 - 10^8 \, s^{-1}$, although the exact nature of the latter process could not be identified from the ESR measurements alone (the possibility of participation of molecular movements has also to be considered). The subsequent X-ray diffraction analysis of the low temperature structure of the A/TCNB crystal [10] revealed the sublattices as neighboring stacks within the elementary cell. The preferential triplet energy transfer is thus either parallel to the stacks or perpendicular to them between the uniformly oriented complexes in the adjacent cells. ESR experiments alone cannot give an unequivocal solution of this problem but the theoretical work of Petelenz and Smith [116] appears to strongly suggest that the energy transport indeed takes place preferentially along the stacks. The authors of the quoted paper calculate the energy transfer integrals both along the stack and perpendicular to it, taking into consideration the CT character of the T_1 excitons (3–10%) and the crystal structure. The calculations yield the value 0.02–0.12 cm^{-1} for the energy transfer integral along the stack and 0.0001–0.02 cm^{-1} perpendicular to it. Thus even with the very small charge transfer character of the triplet excitons in A/TCNB, they are able to move efficiently along the stacks, contrary to the earlier assumptions of Möhwald and Sackmann. The exciton mobility in A/TCNB can thus be considered as quasi one-dimensional.

The problem of the nature of the $D_1 \leftrightarrow D_2$ exchange taking place below the phase transition in A/TCNB was treated further by Möhwald et al. [117]. The authors agreed with the conclusions of Ponte Goncalves concerning the exciton dynamics in this crystal, including the hopping rates estimated by him. Also, they clarified the nature of the above mentioned exchange process in the following way: the ν_{12} exchange frequency is first calculated from the broadening of the individual signals originating from the nonequivalent stacks in the slow exchange regime using Eq. (3.3). This value is then put into Eq. (3.2) which relates the observed shifts of the resonance fields due to exchange with the hopping frequency. Consequently the difference of resonance fields in the absence of exchange effects is calculated according to

$$B^0_{12} = [(B_{12})^2 + 6(\Delta B - \Delta B_0)^2]^{1/2} \qquad (3.4)$$

Ultimately this difference is transformed into the angular difference of anthracene orientations necessary to create the calculated resonance field difference. In this way the exciton dynamics can be separated from the molecular dynamics. It is found that the interstack exciton hopping indeed takes place with a temperature-independent frequency of $\nu_{12} \sim 10^7 \, s^{-1}$. On the other hand, part of the experimental effect of the temperature shift of the individual ESR signals must be attributed to a reorientation of the anthracene molecules below T_c: just below T_c the angle between the long in-plane axes of anthracene (2ϑ) is $10°$ and at

TABLE VII

Intrastack Exciton Mobilities and their Activation Energies (The Mobilities are Either Taken from Indicated References or Calculated from the ESR Linewidths Reported Therein)

Crystal	T (K)	Mobility (s^{-1})	ΔE (cm^{-1})	CT character (%)	Ref.
A/TCNB	300	$\geq 6.4 \times 10^8$	0	4–5	[35,117]
	300	$\sim 2 \times 10^{10}$	–		[118]
	300	$\sim 3 \times 10^9$	–		[122]
A/TCPA	1.2	$\geq 10^8$	–	3–4	
	300	$\geq 10^8$	0		[37]
Ph/TCNQ	300	$\geq 10^8$	0	~ 0	[78]
F/TCNQ	200	$\geq 10^8$	0	~ 0	[94]
B/TCNB	150	5×10^6	–	54	[100]
	300	5×10^7	700		[100]
			250 ± 80		[56]
F/TCNB	100	$\geq 6 \times 10^7$		70	[51]
	300	$\geq 5 \times 10^8$	400 ± 100		[51]
Nd$_8$/TCNB	130	$\geq 3 \times 10^7$	–		[41,110]
	315	$\geq 2.5 \times 10^8$	136	32	[41,110]
P/TCPA	200	$\geq 2.5 \times 10^8$	20 ± 10	30	[44,121]

100 K it is 16°. The activation energy of $700 \, cm^{-1}$ found previously by Ponte Goncalves is most probably linked to the molecular reorientation processes. These processes are further discussed in Section V.

The interpretation of the temperature effects on the ESR spectra of triplet excitons in A/TCNB by Möhwald et al. was subsequently criticized by Park and Reddoch [118]. Instead of the temperature-dependent molecular reorientation below the phase transition temperature, the authors suggested an exchange (hopping) process taking place between four sites. Sites "AI" and "AII" would then represent the regularly oriented anthracene molecules in the respective stacks I and II, while the sites "BI" and "BII" represent then "wrongly" oriented molecules, that is, molecules in stack I oriented in the fashion characteristic for stack II and vice versa. The following parameters are fitted to the experimental data concerning the ESR linewidths and signal splitting in the low temperature phase: intrastack (ν_{11}) and interstack (ν_{12}) hopping rates and the populations of the particular sites. For the mathematical apparatus employed by Park and Reddoch, the reader is referred to the original paper. The results obtained are the following: $\nu_{11} \sim 2 \times 10^{10} \, s^{-1}$ (tempera-ture-independent) $\nu_{12} \sim 10^8 \, s^{-1}$ just below T_c (thermally activated with $\Delta E_a = 1051 \, cm^{-1}$) and the concentration of the "disoriented" molecules as high as 40% just below T_c. In comparison with the results of Ponte Goncalves [35] and Möhwald et al. [117], the interstack hopping rate is similar but the estimated activation energy of the interstack hopping is about 50% higher. Also, the intrastack hopping frequency obtained by Park and Reddoch is 1.5 orders of magnitude larger than the one found by both Ponte Goncalves and Möhwald et al. This would be a major discrepancy provided the model of four-site exchange were valid. How-ever, the final determination of the A/TCNB crystal structure [10] confirms the notion of Möhwald et al. (temperature-dependent reorienta-tion of molecules) rather than the one of Park and Reddoch (a significant temperature-dependent molecular disorder below T_c). It seems therefore that the values given by Ponte Goncalves and Möhwald et al. are closer to reality, although being calculated from Eq. (3.1), they represent lower limits only.

A very interesting approach to the problem of the anisotropy of the excitonic mobility in the A/TCNB crystal was reported by Erdle and Möhwald [119]. These authors took advantage of the similarity of struc-tures of the A/TCNB and P/TCNB crystals. They were able to grow mixed crystals $A_c/P_{1-c}/TCNB$, where c is the mole fraction of anthra-cene. Crystals with $c = 10^{-3}$, 0.25 and 0.75 were used. As the T_1 state of phenanthrene lies more than $2000 \, cm^{-1}$ higher than that of anthracene, the triplet excitation in mixed CT crystals is restricted to the "cages" or domains formed by anthracene. Erdle and Möhwald [120] noticed that the ESR linewidth of a mixed crystal depends on the coefficient c; when $c = 10^{-3}$, the linewidth is typical for a localized A/TCNB triplet (10–16 G depending on the orientation), but with increasing values of c the linewidth is reduced, which means that the triplet excitation becomes mobile. Erdle and Möhwald named these excitations in their mixed crystals "mini-excitons".

The presence of mini-excitons in the mixed crystals $A_c/P_{1-c}/TCNB$ and the dependence of their ESR linewidths on the coefficient c prompted Erdle and Möhwald to construct a statistical model allowing excitons to move preferentially in one or two dimensions, comparing the computed dependence of the linewidths with the experimental results. The following assumptions were made:

- Phenanthrene and anthracene molecules are distributed randomly on the donor sites.
- Every molecule has four nearest neighbors.
- Triplet exciton diffusion takes place only between nearest neighbors.
- Excitation energy is distributed solely over the anthracene moiety.
- All anthracene molecules absorb photons with equal probability.
- ESR absorption lines of localized triplet states have a Gaussian shape with linewidths ΔB_1 which results from an envelope over the unresolved hyperfine structure.
- Statistical orientational disorder has no influence on the ESR linewidth.
- The width ΔB_i of the exciton line of a domain containing i molecules is a factor of $\sqrt{i}$ smaller than that of a localized triplet state, that is, $\Delta B_i = i^{-1/2} \Delta B_1$.
- Excitonic motion is either:
  a. very fast along one crystallographic direction, slow along other directions (one-dimensional propagation);
  b. very fast within one crystallographic plane with no anisotropy of diffusion in this plane (two-dimensional propagation).

The calculations are performed in three steps:

- According to the relative concentration c, the two types of donors are distributed over a quadratic net of dimension 100×100 units, the coordinates given by a numerical random procedure.
- From the data obtained, the size distribution of the anthracene domains in the mixed crystal is calculated.

- ESR absorption $I(B)$ is calculated as a function of the magnetic field B. If $n(i)$ denotes the number of domains containing i anthracene molecules, the lineshape $I(B)$ is given by the following relation:

$$I(B) = \sum_i in(i) i^{-1/2} \Delta B_1^{-1} \exp\left(\frac{-2iB^2}{\Delta B_i^2}\right) \qquad (3.5)$$

The linewidth of this function is defined as the distance between the points of inflection and calculated using as ΔB_1 the width measured for the isolated anthracene trap in the P/TCNB crystal ($c = 10^{-3}$). The ESR linewidth obtained by the above method is shown in Fig. 3.5 as a function of anthracene concentration c for a one-dimensional (solid lines) and two-dimensional (dashed lines) analysis. The circles indicate the experimental points.

It follows from Fig. 3.5 that the one-dimensional model shows a better agreement with the experiment than the two-dimensional one. However, deviations from the purely one-dimensional case are obvious, which permits an estimate of the exciton hopping rate perpendicularly to the preferred direction: $1 \times 10^7 < \nu_{perp} < 4 \times 10^7\ \mathrm{s}^{-1}$. Comparing these values with the hopping rate in the preferred direction ($6 \times 10^8\ \mathrm{s}^{-1}$ [35,117]) yields an anisotropy for the excitonic mobility in A/TCNB of between 15:1 and 60:1.

Erdle and Möhwald were reluctant to attribute the obtained hopping rates to any physical direction in the crystal. However, in view of the

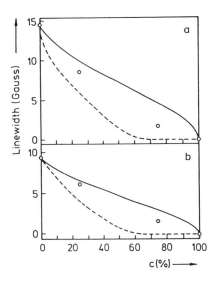

Figure 3.5. ESR linewidth as a function of anthracene concentration c in the mixed crystals $A_c/P_{1-c}/TCNB$ for two canonical orientations (a) $B \parallel x$ (b) $B \parallel y$. The circles are experimental data, the dashed lines are computed for a two-dimensional model, the solid lines for a one-dimensional model (from [119]).

results discussed earlier in this section, the preferred direction of the exciton diffusion can quite safely be identified with the stack axis.

While A/TCNB has been the most widely examined CT system so far, other crystals have also received significant attention regarding their triplet exciton dynamics. In A/TCPA, triplet excitons can be followed experimentally over an unusually broad temperature range 1.2–300 K by two independent magnetic resonance techniques: zf-ODMR and ESR [37], in addition to optical emission. Both the zero field and high field excitonic signals at low temperatures are quite broad in comparison with other CT systems: in zero field $\Delta \nu = 2$ MHz extrapolated to zero rf-power at 1.2 K and in high field $\Delta B = 1.8$–3.8 G at 3.8 K. The lines do not get significantly narrower at higher temperatures which could mean that the mobility of triplet excitons in A/TCPA is lower than in previously discussed A/TCNB. In fact, application of Eq. (3.1) gives a hopping rate of $\geq 10^8 \, s^{-1}$ in A/TCPA, a lower value than in A/TCNB. This is further confirmed by the observed anisotropy of the excitonic ESR linewidth which suggests incomplete averaging of the hyperfine structure. However, the incompletely averaged hfs cannot be the reason for the broad zero field signals since the hyperfine interaction enters the zero field Hamiltonian only in the second order. Another process must be responsible for this phenomenon. The most plausible reason may be found in the energy levels in the A/TCPA crystal.

In A/TCPA, excitation spectroscopy discovered two triplet bands separated by $40 \, cm^{-1}$ [29,30] (see Section II.B.2). The origin of these bands can be seen in the crystal structure which is built of two symmetry-unrelated sublattices in the low temperature phase (<194 K [11]). Within one stack, the complexes belonging to each of the sublattices interchange (Fig. 3.6). If an exciton moves along the stack, as in A/TCNB, it encounters two interchanging inequivalent anthracene molecules. No splitting of the excitonic signals due to different sublattices has been found either in zero field nor in high field. If the exciton motion were restricted to each of the sublattices, one should see a splitting of the ESR signals of up to 7 G, depending on the crystal orientation. Instead, one observes a single pair of broadened signals, both in zero and high field, which means that the exciton visits both orientations of anthracene molecules, but provides little information on the mobility of the excitons.

Upon increasing the temperature both the zero and high field linewidths undergo only minimal changes, so the hopping rate of triplet excitons in A/TCPA seems to be almost temperature-independent, as is the case in A/TCNB outside the phase transition region.

The specific structure of A/TCPA does not allow one to draw definite conclusions about the anisotropy of the exciton motion, although the

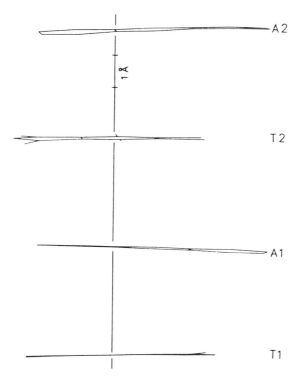

Figure 3.6. Crystal structure of A/TCPA. Only half the unit cell consisting of one stack is shown, with the view perpendicular to the stack axis. Note the two different positions of anthracene (A1 and A2) and the two different positions of TCPA (T1 and T2) within the stack (from [11]).

arguments above tend to point to a one-dimensional motion along the stacks. In the low temperature phase, the anthracene molecules belonging to the symmetry-unrelated sublattices differ in their orientations by no more than 4°. In the absence of any splitting of the ESR signals which could be attributed to this difference, the exciton hopping rate between these two orientations can be estimated as $>5 \times 10^7 \text{ s}^{-1}$. This is, however, not the equivalent of the interstack hopping rate in A/TCNB, as the two orientations alternate within each stack and not in neighboring ones as in A/TCNB. It is instead the lower estimate of the intrastack hopping rate which is in agreement with the previously derived value of $\geq 10^8 \text{ s}^{-1}$. The anthracene molecules belonging to each of the sublattices in different stacks of the unit cell are translationally equivalent, the whole A/TCPA complex is inverted, but anthracene has two in-plane symmetry axes, so the inversion does not influence its translational equivalence. No inter-stack hopping rate can therefore be derived from ESR studies.

The other two systems (of triplet excitons) with a negligible CT character which have been investigated by magnetic resonance techniques are Ph/ and F/TCNQ. In both of them the excitons can be detected over a wide temperature range and the exciton hopping rate ($\geq 10^8$ s^{-1}) was found to be temperature-independent [78,94]. In both cases also the hyperfine structure of the ESR excitonic signals is not completely averaged as shown by a considerable anisotropy of the linewidth. As in A/TCPA, no information on the dimensionality of the exciton movement can be obtained from the ESR data due to the crystal structures, which are in both cases characterized by translational equivalence of the TCNQ molecules on which the triplet excitation is localized.

The crystal structure of P/PMDA provides a possibility to study the dimensionality of triplet energy transfer. The article of Keijzers and Haarer [46] treated the subject of exciton dynamics in this crystal only marginally (no temperature-dependent experiments were performed) but at 130 K the intrastack hopping rate for the excitons estimated according to Eq. (3.1) as $\nu_{11} \geq 2 \times 10^8$ s^{-1} was found to be two orders of magnitude larger than the interstack rate ($\nu_{12} < 1.2 \times 10^6$ s^{-1}). The latter was esti-mated from an experiment depending on orienting the crystal in the magnetic field in such a way that the ESR signals from each of the two sublattices almost coincided. The minimal field difference at which two signals can be recognized without any sign of exchange between them, when cast into frequency units, yields the upper limit of the interstack hopping rate.

In several other CT systems studied by magnetic resonance, the exciton hopping rate was found to be thermally activated as judged from the temperature dependence of the ESR linewidth; for example, the B/TCNB crystal discussed at the beginning of this section. Independent measurements conducted on several samples of B/TCNB from different batches [56] established the dependence of the inhomogeneously broadened linewidth of the immobile species at low temperature on the crystal quality. In the best crystals, the maximal linewidth was found to be 12 G rather than the previously reported 16 G. Also, the activation energy of the narrowing process is dependent on this factor, and was found in the best crystals to be considerably smaller, 250 ± 80 cm^{-1} than the previously reported 700 cm^{-1}.

In the F/TCNB crystal, which is practically isomorphic with B/TCNB, the effect of thermal activation of the excitonic ESR signals was also found [51], and the activation energy estimated as 400 ± 100 cm^{-1}. The exciton hopping rate at room temperature is of the same order of magnitude as in other CT crystals, that is, $\geq 5 \times 10^8\,s^{-1}$, but at low temperature (100 K) it is reduced to $<10^7\,s^{-1}$. In the case of F/TCNB, the origin of the thermal activation of the exciton hopping rate can be looked for in the static orientational disorder of fluorene molecules known to exist in the crystal (such a possibility was already invoked for B/TCNB by Ponte Goncalves in [35]). However, in B/TCNB no disorder at all has been reported by structural examinations [90] and the thermal effect on the linewidth is still present, so this hypothesis does not seem to be plausible. Rather, the self-trapping idea of Möhwald and Sackmann seems more applicable to F/TCNB, particularly in view of the considerable CT character of triplet excitons in this system (70%).

Another CT crystal in which thermal activation of the excitonic ESR linewidth has been reported is Nd_8/TCNB (perdeuterated naphthalene/TCNB) [110]. The exciton hopping rate at 130 K is $\geq 3 \times 10^7\,s^{-1}$ and at 315 K $\geq 2.5 \times 10^8\,s^{-1}$ calculated as usual from Eq. (3.1). The activation energy of the process is $136\,cm^{-1}$, which correlates well with the much smaller CT character of triplet excitons in this crystal ($\sim 25\%$) compared to B/ and F/TCNB.

The effect of thermal activation of the ESR linewidth was restricted to the crystals containing TCNB as an acceptor until it was also found in P/TCPA. In the P/TCPA crystal, triplet excitations are mobile at 250 K as judged by the motionally narrowed linewidth of $\sim 2\,G$. The hopping rate is $\geq 2.5 \times 10^8\,s^{-1}$ as estimated from Eq. (3.1). The ESR linewidth remains unchanged as the temperature is lowered to about 80 K, but upon further cooling increases to about 10 G at 4.2 K [121], a value typical for immobilized (trap) states (Fig. 3.7). The activation energy of this process is small, $20 \pm 10\,cm^{-1}$, estimated using Eq. (3.1), much smaller than that reported for Nd_8/TCNB in spite of the comparable CT character of triplet excitons in both systems (in P/TCPA $\sim 30\%$). It is worth noting that this particular trap is accompanied by a number of other defects (discussed in Section VI) [44], which behave in a very different way upon changing the temperature; their intensities undergo profound changes (thermal trapping and de-trapping), but the linewidths remain approximately stable. This might mean that the trap of a $20\,cm^{-1}$ depth in P/TCPA is a "self-trap".

In view of the above, the early hypothesis of Möhwald and Sackmann [48,100], that self-trapping is the reason for the thermal activation of the excitonic ESR signals, remains an intriguing possibility since most attempts to detect mobile excitons over a wide enough temperature range in crystals with high CT character of the T_1 state have failed. Such investigations, for example, have not been successful in P/PMDA [123] and HMB/TCNB [54] because ESR signals of mobile excitons in these

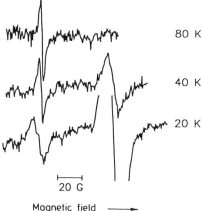

80 K

40 K

20 K

Figure 3.7. Fragment of the triplet ESR spectrum of P/TCPA. Two trap signals are observed: one on the left showing a temperature-dependent narrowing and almost constant intensity (a self-trap), the other of constant width, but strongly temperature-dependent intensity (from [121]).

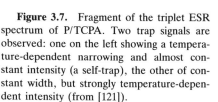

├──────┤
20 G

Magnetic field ──────→

systems disappear on lowering the temperature before any effect on the linewidth can be observed. The reason for this in the case of HMB/ TCNB was given as trapping on a deep (and invisible to ESR) defect in the intermediate range of temperatures [53]. Similarly, mobile triplet excitons of pure (100%) CT character could not be detected by ESR in HMB/TCPA [54] nor in HMB/PMDA [56], most probably due to their very short lifetime and small steady-state concentration.

The concept of self-trapping of triplet excitons in CT crystals was critically discussed by Kozankiewicz and Prochorow [53] from the point of view of optical spectroscopy. In particular, self-trapping is considered incompatible with the presence of P-type delayed fluorescence in B/ TCNB and HMB/TCNB at low temperatures. At this point a word of caution should be expressed concerning the meaning of the terms "mobile" and "immobile" in the case of triplet excitons. P-type delayed fluorescence may appear if an exciton moves on average over at least one complex during its lifetime. Assuming a standard exciton lifetime of 1 ms, a hopping rate of $>10^3$ s^{-1} should in principle be sufficient for the P-type delayed fluorescence to appear, that is, excitons hopping with this rate can be described as "mobile" in a typical stationary optical experiment. On the other hand, the lower limit of the exciton hopping rate as detected by ESR can be derived from Eq. (3.1) by putting $\Delta B = \Delta B_0$, that is, no narrowing, which gives a rate of $\sim 5 \times 10^6$ s^{-1}. In other words, any triplet exciton with a hopping rate falling in the range 10^3–10^6 s^{-1} will be considered as "mobile" in an optical experiment and "immobile" or trapped in ESR. It is our opinion that the triplet excitons of a significant CT character may fall into this category. This assertion gains support from the fact that P-type delayed fluorescence in B/ and HMB/TCNB at low temperatures, although present, is usually much weaker (no quantitative data on its efficiency are known) than in the crystals with zero CT character of triplet excitons. As the T-T annihilation rate depends on the exciton collision probability, the reduced mobility of the excitons in these systems might be one of the reasons for the low yield of the P-type delayed fluorescence.

In summary, the following conclusions concerning the triplet exciton mobility in CT crystals can be formulated:

- In crystals in which the triplet excitons are of neutral (local) character, their hopping rates are temperature-independent and on the order of 10^8–$10^9\,\text{s}^{-1}$.

- In crystals with a significant CT character of the triplet excitons, the hopping rate is reduced at low temperatures and the activation energy of this process correlates at least qualitatively with the CT character. The reason for this may be found in self-trapping.

- The excitons move preferentially along the stacks independent of their CT character; in the two systems which, due to their particular structure, could be studied from the point of view of the anisotropy of the exciton hopping rate (A/TCNB and P/PMDA), the interstack hopping rates were found to be at least 1–2 orders of magnitude lower than the intrastack rates. Triplet excitons in CT crystals can be therefore considered as quasi one-dimensional.

C. Averaging of Fine Structure Tensors

In Section II.C the importance of zfs parameters for characterizing the T_1 state of a CT complex and hence the triplet excitons in a CT crystal was discussed. It was also said there that the zfs parameters are by no means constant in a given crystal, but depend, sometimes significantly, on temperature. As the possible reasons for this, three distinct processes were postulated, namely thermal expansion of the lattice, molecular dynamics and exciton dynamics. In Section III.C we are going to discuss the influence of the last factor (exciton dynamics) on the fine structure tensor (or tensors) and hence also on the apparent CT character of triplet excitons.

Vyas and Ponte Goncalves [98] were the first to postulate triplet exciton dynamics as the possible reason for the temperature changes of zfs parameters in A/TCNB, suggesting a rapid hopping between the lowest (T_1) triplet band and some higher lying triplet band. The presence of the latter is, however, purely speculative and has not yet been confirmed by optical excitation spectroscopy. For this reason, the discussion of Vyas and Ponte Goncalves on A/TCNB is not explored further. The process of inter-band scattering has nevertheless been demonstrated in the case of the A/TCPA crystal, in which the presence of a second triplet band lying $40\,\mathrm{cm}^{-1}$ higher than the lowest one was unequivocally confirmed by excitation spectroscopy. The explanation for this band is due to the specific crystal structure (Section II.B). In the following, we take this system as an example of the influence of inter-band exciton scattering on the fine structure.

The temperature dependence of the zfs parameters of the triplet excitons in A/TCPA is observed both in high field (ESR) and zero field (ODMR). Figure 3.8 presents the temperature dependence of the D parameter obtained from ESR (the changes of the E parameter lie within the ESR experimental error). This suggests that at least two independent processes take place in the crystal, one active below 100 K and the other above 100 K (in the intermediate temperature range both processes are superimposed):

$$D(T) = D_1(T) + D_2(T) \tag{3.6}$$

The first process $D_1(T)$ is identified as the previously discussed (Section III.B) thermally activated hopping of the triplet excitons between the two triplet bands separated by $40\,\mathrm{cm}^{-1}$. It must be assumed that each of the

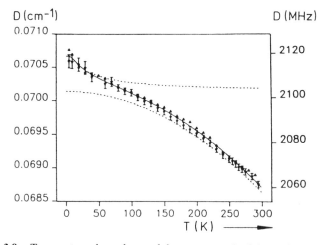

Figure 3.8. Temperature dependence of the parameter D of the excitonic zfs tensor in A/TCPA obtained from ESR spectra. The triangles are two sets of experimental points, the solid line represents a fit using the model discussed in the text, the dashed lines show the two processes contributing to the temperature effect observed (from [56,36]).

bands has a distinct zfs tensor, which is reasonable since the crystal structure consists of two symmetry unrelated sublattices differing slightly by the orientation and configuration of the complexes. The second process $D_2(T)$ is attributed to molecular dynamics and specifically to the librations of the anthracene molecules, and is discussed in Section V.

We may treat the hopping between two triplet bands as a case of a fast exciton exchange between the two magnetically inequivalent sites "1" and "2" having Larmor frequencies ω_1 and ω_2. In the ESR spectrum, a single line appears centered at the frequency,

$$\tilde{\omega} = f_1 \omega_1 + f_2 \omega_2 \tag{3.7}$$

where

$$f_1 = \frac{P_{21}}{P_{12} + P_{21}} \quad \text{and} \quad f_2 = \frac{P_{12}}{P_{12} + P_{21}} \quad (f_1 + f_2 = 1)$$

P_{12} and P_{21} are the probabilities of the jumps "1"→"2" and "2"→"1", f_1 and f_2 are therefore the probabilities of populating the level "1" and "2", respectively. If the two sites are in thermal equilibrium, that is $P_{12}/P_{21} = \exp(-\epsilon/kT)$ where $\epsilon = E_2 - E_1$ is the energy difference of the sites; the observed frequency of the zero field transition ν_{ODMR} is also temperature-dependent, as are the zfs parameters D and E [124]:

$$\tilde{D} = D_1 + (D_2 - D_1) \frac{\exp(-\epsilon/kT)}{1 + \exp(-\epsilon/kT)} \tag{3.8a}$$

and

$$\tilde{E} = E_1 + (E_2 - E_1) \frac{\exp(-\epsilon/kT)}{1 + \exp(-\epsilon/kT)} \tag{3.8b}$$

where D_1, E_1 and D_2, E_2 are the zfs parameters of the excitons moving in the lower and higher band, respectively.

The above equations are valid in the case where the two tensors have different zfs parameters but the same orientation. If the two tensors differ also in orientation, as expected in A/TCPA because of its crystal structure, the eigenvalues of an averaged tensor have to be calculated:

$$\mathbf{F} = f_1 \mathbf{F}_1 + (1 - f_1)\mathbf{F}_2 = \frac{1}{1 + \exp(-\epsilon/kT)} \mathbf{F}_1 + \left(1 - \frac{1}{1 + \exp(-\epsilon/kT)}\right) \mathbf{F}_2 \tag{3.9}$$

The tensor $\mathbf{F}_2$ of the higher lying band may be expressed in terms of the tensor $\mathbf{F}_1$ and the orientation difference between them via perturbation theory:

$$\mathbf{F}_2 = \mathbf{T}^{-1} \begin{bmatrix} X^* & & \\ & Y^* & \\ & & Z^* \end{bmatrix} \mathbf{T}$$

where

$$
\begin{aligned}
X^* &= (D_1 + \delta D)/3 - (E_1 + \delta E) \\
Y^* &= (D_1 + \delta D)/3 + (E_1 + \delta E) \\
Z^* &= 2/3(D_1 + \delta D)
\end{aligned}
\tag{3.10}
$$

and $\delta D = D_2 - D_1$, $\delta E = E_2 - E_1$ and $\mathbf{T}$ is the orthonormalized transformation matrix describing the orientation difference of the two tensors.

The experimental data for A/TCPA used in the fitting procedure are the following: $D_1 = 2126.5\,\text{MHz}$, $E_1 = 241.0\,\text{MHz}$ (these are the excitonic zfs values at 1.2 K), $\epsilon = 40\,\text{cm}^{-1}$ (from the optical spectra), and the matrix $\mathbf{T}$ is known from the crystal structure. The best-fitted values are

$$\delta D(= |D_2| - |D_1|) = -14\,\text{MHz}, \qquad \delta E(= |E_2| - |E_1|) = -3\,\text{MHz}$$

The results of the fits describing the parameters of the zfs tensors corresponding to the two symmetry-unrelated sublattices are presented in Table VIII. The simulated temperature dependencies of the zf-ODMR transition frequencies are shown in Fig. 3.9 together with the experimental results.

TABLE VIII
Zfs Parameters and ODMR Transition Frequencies of the Two Fine Structure Tensors Originating from the Two Triplet Bands in A/TCPA. The Values in the First Column are Those of the Mobile Excitons at 1.2 K (At This Temperature Thermally Activated Jumping Can Be Neglected), While Those in the Second Column Were Calculated from the Temperature Dependencies of the ODMR Frequencies According to Eqs. (3.8)

$D_1 = 2126.5\,\text{MHz}$	$D_2 = 2112.5\,\text{MHz}$
$|E_1| = 241.0\,\text{MHz}$	$|E_2| = 238.0\,\text{MHz}$
$\nu_1^{xz} = 2367.4\,\text{MHz}$	$\nu_2^{xz} = 2350.4\,\text{MHz}$
$\nu_1^{yz} = 1885.5\,\text{MHz}$	$\nu_2^{yz} = 1874.5\,\text{MHz}$
$\nu_1^{xy} = 481.9\,\text{MHz}$	$\nu_2^{xy} = 475.9\,\text{MHz}$

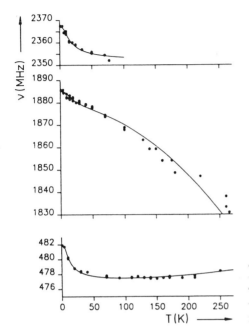

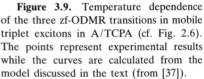

Figure 3.9. Temperature dependence of the three zf-ODMR transitions in mobile triplet excitons in A/TCPA (cf. Fig. 2.6). The points represent experimental results while the curves are calculated from the model discussed in the text (from [37]).

Applying a similar procedure to the high field ESR spectra gives very similar results concerning the D parameter (the poorer accuracy of the high field method does not allow the temperature dependence of E to be determined). Also, while it was predicted that the fine structure principal axes of the averaged (i.e., experimentally determined) tensor change their positions by about 1.5° between 4 and 120 K, such small shifts fall within the ESR experimental error and could not be detected.

In order to conclude this section, we would like to point out that the A/TCPA crystal offers a good example for examining the influence of the triplet exciton motion on the fine structure. Although the thermal expansion of the lattice cannot be taken into account due to the lack of data at low temperatures, the postulated hopping between the two sublattices can well describe the observed temperature changes of the zfs parameters below 100 K. At higher temperatures, the populations of the two triplet bands become practically equal ($kT \gg \epsilon$) so one cannot explain by exciton hopping the observed dependencies above the temperature of the first phase transition. However, the superposition of another effect, namely molecular librations, which become active at $T > 100$ K, allows one to reproduce all the experimental effects seen in Fig. 3.9. This is further discussed in Section V.

D. Coherence Versus Incoherence

The hopping rates of triplet excitons discussed in Section III.B are estimated mostly on the base of ESR results under the a priori assumption of the exciton motion taking place as incoherent hopping between the individual complexes (Eqs. (3.1)–(3.3) are valid only for this case). All the phase memory is lost between the jumps so the motion is incoherent both on a micro and macro scale. The principal reason for the lack of coherent effects in CT crystals is the small width of the triplet band which is usually on the order of a few cm^{-1}. As most experiments on triplet excitons in CT crystals have been run at relatively high temperatures (>50 K), this assumption seems reasonable.

Recently, however, a few studies have been published on triplet excitons in CT crystals at very low temperatures (1.2–4.2 K). It is well known that some one-component molecular crystals exhibit triplet exciton coherence at these conditions. The question of the presence or absence of such effects in the CT crystals under study again becomes important. In the following, we discuss the two CT systems in which mobile triplet excitons can be investigated by magnetic resonance techniques at very low temperatures: A/TCNB and A/TCPA.

The excitonic zf-ODMR signals in A/TCNB were first detected at 1.2 K and identified as such by Steudle and von Schütz [125] while monitoring the delayed fluorescence. The DF-ODMR signals are distinctly different from those obtained while monitoring the phosphorescence (Fig. 3.10) and appear always in doublets split by a few MHz in each of the zero field transitions. The attribution of the signals to mobile excitons is based on the following arguments:

- the very small linewidth (less than 1 MHz);
- the saturation behavior; hole-burning experiments have been unsuccessful, there is no broadening with increasing rf power, no quadrupole satellites can be detected at high rf power, and there is a coupling of the two signals in each doublet.

The appearance of doublets of the excitonic signals was explained by von Schütz et al. [126] as caused by the Davydov splitting due to the two translationally inequivalent A/TCNB complexes in the low temperature elementary cell. The following theoretical model was developed by Schmid and Reineker [127] to account for the experimental results.

In the T_1 state of A/TCNB, the excitation is mainly localized on anthracene. Thus the model considers two differently oriented anthracene molecules in the unit cell which are represented by two differently oriented two-level systems. There are stacks of parallel-oriented mole-

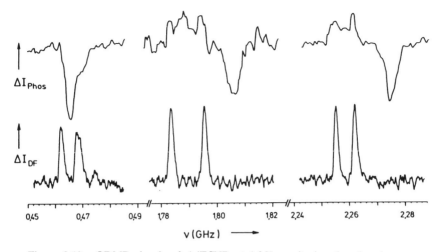

Figure 3.10. ODMR signals of A/TCNB at 1.2 K monitoring the phosphorescence (upper spectrum) and delayed fluorescence (lower spectrum) (from [126]).

cules along the a axis. Molecules in the neighboring stacks in the c' direction are also oriented parallel to each other while those in the b direction differ in orientation (Fig. 3.11). The model assumes a predominantly one-dimensional motion along the stacks while neglecting interactions in the c' direction. The electronic part of the Hamiltonian is given by

$$\mathbf{H}_{el} = \sum_n \sum_{i=1}^{2} \epsilon \mathbf{b}_{ni}^{+} \mathbf{b}_{ni} + \sum_{n \neq n'} \sum_{i=1}^{2} \sum_{i'=1}^{2} J_{n-n',ii'} \mathbf{b}_{ni}^{+} \mathbf{b}_{n'i'} \qquad (3.11)$$

where $\mathbf{b}_{ni}^{+}$ and $\mathbf{b}_{ni}$ are creation and annihilation operators for an exciton-hole pair localized at site i in the nth unit cell, ϵ is the excitation energy of the two-level system and $J_{n-n',ii'}$ is the interaction matrix element of molecule i in the nth unit cell with molecule i' in the n'th unit cell.

The spin Hamiltonian in zero field consists of the two fine structure terms denoted $\mathbf{F}_1$ and $\mathbf{F}_2$ corresponding to the two differently oriented molecules:

$$\mathbf{H}_S = \sum_n \sum_{i=1}^{2} \mathbf{b}_{ni}^{+} \mathbf{b}_{ni} \bar{\mathbf{S}} \cdot \mathbf{F}_i \cdot \bar{\mathbf{S}} \qquad (3.12)$$

Taking into account the translational symmetry of the crystal and only the interaction between nearest neighbors, the total Hamiltonian may be written as

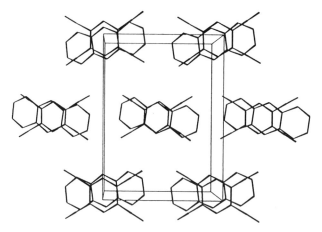

Figure 3.11. Crystal structure of A/TCNB: a view along the stack axis. Two orientations of the complex in the unit cell, differing mostly by a rotation of anthracene molecules are visible (after [10]).

$$\mathbf{H} = \sum (J + A)\mathbf{a}^+_{k+}\mathbf{a}_{k+} + \sum (J - A)\mathbf{a}^+_{k-}\mathbf{a}_{k-} + \bar{\mathbf{S}} \cdot \mathbf{M} \cdot \bar{\mathbf{S}}$$
$$+ \sum \bar{\mathbf{S}} \cdot \mathbf{D} \cdot \bar{\mathbf{S}}(\mathbf{a}^+_{k+}\mathbf{a}_{k+} - \mathbf{a}^+_{k-}\mathbf{a}_{k-}) \qquad (3.13)$$

where

$$J = 2J_a \cos k_a + 2J_c \cos k_c$$
$$A = [(J_0 + J_b \cos k_b)^2 + J_b^2 \sin^2 k_b]^{1/2}$$
$$\mathbf{M} = 1/2(\mathbf{F}_1 + \mathbf{F}_2), \qquad \mathbf{D} = 1/2(\mathbf{F}_1 - \mathbf{F}_2)$$

$\mathbf{a}^+_{k+}$, $\mathbf{a}^+_{k-}$, $\mathbf{a}_{k+}$, $\mathbf{a}_{k-}$ are the creation and annihilation operators depending on the $\bar{k}$ vector derived from the $\mathbf{b}^+_{ni}$ and $\mathbf{b}_{ni}$ operators. Here, $\bar{k}$ is the wave vector of the exciton and J_a and J_c are the interaction matrix elements between nearest neighbors in the a and c directions, respectively. J_0 and J_b describe the interaction between differently oriented nearest neighbors in the b direction in the same and adjacent unit cells, respectively.

Neglecting the fine structure terms, the Hamiltonian (3.13) is diagonal in the electronic operators, that is, two bands of energies are created depending on the components k_i ($i = a,b,c$) of the wave vector. This is the well-known Davydov splitting. Taking into account the influence of the sum and difference tensors $\mathbf{M}$ and $\mathbf{D}$, each Davydov band is split into subbands, which are depicted in Fig. 3.12.a for $k_a = k_c = \pi/2$ as a

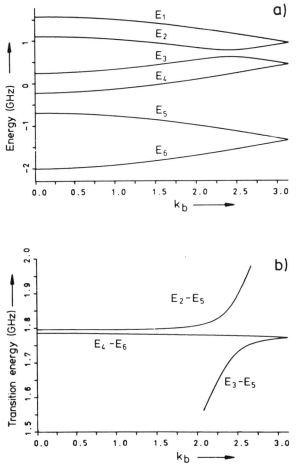

Figure 3.12. (a): Triplet exciton sub-band energies as function of the wave vector k_b for $J_b = 0.011$ cm^{-1}, $D = 2019.5$ MHz and $E = -248$ MHz using the mutual orientation of molecular and crystal axes in A/TCNB as given in [10]. (b) Energy differences for the $D - |E|$ transition as a function of the wave vector k_b. The transition energies are represented only for transition probabilities larger than 0.02 (from [126]).

function of k_b. The best-fitted parameters are $J_0 = J_b = 0.011$ cm^{-1}, $D = 2019.5$ MHz, $E = -248$ MHz. The mutual orientation of the anthracene molecules known from the structural data was used in the calculations.

Figure 3.12.b shows the energy difference between the sublevels for the $D - |E|$ transition. Figure 3.13 depicts the calculated ODMR spectrum, which was obtained by adding the transition probabilities of all k

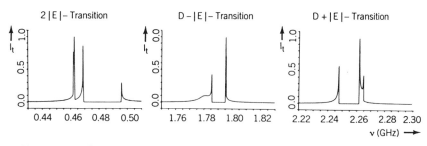

Figure 3.13. Calculated excitonic ODMR lines in A/TCNB. The histograms are given for infinite coherence lengths of the excitons (from [126]).

values whose transition energy lies in a certain interval ΔE. One may recognize the same splitting which appears in the experimental spectra, showing that the model of triplet excitons moving coherently along the stacks, with a small interaction perpendicular to them, is valid for A/TCNB, at least at 1.2 K. The discrepancy between the model and the experiment lies in the shape of the signals; while the model predicts asymmetrical lines, the experiment yields almost perfect Gaussians. The reason for this may be looked for in the inhomogeneous broadening caused by crystal imperfections, impurities, etc., as the theoretical curves were calculated assuming infinite coherence length. Thus the exciton motion may be microscopically coherent while macroscopically incoherent, in analogy to certain other one-dimensional crystals like tetrachlorobenzene [15,128].

The presence of coherence in the excitons in A/TCNB was also postulated by Park and Reddoch [129]. These authors reported the ESR observation of mobile triplet excitons in addition to the ubiquitous trap signals at temperatures as low as 4.2 K (in all previous reports, excitonic ESR signals disappeared under the trap signals at $T < 25$ K). Park and Reddoch oriented their crystals in the magnetic field in such a way that the two pairs of the low temperature excitonic signals were only a few Gauss apart. On lowering the temperature from 104 to 25 K, the one pair of signals, initially observed at this orientation due to interstack motion, splits into two pairs, just as expected. However, below 25 K the signals merge again to yield one pair at 4.2 K for this particular orientation. The authors interpreted this result as an increase of the interstack hopping rate from $\sim 10^7 \, \text{s}^{-1}$ to $\sim 10^8 \, \text{s}^{-1}$ on lowering the temperature, and this in turn as caused by the onset of coherence in the exciton motion. One should be aware, however, that the crystal structure of A/TCNB was followed down to only 119 K [10] and found to be temperature-dependent. It is quite conceivable that the structure of A/TCNB undergoes further modifications between 119 and 4.2 K and this can perhaps account for the effects observed by Park and Reddoch as well.

The interpretation of von Schütz et al. was critically discussed by Bos and Schmidt [130]. These authors employed electron spin echo (ESE) and other pulsed resonance methods together with time-resolved optical excitation to study the triplet species in A/TCNB at low temperature (1.2 K). Only one zero field transition ($D + |E|$) was examined. Under these conditions, a structured signal appears at a frequency very similar to that reported by von Schütz et al. for the doublet excitonic signals. This undoubtedly originates from trapped excitons and the structure can be attributed to a discrete distribution of shallow traps of very similar zfs parameters. Bos and Schmidt discovered a rapid communication between these traps taking place even at 1.2 K. This communication proceeds directly between the traps and not via the triplet band. No mobile exciton signals could be detected at 1.2 K.

Bos and Schmidt claimed that the traps detected by them in the ESE spectra are also responsible for the double DF-ODMR signals detected by von Schütz et al. [125,126] which were apparently wrongly attributed to mobile excitons. This would make the entire discussion presented earlier on the coherence effects in A/TCNB groundless. Our opinion is that the conclusions of Bos and Schmidt are somewhat too far-reaching. Most importantly, the experimental techniques employed by the two groups were entirely different. Bos and Schmidt used direct microwave detection of the resonance while von Schütz et al. used an optical method. The latter authors detected the delayed fluorescence resulting from the annihilation of mobile and localized states. In that case, only a small percentage of all triplet states is observed. ESR, however, is governed by the high concentration of localized triplet states, as has been also shown by optical nuclear polarization and nuclear relaxation measurements (see Section III.E). Secondly, the T_1 state of A/TCNB is practically anthracene-like and the numerous traps discovered by Steudle [131] do not differ much in CT character having zfs parameters very close to anthracene, and neither do the mobile excitons. Zfs parameters alone do not allow one to distinguish between the mobile and immobilized species. In any case, the matter has not been finally settled yet.

The other CT system in which mobile triplet excitons are readily detected at the temperature range where coherence effects may take place is A/TCPA. Despite the relative broadness of both zf-ODMR and ESR signals (see Section III.B) there can be no serious doubt that the signals originate from mobile excitons and not from trapped ones; they are almost perfectly Lorentzian and homogeneously broadened as shown by saturation experiments, and also by time-resolved ESR [37]. In the latter case, the recorded decay time of the free induction decay (FID), 186 ns, agrees remarkably well with the observed cw-ODMR linewidth of 1.8 MHz, proving that this linewidth is indeed homogeneous. However, the perfect Lorentzian lineshape of the ODMR and ESR signals, as well as the observed large linewidths, do not agree with the parameters of the coherent triplet excitons observed in other molecular crystals such as tetrachlorobenzene [128]. Neither do they agree with the width or shape of the excitonic signals found in the previously discussed A/TCNB crystal. Finally, the fact that neither the lineshape nor the linewidth undergo significant changes with the temperature point to pronounced differences between A/TCPA and other systems. It seems therefore that triplet excitons in A/TCPA move incoherently even at a temperature as low as 1.2 K. The reason for the loss of coherence at such a low temperature may be found in the scattering of the excitons between the two triplet bands separated by 40 cm^{-1} and originating from the two symmetry-unrelated sublattices. The same phenomenon is responsible for the increased linewidth discussed in Section III.B, and for the temperature dependence of the zfs parameters discussed in Section III.C.

We are aware of no data concerning the problem of triplet exciton coherence in CT crystals other than the two systems discussed above. For the case of these crystals in which the triplet excitons have a significant CT character, and thus are characterized by a strong exciton-phonon coupling, one may expect that the coherence is already destroyed at a very low temperature. Hence the exciton propagation in the crystal must be described by the incoherent hopping between individual complexes, giving one reason for the very small expected triplet bandwidth.

E. Interaction of Mobile Excitons with Nuclear Spins

1. Optical Nuclear Polarization

Another consequence of the mobility of triplet excitons is their interaction with the nuclear spins in the molecules visited during their lifetime. This causes a relaxation of the nuclear spins and leads to a disturbance of the equilibrium populations of the nuclear spin levels, a phenomenon known as optical nuclear polarization (ONP). The role of the mobile triplet excitons in causing ONP was recognized only after considering some other possible mechanisms, as presented below.

An ONP experiment, which is in principle a version of NMR, has been used to study the properties of many aromatic molecules. According to our knowledge it has been applied with success to only one CT crystal, namely A/TCNB, in which it yields valuable information on the triplet exciton dynamics and in particular complements the results pertaining to the exciton hopping rates obtained by ESR [122, 132, 133]. The principles and theory of ONP are extensively treated in [134], so only a brief outline is presented below in order to facilitate the understanding of the results reported in the summarized papers.

A typical ONP experiment can be divided into four consecutive steps:

- absorption of light leading to the population of the S_1 state;
- intersystem crossing to the T_1 state creating the optical (electron) spin polarization (OSP, discussed in Section IV);
- coupling of the electronic and nuclear spins by hyperfine interaction, OSP is transferred to the nuclear reservoir creating the ONP;
- decay to the S_0 state with the general preservation of the ONP as electronic transitions do not affect it.

This cycle is performed repeatedly in the experiment and the basic observable is the bulk polarization p defined as

$$p = \frac{n_\alpha - n_\beta}{\sum_\lambda n_\lambda} \tag{3.14}$$

n_λ ($\lambda = \alpha, \beta$) are the populations of the nuclear Zeeman levels. Because the cycles are repeated on a time-scale short in comparison with the relaxation times, one observes an exponentially increasing value of p,

$$p(t) = p_0[1 - \exp(-t/T_{1L})] \tag{3.15}$$

T_{1L} is the nuclear spin polarization time. The initial slope of this function, $dp(0)/dt$, called the initial polarization rate, is the measure of the number of spins polarized per excitation cycle. It is the most often used observable in an ONP experiment and is generally a function of the magnetic field, crystal orientation, and temperature.

The observed room temperature initial polarization rates as functions of the magnetic field in A/TCNB are shown in Fig. 3.14. They form very broad dependencies (linewidths on the order of thousands of Gauss) with a few very narrow peaks at about 366 and 711 G.

In order to interpret these ONP patterns, one has to consider the two possible mechanisms of the hyperfine coupling which may in principle lead to ONP. They depend on the time behavior of the hyperfine interaction Hamiltonian,

$$\mathbf{H}_{HF}(t) = \bar{\mathbf{S}} \cdot \mathbf{A}_{hfs}(t) \cdot \bar{\mathbf{I}} \tag{3.16}$$

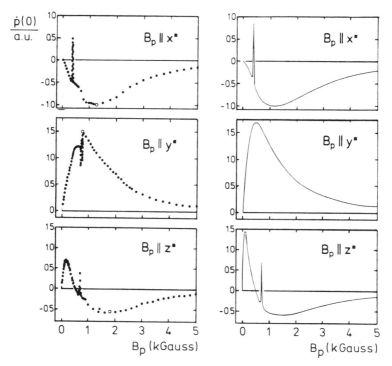

Figure 3.14. Initial ONP rate $dp(0)/dt$ of A/TCNB at $T = 300\,\text{K}$ as a function of the polarization field B_p. Left part: normalized experimental curves for B_p along the three zfs axes. Right part: corresponding model calculations (from [122]).

where $\mathbf{A}_{\text{hfs}}(t) = \mathbf{A}_{\text{av}} + \mathbf{A}(t)$, that is, $\mathbf{A}_{\text{hfs}}(t)$ is divided into a static part $\mathbf{A}_{\text{av}}$ and a fluctuating part $\mathbf{A}(t)$ which vanishes when averaged in time.

1. $\mathbf{A}(t) = 0$, $\mathbf{A}_{\text{av}} \neq 0$. The coupling of an electronic spin S to a nuclear spin I is caused by the static hyperfine interaction. This causes a mixing of the pure spin states and transfers the population differences of the electronic spins to nuclear spins. A perturbation treatment shows that the mixing is proportional to $A/\Delta E$ (A is the hfs constant, ΔE is the energy separation of the sublevels). The dependence of ΔE on the magnetic field causes the corresponding change in the ONP magnitude with the strongest effects observed near level crossing fields B_{LC} where ΔE reaches its minimum. A more detailed consideration shows that the level crossing region is changed to a regime of hyperfine-induced level anti-crossing (LAC). The ONP field dependence is given by a superposition of two Lorentzian lines of opposite sign centered at the LAC fields:

$$B_{\text{LAC}} = B_{\text{LC}} \pm A_{\parallel}/2g\beta_e \qquad (3.17)$$

$A_{\parallel}$ is the hfs component parallel to the magnetic field. The LAC mechanism is not compatible with the experimental curves shown in Fig. 3.14. The Lorentzian curves predicted from the theory can in principle be broadened by electronic spin relaxation, but the fit of corresponding parameters to the experimental curves yields results that are quite unrealistic. One has to look therefore for another mechanism that could be responsible for the creation of ONP in A/TCNB.

2. $\mathbf{A}(t) \neq 0$, $\mathbf{A}_{\text{av}} = 0$. The time average of the hyperfine interaction is zero but the fluctuating hyperfine terms may induce simultaneous electronic and nuclear spin transitions [135]. This phenomenon is analogous to the well-known Overhauser effect. With only scalar coupling considered, the ONP signal is proportional to the number of transitions W_{-+} with $\mathbf{S}^-\mathbf{I}^+$ character minus W_{+-} for $\mathbf{S}^+\mathbf{I}^-$ transitions. This means that for any pair of electronic spin states $|m\rangle$ and $|m'\rangle$,

$$W_{-+}^{(mm')} - W_{+-}^{(mm')} \propto J(\omega_{mm'})\mathbf{A}^2$$
$$\times (|\langle m|\mathbf{S}^+|m'\rangle|^2 - |\langle m|\mathbf{S}^-|m'\rangle|^2)(N_m - N_{m'}) \qquad (3.18)$$

N_i is the population of $|i\rangle$, $J(\omega_{mm'})$ the spectral density of the fluctuation at the transition frequency $\omega_{mm'}$. The initial polarization rate is then given by

$$dp(0)/dt \propto \mathbf{A}^2 \sum J(\omega_{mm'})(N_m - N_{m'})$$
$$\times (|\langle m|\mathbf{S}^+|m'\rangle|^2 - |\langle m|\mathbf{S}^-|m'\rangle|^2) \qquad (3.19)$$

In sufficiently high magnetic field, zfs can be neglected and this equation is reduced to a much simpler form [136]:

$$dp(0)/dt \propto 1/2\mathbf{A}^2 \langle \mathbf{S}^z \rangle J(\Omega) \qquad (3.20)$$

$\langle \mathbf{S}^z \rangle$ is the optical electron spin polarization and $\Omega = \omega_s - \omega_I \simeq \omega_s$ the electronic Larmor frequency. Both $\langle \mathbf{S}^z \rangle$ and $J(\Omega)$ depend in a characteristic manner on the static field B_p. Therefore from the ONP field dependence, one obtains information on the mechanism of processes that determine the relaxation rates. The broad ONP signals shown in Fig. 3.14 can be interpreted in terms of the relaxation caused by fluctuating hyperfine interactions described above.

The final proof, however, of the dominant role played by hyperfine interaction fluctuations caused by the mobility of triplet excitons was delivered by an experiment called radiofrequency-induced ONP (rf-ONP). It is described in detail in [137] and depends on an observation of ONP while saturating the rf transitions, usually in an external magnetic field. Rf-ONP yields spectra that are equivalent to ESR. Without going into details, the shape of a rf-ONP signal is totally different for the two possible mechanisms of hyperfine interactions discussed above. In the case of static interactions, one expects signals that are combinations of a positive and a negative part, and thus resemble the first derivative signals observed in ESR. If the fluctuating hyperfine interactions dominate, the rf-ONP signals should be absorption-like and Gaussian-shaped. The actual rf-ONP spectra of A/TCNB are shown in Fig. 3.15. There is no doubt that the observed lines are symmetrical and Gaussian-shaped and so the role of the mobility of triplet excitons in A/TCNB in causing the hyperfine interactions to fluctuate in time and thereby causing the observed ONP pattern is confirmed.

The parameters characterizing the relaxation caused by fluctuating hyperfine interactions are: triplet exciton hopping time τ and relative coupling strength $\langle A^2 \rangle / K$ (A is the hyperfine coupling constant, K is the total triplet decay rate). In order to obtain the numerical values of these parameters in A/TCNB, a model [138] was used taking into account the selectivity of the population and depopulation of the triplet sublevels using the values for anthracene [139]. Also, for the sake of simplicity, the triplet exciton mobility was considered three-dimensional. The results of the fits are shown in Fig. 3.14 beside the experimental results. The

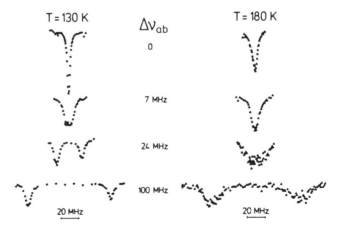

Figure 3.15. Rf-ONP spectra of A/TCNB at four different orientations of the crystal with respect to the magnetic field expressed by the separation of the magnetically inequivalent signals in frequency units ($\Delta \nu_{ab}$). Exchange effects are prominent at 180 K (from [133]).

best-fitted exciton hopping time is 3×10^{-10} s which corresponds to a hopping rate of 3×10^{9} s^{-1} and a relative coupling strength $\langle A^2 \rangle / K$ of 5×10^{9} s^{-1}. The ONP experiments yield a hopping rate of triplet excitons in A/TCNB which falls exactly between the values reported by Ponte Goncalves and Möhwald et al. on the one hand ($\sim 6 \times 10^8$ s^{-1}) and Park and Reddoch on the other ($\sim 10^{10}$ s^{-1}). The interstack hopping frequency could be estimated from the rf-ONP experiment shown in Fig. 3.15 to be $\sim 1.5 \times 10^7$ s^{-1} at 130 K and $\sim 5 \times 10^7$ s^{-1} at 180 K, also in agreement with the ESR results. No additional information was obtained on the dimensionality of the exciton motion.

2. Triplet Exciton Dynamics via Nuclear Relaxation Mechanisms

Although the absolute value of the nuclear spin-lattice relaxation rates in molecular crystals in the presence of light excitation often cannot be evaluated with respect to the exciton dynamics, the frequency (or field) dependence of T_{1L}^{-1} (nuclear relaxation rate in presence of light) is an excellent measure of the correlation times of the exciton–nuclear interaction. Such experiments were reported in [122,140].

Normally, NMR experiments as a function of frequency are restricted to roughly 1.5 orders of magnitude (~ 10–500 MHz). The lower limit is due to dead time and sensitivity reasons, the upper limit to the maximum field available.

There is a striking feature in most organic crystals, however, namely that the nuclear relaxation times are very long in the dark, that is, without illumination, typically $\sim 10^3$ s, if no mobile side groups such as CH_3 are present [141]. This makes the procedure of field cycling applicable [142]. In the dark, the field is switched slowly from the detection value to the relaxation field of interest, B_{relax}; then the light is switched on for a given time t_{light} and the nuclei relax. After switching the light off, the field is switched rapidly to the detection value and the magnetization is measured as a function of t_{light}, thus allowing the evaluation of T_{1L}.

As the geometry and intensity of the light excitation is unaffected by this procedure, T_{1L}^{-1}, the light-induced relaxation rate of the nuclei (usually protons), can be measured at very low fields, extending the range of detection considerably, for example, to 10^{-2}–10^5 G. The result of such a measurement is shown in Fig. 3.16. T_{1L}, the longitudinal nuclear spin relaxation time in the presence of light which is also the rise time of the ONP $p(t)$ (see preceding section), is also determined by the exciton-induced fluctuation of the hyperfine interactions. Consequently, the field dependence of T_{1L} is closely connected with the exciton mobility [136].

To evaluate the exciton hopping time τ from the T_{1L} data, it is

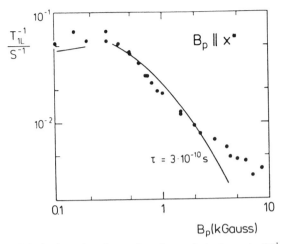

Figure 3.16. Spin lattice relaxation and nuclear polarization rate T_{1L}^{-1} of A/TCNB at $T = 300$ K as a function of the magnetic field. The solid line is calculated for isotropic exciton hopping with a hopping time of 3×10^{-10} s (from [122]).

sufficient to consider the high field approximation, analogously to Eq. (3.20) [136] (that means we neglect the zero field splittings):

$$T_{1L}^{-1} = A^2[\bar{S}(\bar{S} + 1) - \langle (S^z)^2 \rangle]J(\Omega) \qquad (3.21)$$

The term $\mathbf{S}(S + 1) - \langle (S^z)^2 \rangle$ shows only a minor dependence on B_p so $T_{1L}^{-1} \propto J(\Omega)$. Figure 3.16 shows T_{1L}^{-1} as a function of B_p. For fields below 400 G T_{1L}^{-1} is almost constant. Above 500 G, T_{1L}^{-1} begins to decrease monotonically and approaches a constant level in very high fields that is determined by the dark rate T_{1D}^{-1} for relaxation in the absence of light irradiation. When using the spectral densities given in [143], we can calculate T_{1L}^{-1} for one-, two- and three-dimensional exciton hopping. However, the resulting differences for the spectral density are small and cannot be discriminated within the limits of the experimental accuracy. The dispersion curve shows an almost identical interval of strong decrease for all cases of different anisotropy. The characteristic frequency position with $\omega_p \tau \simeq 1$ ($\omega_p = \gamma B_p$) is closely connected with the largest elements of the diffusion tensor $\mathbf{D} \simeq 1/\tau$. Therefore the resulting determination of τ is practically independent of the anisotropy. The best fit of the theory to the data is shown in Fig. 3.16 by the solid curve. It renders a value $\tau = 3 \times 10^{-10}$ s which agrees well with the ONP analysis discussed in the previous section and the results on tetrachlorobenzene [144].

F.   Interaction with Doublet States

A triplet exciton moving in the crystal may encounter on its way a charge carrier trapped on some defect and characterized by spin $= 1/2$ (a doublet). We do not discuss here the role of triplet excitons in creating charge carriers, as it is one of the consequences of the triplet-triplet exciton annihilation which is discussed in Section IV.C.5. Rather, an interaction with an already present charge carrier is discussed, a phenomenon observed in the A/PMDA crystal by Ziegler and Karl by means of an experiment called triplet-doublet magnetic resonance (TDMR) [93]. TDMR is a high field resonance technique depending on a measurement of the photocurrent as a function of crystal orientation relative to the magnetic field. The observed modulation of the photocurrent is due to the following phenomena.

A triplet exciton (T) moving in the crystal encounters on its way a charge carrier trapped on some defect (D_{tr}). The defect can be for example a disoriented molecule and the trapped carrier is a molecular radical ion characterized by spin $\sigma = 1/2$ (a doublet). As a result of triplet–doublet interaction, a transient pair $(T \ldots D)$ is created, which can either dissociate back into the exciton and the charge carrier or decay into the molecule in the ground electronic state (S_0) and a thermally excited charge carrier D^*. The latter relaxes into the conduction band and contributes to the photocurrent:

$$T + D_{tr} \rightarrow (T \ldots D) \rightarrow S_0 + D^*$$

The pair state Hamiltonian in the TD problem is written as the sum of the Hamiltonians of the isolated triplet and doublet states:

$$\mathbf{H}_{TD} = g_T \beta \bar{\mathbf{B}}_0 \cdot \bar{\mathbf{S}} + \bar{\mathbf{S}} \cdot \mathbf{F} \cdot \bar{\mathbf{S}} + g_D \beta \bar{\mathbf{B}}_0 \cdot \bar{\boldsymbol{\sigma}} \qquad (3.22)$$

g_T, g_D = g-factors of the triplet and doublet state, respectively.

The result of this operator acting on the pair state wave function is shown in Fig. 3.17 for the case of $g_T = g_D$. Of the six spin sublevels created, two are of pure quartet character (Q) and do not contribute to the photocurrent. The other four have mixed doublet-quartet character (D/Q). If the energy difference between these levels (indicated by arrows) is reduced to zero by a proper orientation of the crystal relative to the magnetic field, the two now degenerate pair states turn into one pure doublet and one pure quartet state. Thus the number of states carrying doublet character and contributing to the photocurrent is reduced from four to two. As a consequence, the photocurrent is mod-

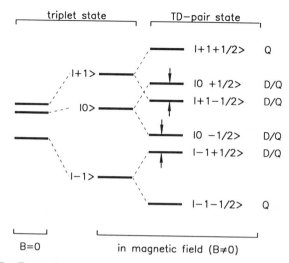

Figure 3.17. Energy level diagram of triplet state and triplet-doublet pair levels in a magnetic field. D, doublet; Q, quartet character (after [93]).

ulated when the crystal is rotated in the magnetic field and characteristic minima (also called "resonances") appear (Fig. 3.18). From the rotational patterns of the photocurrent peaks, all the information relevant to the zfs tensor of the triplet excitons can be extracted. In the case of the excitons in the A/PMDA crystal, the zfs parameters determined by

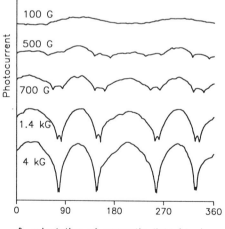

Figure 3.18. Anisotropy of photocurrent in A/PMDA when the magnetic field is rotated in a fixed crystal plane for the field values indicated (from [93]).

TDMR are the following: $D = \pm 0.0665(5)$, $E = \mp 0.00765(1)$ cm^{-1}. The principal axes of the tensor agree within the experimental error (less than 1°) with the symmetry axes of the anthracene molecules in the triclinic A/PMDA crystal. These results stand in very good agreement with those obtained by ESR [92] and show that TDMR can be an alternative experimental method for determining the magnetic properties of mobile triplet excitons in CT crystals.

IV. KINETICS OF TRIPLET EXCITONS IN CT CRYSTALS

A. General

In Section IV, we intend to discuss the influence of the triplet exciton mobility on the excitonic lifetime and consequently the population of the excitonic band, as well as of individual triplet spin sublevels. Among the processes that appear as a consequence of increased excitonic mobility and determine the population of the band, the most prominent role is played by trapping on defects, interaction with defects already populated by other excitons (heterogeneous triplet-triplet annihilation), and interaction with other mobile excitons (homogeneous T–T annihilation). These phenomena are discussed in Section IV.C.

Before we embark on this discussion, however, we want to devote Section IV.B to kinetic processes characteristic of the T_1 state itself, which are not sensitive to the excitonic mobility, in analogy to the discussion of fine structure in Section II.

Finally, Section IV.D is devoted to a review of kinetic schemes in CT crystals which differ in the CT character of the T_1 state.

B. Kinetics of Isolated Triplet States

1. Triplet Sublevel Population Rates

As is well known, most triplet states are populated from singlet states by intersystem crossing (ISC). From the theory of ISC, it follows that this spin-forbidden process becomes partially allowed by admixture of some singlet character into the triplet state and vice versa, which occurs predominantly by spin-orbit coupling (SOC). The SOC operator mixes only those singlet and triplet states that have identical symmetry. The symmetry of a singlet state is fully determined by its orbital symmetry, whereas the symmetry of a triplet state is given by a product of the symmetry of the orbital part of the triplet wave function $^3\Phi_i$ and the respective triplet spin function, which is different for each of the τ_x, τ_y and τ_z states. The admixture of singlet character to the full T_1 wave function is significant only for the triplet sublevel for which

$$\Gamma_{T_n} = \Gamma_{3_{\Psi_i}} \otimes \Gamma_\tau = \Gamma_S$$

where the Γ are the irreducible representations of the molecular point groups that transform like the individual functions.

As a result, for aromatic molecules characterized by low symmetry and hence also for CT complexes formed from them, one (or less frequently two) of the triplet sublevel population rate constants is much larger than the others. The triplet sublevels are thus populated with very unequal efficiencies. An applied high external magnetic field mixes the zero field population rate constants S_i ($i = x, y, z$) to yield high field constants S_u ($u = 1, 0, -1$). Consequently one of the high field constants is usually either smaller or larger than the others, so that the high field triplet sublevels $|-1\rangle$, $|0\rangle$ and $|+1\rangle$ are also unevenly populated. Spin lattice relaxation (SLR) attempts to bring the population differences created as a result of the ISC transition back to Boltzmann equilibrium. If SLR is fast in comparison with the triplet lifetime, thermal equilibrium is established. If not, the steady-state populations of the triplet spin sublevels deviate strongly from Boltzmann equilibrium, a phenomenon known as optical spin (or electron) polarization (OSP or OEP or spin alignment; for a review see [145]). The triplet sublevel population and depopulation processes both in zero and high magnetic field are shown in Fig. 4.1.

The absolute values of the triplet population rate constants S_i can be determined directly by a variety of ODMR methods. There is a substan-

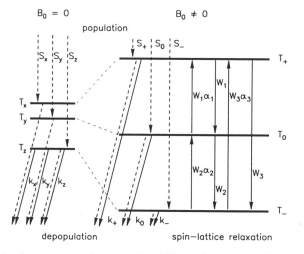

Figure 4.1. Rate constants for population (S_i) and depopulation (k_i) in zero and high magnetic field, and for spin lattice relaxation (W_i) in high field. $\alpha_i = \exp(-\Delta E_i/kT)$; ΔE_i is the energy difference between Zeeman levels (after [145]).

tial amount of literature data available, resulting mostly from experiments on CT complexes isolated in glasses or in some cases isolated defects in crystals. In our opinion, these results do not bring much insight into the general properties of CT crystals and we will not therefore pursue this discussion. Rather, we concentrate on the number of stationary ESR experiments involving short-lived triplet species, which depend on an observation of OSP to extract relative values of the population rate constants. These experiments on crystals have brought interesting and often unexpected results concerning the validity of certain approximations used in describing wave functions of CT complexes.

OSP manifests itself in an ESR spectrum of a single crystal by a reduced intensity of one of the two generally expected absorption signals, or even by a change of its phase, that is, one of the signals becoming emissive. The triplet ESR spectrum of the Ph/TCNQ crystal shown in Fig. 2.8.a shows typical OSP. Because the high field triplet sublevel population rates are functions of the zero field rates and the strength and orientation of the magnetic field, the intensities of the observed ESR signals vary when the crystal is rotated in the field. Also, the phases of the signals may exchange at certain crystal orientations, yielding characteristic angular OSP patterns. These patterns can give insight into the population mechanisms as shown below.

It should be mentioned at this point that recently a new phenomenon has been observed as a source of OSP in some CT crystals: the excited singlet state fission into two triplet states [39]. This mechanism is efficient in the case where the lowest triplet state energy is very small, on the order of half the energy of the lowest excited singlet state, as shown in [39] for the F/TCNQ and Ph/TCNQ crystals. Below, however, the more common mechanism of the creation of OSP is discussed.

The CT crystal P/PMDA was already discussed from the point of view of its zfs tensor and CT character (Section II.C.3). It was found by Keijzers and Haarer [46] that the triplet zfs tensor in this crystal can be satisfactorily described by a linear combination of the zero order local triplet state of phenanthrene and a zero order CT state. Keijzers et al. [146], however, observed in a subsequent paper that the OSP pattern of triplet excitons in this CT crystal cannot be explained by a wave function constructed from such a combination. This is so because both the local phenanthrene and CT triplets are $\pi\pi^*$ states for which it is known that one- and two-center SOC integrals vanish by symmetry [147]; hence their contributions to the ISC processes are minimal. Therefore, more complete triplet wave functions of both triplet and singlet states of the complex are necessary to account for the observations, and these have to include for symmetry reasons the donor and acceptor states of $\sigma\pi^*$ and

$n\pi^*$ character as well. In the case of phenanthrene and PMDA, these states can only mix with the $\pi\pi^*$ states by vibronic coupling. In other words, the two-electron triplet wave function $^3\Psi(1,2)$ previously considered as a linear combination of the zero order CT state and only one locally excited state of either a donor or acceptor (Eqs. (1.6) and (1.7)) has to be expanded by addition of the following extra terms:

$$\sum c_{\sigma D}{}^3|D_\sigma(1)D_{\pi^*}(2)| + \sum c_{\pi D}{}^3|D_\pi(1)D_{\pi^*}(2)|$$
$$+ \sum c_{nA}{}^3|A_n(1)A_{\pi^*}(2)| + \sum c_{\pi A}{}^3|A_\pi(1)A_{\pi^*}(2)| \qquad (4.1)$$

(the summations are over all the excited states of the symmetry concerned). The CT singlet wave function $^1\Psi(1,2)$ can be represented in a similar way.

Calculation of the ISC transition probability from the S_1 to the T_1 state of the CT complex requires evaluating the following matrix element:

$$\langle\,^1\Psi(1,2)|H_{\mathrm{LS}}|\,^3\Psi(1,2)\rangle \qquad (4.2)$$

where H_{LS} is the spin orbit Hamiltonian.

Following the way Keijzers et al. solved this problem in detail is beyond the scope of this paper, hence the reader is referred to the original paper. The general outline of their work is as follows:

- After considering the mutual energy position of the $\sigma\pi^*$ and $n\pi^*$ states in both donor and acceptor, only one of the original 36 matrix elements is found to be significant, namely the one which couples the $\pi\pi^*$ states with the $n\pi^*$ states of PMDA.

- Theoretical calculation of the energy levels of the ground and low lying excited states of the PMDA molecule [148] shows that they are both of $n\pi^*$ character. In the triplet manifold, however, the $\pi\pi^*$ state lies very close to the $n\pi^*$ state, thus allowing an efficient vibronic coupling of the $\pi\pi^*$ to the $n\pi^*$ triplet. The main ISC channel in the P/PMDA complex thus leads to a $n\pi^*$ triplet of PMDA which is in turn mixed with the $\pi\pi^*$ triplet.

- After considering the ratio between the p_x and p_y LCAO coefficients in the n orbital of the carbonyl group in PMDA, which mostly contributes to the $n\pi^*$ state, Keijzers et al. [146] estimate the ratio at which the τ_x and τ_y zero field spin functions of PMDA participate in the ISC process, and eventually obtain the relative population rate constants of the triplet sublevels $S_y : S_x = 1.79 : 1$ (S_z is taken as zero in analogy with other aromatic molecules).

This important result was subsequently used to account for the OSP pattern of triplet excitons in the P/PMDA crystal shown in Fig. 4.2. First, high field population rate constants are calculated from the formula

$$S_u \, (u = 1, 0, -1) = \sum |c_{iu}^2| S_i \, (i = x, y, x) \tag{4.3}$$

(the c_{iu} coefficients depend on the crystal orientation in the magnetic field and can be calculated for a given orientation by diagonalization of the spin Hamiltonian containing zero field and Zeeman terms). The steady-state populations N_1, N_0 and N_{-1} of the high field spin sublevels $|1\rangle$, $|0\rangle$ and $|-1\rangle$ are then calculated by solving a set of three coupled linear differential equations (4.4) under steady-state conditions. The five parameters introduced into the equations are: the high field population rates S_1, S_0 and S_{-1}, the triplet lifetime (as determined from the delayed fluorescence decay time) and the spin lattice relaxation time T_1 (determined from ESR saturation measurements).

$$dN_1/dt = 0 = (-k - 2W)N_1 + W\alpha_1 N_0 + W\alpha_2 N_{-1} + S_1 N_S \tag{4.4a}$$

$$dN_0/dt = 0 = WN_1 + [-k - W(1 + \alpha_1)]N_0 + W\alpha_1 N_{-1} + S_0 N_S \tag{4.4b}$$

$$dN_{-1}/dt = 0 = WN_1 + WN_0 + [-k - W(\alpha_1 + \alpha_2)]N_{-1} + S_{-1} N_S \tag{4.4c}$$

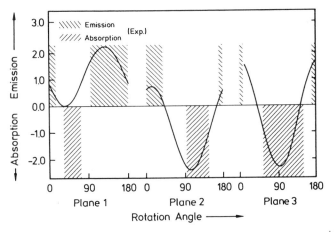

Figure 4.2. Angular dependence of the observed spin polarization of the $|1\rangle$ to $|0\rangle$ ESR transition in A/PMDA for rotations around three mutually perpendicular axes. The differently shaded areas correspond to regions in which absorption or emission lines are observed. The solid lines are calculated assuming $S_y : S_x = 1.79 : 1$ (from [146]).

α_1 and α_2 are the Boltzmann factors $\exp(-\Delta E/kT)$ for the $\Delta m_s = \pm 1$ and ± 2 transitions, respectively, and N_S is the singlet population.

The following assumptions were made in [146] before deriving these equations:

   i. the SLR rates W_1, W_2 and W_3 (Fig. 4.1) are all equal; $W_1 = W_2 = W_3 = W = 1/T_1$;

  ii. the decay rates from the spin sublevels are also equal (spin-independent depopulation is dominated by the exciton annihilation mechanism), $k_1 = k_0 = k_{-1} = k = 1/\xi$, ξ is the triplet lifetime;

 iii. the microwaves do not influence the sublevel populations (no-saturation condition).

The results of solving Eqs. (4.4) for the steady state relative populations, for example, $(N_1 - N_0)/N_S$ for the $|0\rangle \rightarrow |+1\rangle$ transition are shown by the solid lines in Fig. 4.2. This figure also shows the OSP patterns expected assuming the exclusive population of one of the triplet sublevels. The agreement between the calculated relative population rate constants and the experimentally obtained OSP pattern is very good, particularly since no adjustable parameters were used.

The most important conclusion of this procedure is the following. Although in most cases the fine structure (and also, as shown in Section VI, the hyperfine structure) tensor of a CT complex can be satisfactorily described by a wave function formed from a linear combination of only two zero-order states (the local triplet of either donor or acceptor, and the CT state), the efficiency of ISC processes can be explained only after taking into consideration contributions from other zero order states as well (in the case of P/PMDA the singlet and triplet states of the PMDA molecule).

The problem of the participation of the acceptor in the ISC processes in various CT crystals has subsequently been treated by Pasimeni, Corvaja and co-workers. The OSP pattern of the room temperature triplet ESR spectrum of A/PMDA crystal was reported by Montanari et al. [149]. The zfs tensor of A/PMDA can be adequately described using a triplet wave function represented by a local anthracene function with a minimal admixture of the zero order CT state ($c_1^2 \leq 5\%$). Nevertheless the OSP pattern cannot be explained by the triplet population rates of anthracene. A procedure analogous to that of Keijzers et al. [146], but depending on fitting the kinetic constants instead of calculating them theoretically, yields the relative population rates of the triplet sublevels $S_x : S_y : S_z = 1 : 0.4 : 0.15$. These values are similar to those reported in [146] for P/PMDA, except that the S_x and S_y values are interchanged. The reason for this difference is that in A/PMDA the Z principal value of the zfs tensor is taken as negative while in P/PMDA [146] it is taken as positive (as stated previously, the absolute values of zfs parameters are not available from standard ESR experiments). After accounting for this difference, the relative population constants of triplet excitons in A/PMDA are very similar to those of P/PMDA and confirm the notion that the acceptor molecule plays the dominant role in the ISC processes, even if the triplet state is almost completely localized on the donor.

In the CT crystal A/TCNB, the OSP pattern of triplet ESR spectra likewise cannot be explained on the basis of the population rate constants characteristic for anthracene ($S_x : S_y : S_z = 1 : 1 : 0.14$ [139]), whereas those reported for TCNB ($1 : 0.32 : 0.16$ [150]) nicely reproduce the OSP pattern of the complex [103]. The best-fit procedure yields the ratio $1 : 0.30 : 0.25$ for A/TCNB, only slightly different from that of TCNB. In the A/TCNB crystal, the triplet wave function is similar to A/PMDA localized predominantly on anthracene ($c_1^2 \lesssim 5\%$) but the ISC seems to be dominated by the very small admixture of the TCNB wave function both in the S_1 and T_1 states. This situation was found also in other CT crystals built with TCNB as an acceptor: B/TCNB [151], N/TCNB [105], tS/TCNB (trans-stilbene/TCNB) [152] and DPA/TCNB [50], independent of the degree of localization of the electron spin.

Corvaja and Pasimeni [151] observed in the OSP patterns of triplet ESR spectra of the B/TCNB crystal that the inversion points (the angles at which the absorptive ESR signals become emissive and vice versa) are not placed symmetrically with regard to the zfs principal axes. They interpreted this effect in analogy with the situation reported by Schadee et al. [153] for tetramethylpyrazine (TMP) in a single crystal of durene. In TMP, the zfs principal axes are rotated from the molecular axes by the crystal field. As the dominant role in the ISC is played by the n orbitals of the two nitrogen atoms in the TMP molecule, both of which lie in the molecular symmetry plane but not in the zfs principal plane, the spin momentum is not quantized in a zfs principal plane and the triplet is generated in a superposition of the states τ_z and τ_x. In other words, the triplet sublevels are no longer populated independently of each other. Consequently the high field population rates S_u are given by

$$S (u = 1, 0, -1) = \sum_{i,j} |c_u^i c_u^j| (S_i S_j)^{1/2} \qquad (4.5)$$

which differs from (4.3) by the presence of the mixed (coherence) terms.

The above idea was applied to B/TCNB by Corvaja and Pasimeni [151] because in TCNB the C≡N bond axes do not correspond with the molecular symmetry axes (which in turn nearly coincide with the zfs principal axes). In such a case it may be expected that the ISC leads to a triplet with the spin momentum quantized principally in a plane perpendicular to the C≡N bond. However, in the TCNB molecule there are four cyano groups which are placed symmetrically with respect to the molecular axes, and thus the effect is canceled (the four groups are pairwise equivalent). To account for the observed OSP asymmetries, the two pairs of cyano groups must be inequivalent, which can happen if the TCNB molecule is rotated with respect to biphenyl (and the zfs principal axes system). The best-fit procedure yields a rotation angle of 30°. In attempting to explain the asymmetry of the OSP pattern, Corvaja and Pasimeni concluded that TCNB must be rotated in the S_1 state by this angle, which is very large. Asymmetries in the OSP patterns were also found in other CT crystals containing TCNB, as well as other acceptors, for example, in the previously mentioned A/PMDA crystal [149]. In A/PMDA, ISC to a superposition of spin states was also proposed, as the directions of the C=O bonds in PMDA (the carbonyl groups are mainly responsible for the ISC) do not correspond with the zfs principal axes. A rotation angle of PMDA by 4° is necessary to account for the observed effect, a value both realistic and found indeed in the crystal structure [154].

The results of Keijzers et al. and Corvaja, Pasimeni et al. found conformation in an independently performed nonstationary ESR experiment on the triplet state of the N/TCPA crystal by Yu and Lin [42]. Fitting the observed responses of ESR signals to switching the exciting light on and off, the authors were able to extract the relative population rates of the triplet sublevels $S_x : S_y : S_z \simeq 1:1:0$. This result is different from the values characteristic of naphthalene ($S_x > S_y$, determined also by Yu and Lin) in which the triplet wave function of the N/TCPA complex is mostly localized, and suggests that in this case also the ISC process must be dominated by a higher-lying $n\pi^*$ state of the acceptor molecule TCPA.

2. Triplet State Depopulation Processes

The most obvious process leading to depopulation of a triplet state is the direct $S_0 \leftarrow T_1$ decay, either radiative (phosphorescence) or nonradiative. In fact, phosphorescence is commonly observed for CT triplet states of CT complexes isolated either in low temperature glasses or as defects in CT crystals. There have been no systematic studies of phosphorescence in CT crystals until now and we attempt to discuss this problem in Section VI together with the nature of the defects. As for the nonradiative decay of the T_1 state, it has not been studied so far in CT complexes to the best of our knowledge. The few reports available concern rather the nonradiative decay of the S_1 state [155]. It should be mentioned that both radiative and nonradiative decays of the triplet state are spin-selective, like the ISC processes and can therefore lead to optical spin polarization as observed in ESR or ODMR. In most cases, both ISC and direct decay contribute to the OSP.

Another process potentially causing depletion of the triplet state population is the thermally activated (E-type) delayed fluorescence. This phenomenon may take place if the singlet-triplet energy gap is small in comparison with the Boltzmann factor, and it has in fact been found in many molecules, both in the crystalline phase as well as in solutions [156]. In this process, a molecule in the triplet state can receive thermal energy from the environment and undergo a reversed ISC process to the S_1 state where it consequently emits fluorescence. The characteristic features of E-type DF are: a linear dependence of the intensity on the intensity of the exciting light (in the low-intensity regime), since it is a monomolecular process; a thermal activation of the intensity; and a spectral similarity to the prompt fluorescence.

Since, as was shown in Section II.B.1, certain CT complexes have a significantly reduced singlet-triplet energy gap in comparison with most one-component aromatic molecules, it may be expected that they will exhibit E-type delayed fluorescence. Indeed this kind of delayed emission was first reported in low temperature glasses by Beens and Weller [157]. The first observation in a CT crystal followed in an article by Möhwald and Sackmann [100] on B/TCNB. In B/TCNB, the intensity of the delayed fluorescence is strongly temperature-dependent in the temperature range 80–350 K with an activation energy $\Delta E_a = 500 \pm 200 \, \mathrm{cm}^{-1}$. It is also linearly dependent on the exciting light intensity, which in the regime of weak excitation confirms the monomolecular character of the emission. Because of its characteristics, the delayed fluorescence is identified as E-type (originating from a thermally activated $S_1 \leftarrow T_1$ process). A more recent paper on B/TCNB by Corvaja et al. [47] confirmed the observation of E-type delayed fluorescence in this crystal, although it also downplayed its role in the triplet kinetics. According to these authors, E-type DF begins to play a significant role only at $T > 250$ K (Fig. 4.3), although it is never a dominating depopulation mechanism. The appearance of E-type DF can be directly linked to the relatively small value ($1750 \, \mathrm{cm}^{-1}$) of the singlet-triplet energy gap and thus also to the considerable (50%) CT character of the lowest triplet state.

The importance of the E-type delayed fluorescence as a means of monomolecular depopulation becomes more pronounced in the case of crystals in which the triplet state has a larger CT character than in B/TCNB. In such systems, the S_1–T_1 energy gap is reduced significantly. The P/PMDA crystal ($c_1^2 = 76\%$) was investigated by Kozankiewicz [45]

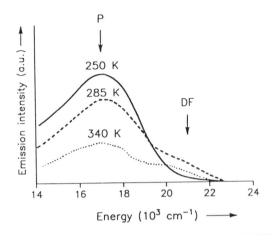

Figure 4.3.   Delayed fluorescence and trap phosphorescence in B/TCNB above 250 K (from [47]).

from the point of view of the exciton kinetics. Above 60 K, the delayed fluorescence is clearly of E-type character ($I_{DF} \sim I_{exc}$), increasing with the temperature and dramatically shortening the lifetime of the triplet species responsible for it from 2.6 ms at 100 K to 12 μs at 280 K (Fig. 4.4). The activation energy of the latter process (600 ± 30 cm^{-1}) agrees very well with the S_1–T_1 energy difference obtained from the spectral positions of DF and phosphorescence.

E-type DF was also observed in the HMB/TCNB crystal (CT character of the triplet state about 90%). In fact, two different kinds of DF were found. At $60 < T < 200$ K, the origin of DF lies in the thermal activation of a defect to a singlet trap correlated with it. This is apparent in Fig. 4.5,

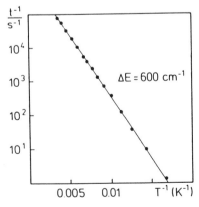

Figure 4.4.   Inverse temperature dependence of the decay time of E-type delayed fluorescence in P/PMDA. The solid line is drawn using an activation energy of 600 cm^{-1} (from [45]).

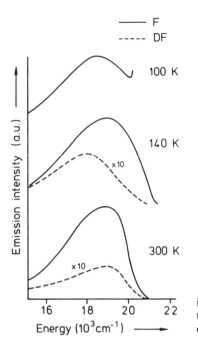

Figure 4.5. Delayed and prompt fluorescence in HMB/TCNB at 300 and 140 K with the excitation over the S_1 band, and at 100 K with the excitation under the S_1 band (from [53]).

which shows that the delayed fluorescence at 140 K is spectrally different from the prompt one (originating from the S_1 band). Also, E-type DF is spectrally similar to the prompt fluorescence observed when exciting the crystal under the S_1 band, that is, directly into the singlet trap. Above 200 K, another kind of E-type DF was identified, originating from thermal activation of mobile triplet excitons to the S_1 band. It effectively shortens the lifetime of triplet species responsible for it (in this case mobile excitons), and the activation energy of this process, $\Delta E_a = 1100 \ cm^{-1}$, is equal to the singlet-triplet energy gap, exactly as in P/PMDA.

The two most extreme cases of the efficiency of the thermal depopulation of the triplet band in CT crystals are HMB/PMDA [158] and HMB/TCPA [159], both having a practically ionic character of triplet excitons and a very small $S_1–T_1$ energy gap. E-type delayed fluorescence appears in both crystals at 30 K (HMB/PMDA) and 40 K (HMB/TCPA) and completely dominates the depopulation mechanisms at higher temperatures. It also drastically shortens the triplet lifetime in HMB/TCPA to as little as 10 μs at 150 K. This very short lifetime of the triplet species may be the reason for the failure of ESR investigations of these two crystals at higher temperatures [54]. The activation energy of this process is $400 \pm 30\,cm^{-1}$ in HMB/TCPA, which is in apparent disagreement with the low temperature stationary emission spectra of this system showing fluorescence and phosphorescence onsets at the same energy [160]. Kozankiewicz and Prochorow explained this discrepancy by the possibility of different configurations of the CT complex in the ground and excited triplet states, providing a Franck–Condon mechanism for reducing the apparent singlet-triplet energy gap.

In view of the articles discussed above, it may be stated that E-type delayed fluorescence plays an important role as a depopulation mechanism of the triplet state in CT crystals with a high CT character of the T_1 state, due to the small S_1–T_1 energy gap. This results in a shortening of the lifetime of the triplet species, from which the singlet triplet energy gap can be determined with an accuracy often better than from the spectral position of the singlet and triplet emission.

C. Kinetics of Mobile Triplet Excitons

1. Resonant Band Phosphorescence

The observation of a resonant band phosphorescence in CT crystals has been reported to our best knowledge in only two systems, A/TCPA [11] and N/TCPA [161]. In both cases, band phosphorescence is extremely weak and observable only over a very limited temperature range around 4.2 K. This is due to depopulation by a variety of other physical processes related to the mobility of excitons, which is discussed in following sections. In any case, triplet band phosphorescence is the least significant band depopulation process and as such is usually neglected in the kinetic schemes of CT crystals.

2. Trapping

A direct result of the excitonic mobility is the possibility of an exciton becoming trapped on a defect whose energy is lower than that of the band. This process effectively depopulates the triplet band. It can be studied by a variety of experimental methods, with the stationary optical emission measurement being the easiest if somewhat rudimentary approach. A simple comparison of the relative intensities of trap and exciton delayed emissions as functions of temperature yields some information on the efficiency of trapping. A more suitable method is time-resolved optical emission spectroscopy. The characteristic trapping time can be derived either from the onset time of trap fluorescence, or, alternatively, from the decay time of the exciton emission, be it phosphorescence or delayed fluorescence. In A/TCNB at 1.2 K, it appears that most (98%) of the delayed fluorescence decays within 100 μs after the excitation pulse. At 4.2 K this short decay constitutes 70% of the total decay of DF (Fig. 4.6a). The origin of the longer decay component is discussed in the next section on triplet-triplet annihilation. The short-lived component represents the decay of the delayed fluorescence due to trapping. Its very high participation means that at 4.2 K and below, trapping is the dominant depopulation mechanism of mobile triplet excitons. On elevating the temperature, its significance decreases in favor of T-T annihilation processes, and above 25 K trapping is negligible (Fig. 4.6.b). This is a general phenomenon in CT crystals, although the temperature at which trapping becomes unimportant in the kinetic scheme varies from one CT system to another. The trapping time (<100 μs) seems to be characteristic not only for A/TCNB, but also for other CT crystals, for instance A/TCPA [36]. It appears, however, that the importance of trapping in A/TCPA is smaller than in A/TCNB. This is further confirmed by ESR and ODMR at very low temperatures which readily reveal the presence of mobile excitons along with trap signals of comparable intensities [36,37].

A time-dependent magnetic resonance experiment (electron spin echo combined with time-resolved optical excitation) was performed on A/TCNB by Bos and Schmidt [85]. Their report is discussed in a more comprehensive way in Section VI. At this point it is necessary to mention that it revealed a characteristic trapping time of mobile excitons on the order of a few microseconds, essentially in agreement with the results of Steudle and von Schütz [125,131].

An interesting series of time-resolved optical experiments on the N/TCPA crystal was reported by Avdeenko et al. [161,162] and Karachevtsev [163] regarding the question of trapping efficiency in one-dimensional crystals. The trapping time of triplet excitons in N/TCPA was estimated from the decay time of the band phosphorescence to be shorter than 1 μs. Interestingly, on raising the temperature from 1.4 to 4.2 K the band phosphorescence intensity decreases and so does the trap phosphorescence intensity (Fig. 4.7). This is quite surprising since one would expect a thermal detrapping effect to result in an increase of the band emission intensity. Karachevtsev [163] developed a model based on the model of Dlott et al. [164] describing the transport properties of one-dimensional excitons in a presence of impurities. According to this model, an exciton can find itself immediately after its creation in a one-dimensional cage limited on both ends by impurities with a higher energy than that of the band, forming effective barriers to the motion. If the concentration of such barriers is much higher than that of the traps, there can be two kinds of one-dimensional cage, those containing at least one trap, and those containing none. In the first kind of cage, an exciton

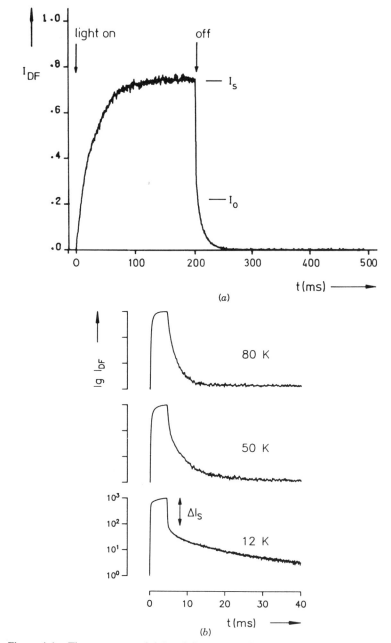

Figure 4.6. Time response of delayed fluorescence in A/TCNB to an excitation pulse (a) at 4.2 K (from [131]) and (b) at higher temperatures (from [91]). $\Delta I_S = I_S - I_0$ is the fast (<100 μs) decay component originating from exciton trapping.

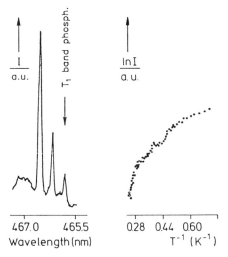

Figure 4.7. The resonant band phosphorescence of N/TCPA along with trap emission (left) and the temperature dependence of its intensity (right, from [161]).

is trapped very quickly, hence its lifetime is $< 1 \ \mu s$. In a "pure" cage, trapping does not take place and the "fast" excitonic lifetime derived directly from the decay of the band phosphorescence is determined by T-T annihilation processes (delayed fluorescence is also present in N/ TCPA although it was not an object of kinetic studies). The cage model helps one understand the temperature behavior of the band phosphorescence; on raising the temperature, an exciton from a "pure" cage can overcome a barrier and find itself in a trap-containing cage, where it almost immediately becomes trapped, causing the band phosphorescence intensity to decrease. This "screened defect" model may also explain the fact why in many CT crystals, mobile triplet excitons can be found by either optical or magnetic resonance methods despite the presence of a significant concentration of traps, usually much higher than in most one-component aromatic crystals (Section VI).

As for the nature of the defects forming the boundaries of the cages, naturally abundant deuterium-substituted compounds may be taken into consideration. It is known from high resolution excitation spectroscopy studies of anthracene, both protonated and selectively deuterated [165] that the deuterated molecules have a triplet energy up to $10 \ cm^{-1}$ higher than that of the protonated ones.

It should be noted that other well-known trapping models developed for the standard one-dimensional molecular crystal, tetrachlorobenzene (TCB) [166,167] have not been tested on CT crystals to our best knowledge. The different temperature behavior of the band/trap phosphorescence intensity ratio in N/TCPA in comparison to TCB points out, however, significant differences between single-component crystals and CT crystals. These differences with respect to the nature and number of defects are discussed in Section VI.

3. Triplet-Triplet Exciton Annihilation: Optical Studies

The significant mobility of triplet excitons in any crystal results in a finite probability of two such excitons meeting at a given crystal site and forming a short-lived triplet pair [168]. The T-T transient pair may either dissociate back with a rate constant k_{-1}, decay into an excited triplet and a ground singlet, or decay into an excited singlet and a ground singlet with a rate constant k_2 (the only pathway giving rise to delayed fluorescence), according to the following schema:

$$T + T \underset{k_{-1}}{\overset{k_1}{\rightleftarrows}} [T \ldots T] \begin{array}{c} \nearrow S^* + S_0 \\ \searrow T^* + S_0 \end{array}$$

The decay of such a pair can proceed through several pathways, most important of which is energy relaxation to the S_1 state and consequent emission of P-type delayed fluorescence. Delayed fluorescence can be of two different natures, heterogeneous and homogeneous. The difference between the two lies in the nature of the transient triplet pair; in the heterogeneous case, it consists of one trapped and one mobile exciton, while in the homogeneous case it is formed by two mobile triplet species. The presence of any bimolecular processes can be in principle detected by the quadratic dependence of the delayed emission intensity on the exciting light intensity when depopulation is dominated by monomolecular decay (weak excitation regime), $I_{DF} \sim I_{exc}^n$, $n = 2$. In practice, this dependence is usually less than quadratic ($1 < n < 2$).

The first observation of P-type delayed fluorescence in a CT crystal dates back to the article of Möhwald and Sackmann on A/TCNB [48]. They discovered a bimolecular decay in the temperature range 80–350 K, with the coefficient n between 1.2 and 2. The decay time was 0.28 ms, yielding an excitonic lifetime of $2 \times 0.28 = 0.56$ ms. Importantly, the temperature dependence of this delayed emission was very weak, pointing to a difference with the previously discussed E-type delayed fluorescence.

The mechanisms of the processes causing delayed fluorescence in A/TCNB were further studied over an extended temperature range down to 1.2 K by Steudle [131], Steudle et al. [34] and Sauter [91]. At the lowest temperature, the delayed fluorescence is relatively strong with either indirect excitation via the S_1 band or direct excitation by $T_1 \leftarrow S_0$ absorption. This means that even at 1.2 K, trapping is not efficient enough to depopulate the band completely (at this temperature thermal de-trapping is inefficient and a trapped exciton decays directly to the ground state). The dominant role of trapping at 1.2 K is, however, clear from time-resolved experiments discussed in the previous section. P-type delayed fluorescence at 4.2 K shown in Fig. 4.6 was determined to be heterogeneous by comparison with the theory developed by Ern [169]. The corresponding formulae for the onset and decay of delayed fluorescence are

$$I_{DF}^{on}(t) = K(1 - e^{-at})(1 - e^{-\beta t}) + \rho(1 - e^{-\beta t})^2 \qquad (4.6)$$

$$I_{DF}^{off}(t) = K(e^{-at} + \rho e^{-2\beta t}) \qquad (4.7)$$

where

$$K = 1/2\alpha^2 I_{exc}^2 \gamma'' k_{ET}(k_E + k_{ET})^{-1}(k_E k_T + k_E k_{TE} + k_T k_{ET})^{-1}$$

$$\rho = k_{ET} k_{TE}[k_T(k_E + k_{ET}) + k_E k_{TE}]^{-1}$$

$$\beta = k_T + k_{TE} k_E(k_E + k_{ET})^{-1}$$

$$a = k_E + k_{ET}$$

and α is the triplet absorption constant, I_{exc} is the exciting light intensity, k_{TE} is the rate constant for de-trapping, k_{ET} is the rate constant for trapping, k_E is the rate constant for exciton decay to the ground state, k_T is the rate constant for trap decay to the ground state, and γ'' is the heterofusion annihilation rate.

A fit of the parameters appearing in Eqs. (4.6) and (4.7) to the experimental curve at 4.2 K in Fig. 4.6.a yields

$$\beta = 12 \pm 1.5 \, s^{-1} \, (1/\beta = 80 \, ms)$$

$$a \geq 10^4 \, s^{-1} \, (1/a \leq 100 \, \mu s)$$

The very short decay time can be identified with the time necessary for an exciton to become trapped on a defect. Thus the lifetime of a mobile exciton at 4.2 K is shorter than 100 μs due to trapping. The longer component, which amounts to 2% of the delayed fluorescence at 1.2 K and 30% at 4.2 K can be readily correlated with the lifetime of the trapped exciton taking part in the fusion process (77 ms from phosphorescence measurements). After the excitation is turned off, most of the DF decays within 100 μs, the energy of the triplet excitons being transferred to defects, where it remains for about 80 ms. During these 80 ms, there is a finite probability of thermal de-trapping back into the triplet band, the excitation becoming mobile once again, and contributing to the DF. The temperature dependence of the long decay component of the DF requires the presence of shallow defects, approximately 3 cm^{-1} deep. Such defects have indeed been detected in A/TCNB [131].

The dominance of the trapping component in DF points out the heterogeneous nature of fusion at low temperatures. On elevating the temperature (Fig. 4.6.b), the contribution of the microsecond component decreases and above 25 K, it is absent from the decay of delayed fluorescence. Above 25 K, the concentration of mobile excitons increases and homogeneous fusion becomes dominant, causing a decrease in the lifetime of DF, which at 80 K has a decay time of 1.5 ± 1 ms. This lifetime confirms the earlier report of Möhwald and Sackmann [48] that the delayed fluorescence at higher temperatures has a homogeneous character.

Similar properties of P-type delayed fluorescence were found in the A/TCPA crystal. This was initially examined by Kozankiewicz [170] and Kozankiewicz et al. [171] using stationary optical experiments over the 1.7–300 K temperature range. At low temperatures (1.7–35 K), at which DF competes with trapping, it is of heterogeneous nature. Above about 50 K, it becomes homogeneous as in A/TCNB. The delayed fluorescence in the A/TCPA crystal has subsequently become an object of time-dependent studies by Mühle [36]. Its behavior upon pulsed excitation is similar to that in A/TCNB; there is a short component on the order of microseconds and a longer one in the millisecond range (Fig. 4.8). The

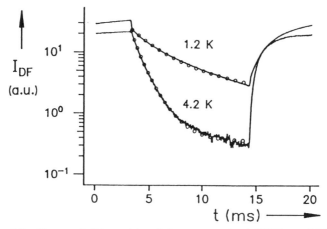

Figure 4.8. Decay of P-type delayed fluorescence in A/TCPA at 4.2 K after an excitation pulse. Only the millisecond component was detected (circles) (from [36]).

millisecond component turns out to be of complex nature, as it can not be fitted to either the formula describing monomolecular or bimolecular decay. A perfect fit is obtained only after assuming a biexponential decay with two different constants, one characteristic for monomolecular processes and the other for bimolecular ones:

$$n_{\mathrm{T}}(t) = \frac{n_0 e^{-\beta t}}{(N_0 \gamma / \beta)(1 - e^{-\beta t}) + 1} \tag{4.8}$$

$(I_{\mathrm{DF}} \sim n_{\mathrm{T}}(t)^2)$, β is the monomolecular decay constant, γ is the T-T annihilation constant, and $n_{\mathrm{T}}(t)$ is the triplet exciton concentration. β represents all monomolecular depopulation processes, one of which is the resonance band phosphorescence found in A/TCPA but not in A/TCNB.

P-type delayed fluorescence has also been found in almost all the CT crystals studied. In most cases, however, its intensity is much lower in comparison to trap emission than in the two systems discussed above, particularly at low temperatures. This is especially true in crystals with a high CT character of the triplet state. The previously discussed idea of self-trapping may be a possible cause. In the extreme cases of 100% CT character (HMB/PMDA and HMB/TCPA), no P-type delayed fluorescence has ever been found to the best of our knowledge.

4. Mechanisms of Triplet-Triplet Annihilation: ODMR studies

The optical experiments described in Sections IV.C.2 and IV.C.3 yield valuable information on the relative importance of the three dominant depopulation processes in CT crystals (T-T annihilation, $S_1 \leftarrow T_1$ thermal activation, and trapping). Regarding the T-T annihilation, however, usually only the total annihilation rate can be obtained from the optical experiments with a limited insight into the nature of the process. In contrast, magnetic resonance experiments, and particularly delayed fluorescence-detected ODMR (DF-ODMR), provide a method for following the T-T annihilation and supplying information on the nature and lifetime of the transient triplet pair, the precursor of the delayed fluorescence. These experiments are described below, and although the high field experiments were chronologically earlier, the zero field results are discussed first.

The first zero field DF-ODMR experiment on triplet excitons in a CT crystal involved A/TCNB and was reported by Steudle and von Schütz [125]. A more comprehensive report followed [72]. Part of the spectrum at 1.2 K (one zero field transition only) is shown in Fig. 4.9. The doublet structure is due to Davydov splitting which was discussed in Section III.

The standard experiment involved a chopper to separate the excitation

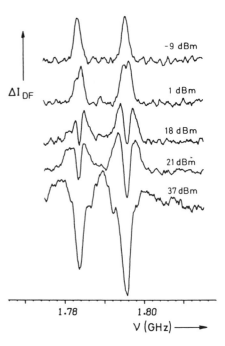

Figure 4.9. Slow passage DF-ODMR signal in A/TCNB at 1.2 K for different RF power levels (experimental conditions as in scheme (i) in the text (from [72]).

and detection (dead time 200 μs) and a quasi-stationary rf field swept very slowly (2 MHz/s). At low rf power level, all the signals were positive, at intermediate levels a negative dip appeared in each signal, and at high power levels the signals changed phase becoming all negative. Apart from the standard cw experiment, time-correlated techniques were also applied, as shown in Fig. 4.10. Their common feature was that the microwaves were pulsed and the pulses either covered the optical excitation time window or the detection window or both. The results of these experiments can be summarized as follows:

   i. The rf pulse overlaps the optical detection period; signals are always positive and increase with increasing rf power.

  ii. The rf pulse overlaps the optical excitation period; signals are always negative and increase in absolute value with increasing rf power.

 iii. The rf pulse overlaps both excitation and detection period; essentially the same situation as in the basic cw experiment (with increasing rf power a negative dip on positive signals appears and subsequently all the signals become negative). Also, at low rf power the dip and even the phase reversal may be achieved by increasing the overlap of the rf pulse with the optical excitation period.

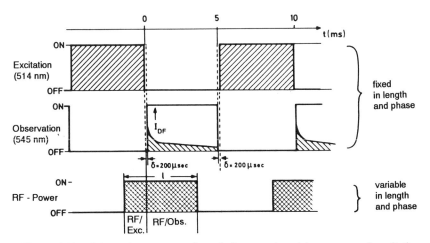

Figure 4.10. Schematic representation of the experimental sequence of excitation, observation and RF power pulses in the time-resolved ODMR experiment on A/TCNB described in the text. The delayed fluorescence intensity I_{DF} is monitored 200 μs after switching off the excitation and blocked 200 μs before the onset of the laser. The length and amplitude of the RF irradiation as well as the overlap with the excitation (RF/exc) and observation (RF/obs) intervals can be varied continuously (after [72]).

The following model was developed by von Schütz et al. [72] to explain the experimental observations outlined above. First, one has to consider the energy levels of the transient T-T pair in zero field. The spin Hamiltonian of the pair is the sum of individual triplet Hamiltonians,

$$\mathbf{H}_{pair} = \mathbf{H}(1) + \mathbf{H}(2) \qquad (4.9)$$

As a result of this operator acting on the spin wave functions τ_{ij} ($i, j = x, y, z$), nine eigenvalues are obtained corresponding to nine energy levels, described by

- 1 singlet function;
- 3 triplet functions; and
- 5 quintet functions.

In the real situation an additional term must be added to the Hamiltonian (4.9) to account for the dipolar spin–spin interaction of the two $S = 1$ spins. This interaction mixes the singlet and quintet levels of the pair so that eventually there are:

- 3 mixed singlet-quintet levels SQ_i described by $|\tau_{ii}\rangle$;
- 3 triplet levels T_i described by $1/\sqrt{2}(|\tau_{ij}\rangle - |\tau_{ji}\rangle)$;
- 3 quintet levels Q_i described by $1/\sqrt{2}(|\tau_{ij}\rangle + |\tau_{ji}\rangle)$.

This situation is illustrated by Fig. 4.11. According to the theory of Johnson and Merrifield [172], only the levels characterized by a nonzero singlet amplitude of their wave functions contribute to the delayed fluorescence. As each of the SQ_i levels has a singlet amplitude of $1/3$, the overall annihilation rate of the process leading to the S_1 state is

$$\gamma_S = \frac{1}{3} k_1 \sum_i w_{ii} \frac{k_2}{k_{-1} + 1/3k_2} \qquad (4.10)$$

k_1 is the T-T collision rate, k_{-1} is the backward dissociation rate, k_2 is the singlet annihilation rate, and w_{ii} is the probability of two τ_i states colliding which depends on the population of the τ_i level.

Since only the τ_{ii} states contribute to the delayed fluorescence, the concentration n_{ii} of the τ_{ii} pairs has to be calculated under given conditions in order to account for the observed changes of DF-ODMR signals. This can be done by solving the set of coupled differential equations:

$$\frac{\delta n_{ii}}{\delta t} = -n_{ii}\left(\frac{2}{\xi_f} + \frac{1}{\xi_{ii}}\right) + \alpha P(n_{ij} - n_{ii}) + \beta n_i^2 \qquad (4.11a)$$

$$\frac{\delta n_{ij}}{\delta t} = \frac{-2n_{ij}}{\xi_f} + \alpha P(n_{ii} - n_{ij}) + \beta n_i n_j \qquad (4.11b)$$

where ξ_f is the exciton lifetime, ξ_{ii} is the lifetime of the τ_{ii} pair.

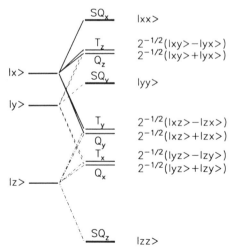

Figure 4.11. Energy levels and spin functions of a transient T–T pair in zero field. The singlet-containing levels contributing to the delayed fluorescence are marked by broader lines.

Similarly, the concentration of the single exciton population n_i is

$$\frac{\delta n_i}{\delta t} = -\frac{n_i}{\xi_f} - \alpha P(n_i - n_j) + \gamma J - \beta(n_i^2 + n_i n_j) \tag{4.12a}$$

$$\frac{\delta n_j}{\delta t} = -\frac{n_j}{\xi_f} - \alpha P(n_j - n_i) + \gamma J - \beta(n_j^2 + n_i n_j) \tag{4.12b}$$

P is the microwave field power, $\alpha P \equiv R_{i \to j} \times |V_{ij}|^2 \hbar^{-1} \rho(E_i - E_j)$ is the rate of microwave-induced transition as given by the golden rule, γJ describes the triplet population (taken as equal for each sublevel) and β is a constant proportional to the hopping time.

The intensity of delayed fluorescence is proportional to the concentration of the pair states n_{ii},

$$I_{DF} \sim n_{ii}/\xi_{ii} \tag{4.13}$$

The DF-ODMR signal ΔI_{DF} is given by the difference of the DF with and without rf power,

$$\Delta I_{DF} = I_{DF} - I_{DF}^0 \sim (n_{ii} - n_{ii}^0)/\xi_{ii} \tag{4.14}$$

In the case of pulsed excitation at a time $t < 0$ followed by observation in a time interval $0 < \lambda < t$, n_{ii} is time-dependent and the stationary values of Eq. (4.14) are replaced by

$$\Delta I_{DF} = \frac{1}{\xi_{ii} t} \int_0^t [n_{ii}(\lambda) - n_{ii}^0(\lambda)]\, d\lambda \tag{4.15}$$

For the true stationary situation (continuous excitation and microwave power) which was not possible in the experiment because of technical difficulties (the excitation light had to be chopped to cut out the prompt fluorescence), it can be shown that

$$\Delta I_{DF} \sim -\Delta N = \frac{-4\alpha P G[\xi_1^0 \xi_2^0 - (\xi_1^0 + \xi_2^0)^2]}{1 + \alpha P(\xi_1^0 + \xi_2^0)} \tag{4.16}$$

where $N = n_{ii} + n_{ij}$, $G = \beta(\gamma \xi_1 J)^2$, $1/\xi_1 = (2/\xi_f) + (1/\xi_{ii}) + \alpha P$, and $1/\xi_2 = 2/\xi_f + \alpha P$. From Eq. (4.16) it can be shown that in stationary conditions for all rf power levels, ΔI_{DF} is positive, that is, the DF-ODMR signals are positive.

For the experimental conditions (ii) described on page 270, the observation takes place in the time period $0 < t < t_0$ and the excitation in the period $-\infty \le t \le 0$. It follows from (4.16) that the number of pair states n_{ii} at $t = 0$ is larger in the system with zero rf field ($P = 0$) than in the system with the rf applied, and thus the delayed fluorescence $I_{DF} <$ I_{DF}^0 for any rf power level, that is, the DF-ODMR signals are always negative. This behavior can be easily understood in a qualitative manner; after or during the light excitaton all pairs having singlet character decay within a very short time. The remaining states have predominantly triplet or quintet character. Applying rf before the observation transfers some of these remaining pairs to states having singlet character, which decay before detection. Thus the DF intensity decreases with the rf on during the excitation period.

For the experimental conditions (i), in the observation interval the initial conditions are the same for both the $P = 0$ and $P \ne 0$ cases. The main part of the rf power is absorbed by the n_{ij} states to yield the n_{ii} pairs as in the stationary experiment, so the DF-ODMR signal is always positive.

Case (iii) is the same as with stationary microwave field. The excitation period is $-\infty < t < 0$, the observation time $0 < t < \xi_0$. ΔI_{DF} has to be computed according to Eq. (4.16). For low rf fields ($\alpha P \xi_{ii} \ll 1$), one obtains

$$\Delta I_{DF} \simeq \frac{1}{2}\, \nu \xi_f (1 - e^{-2t/\xi_f}) \left(\frac{1 + 2\alpha P \xi_f}{1 + \alpha P \xi_f / 2} - \alpha P \xi_f - 1 \right) \qquad (4.17)$$

with $\nu \equiv \beta(n_i(0))^2 \simeq \beta(n_j(0))^2$.

ΔI_{DF} increases with increasing rf power and is at maximum when $\alpha P \xi_f / 2 \simeq 0.225$, after which the signal starts to decrease with increasing rf power. This leads to a dip appearing in the ODMR signal. For a homogeneous line, this dip first appears (as observed) in the center of the resonance (at ω_0), if one calculates the effective field as a function of $(\omega_{RF} - \omega_0)$. In other words, the effective field is lower in the wings, and saturation is achieved there only when higher rf power is applied. If the overall rf field is strong enough, the total ODMR signal becomes zero. This can be interpreted to mean that in a strong rf field, the number of n_{ii} and n_{ij} pair states become equal, and the transitions between them do not contribute to a change of delayed fluorescence. A reversal of the ODMR signal phases can be also understood qualitatively from the above model if additional processes which generate singlet excitons, like the pair dissociation, are taken into consideration.

The peculiarities of the rf power dependencies of DF-ODMR signals provoke a very intriguing question: do the transitions observed by DF-ODMR indeed take place within the transient T-T pair, or do they perhaps occur within single excitons before they collide and form the T-T pair? The second case is also conceivable since the transitions within the single excitons influence the populations of the n_i and n_j states, thus influencing the populations of the n_{ij} and n_{ii} pair states. Figure 4.11 shows that the frequencies at which the transitions appear are exactly the same for single excitons and for a transient pair. The appearance of the dip in the spectrum may lead to an attractive explanation that two different species are responsible for the microwave absorption: a long-lived one easily saturating at low rf levels (single exciton) and a much shorter lived transient species (T-T pair) absorbing at much higher rf power levels. Such an interpretation was suggested by Steudle [131]. In fact, the rf power dependence of the DF-ODMR signals in A/TCNB yields a lifetime of 10^{-7} s, orders of magnitude shorter than the single exciton lifetime of about 7×10^{-2} s, although still at least 2 orders of magnitude longer than the collision time expected from the hopping frequency of the excitons ($\sim 10^9$–10^{10} s^{-1}). Von Schütz et al. [72], however, on the basis of the above model, came to the conclusion that the observed behavior of the ODMR signals cannot be accounted for by transitions within single excitons. In particular, the saturation effect would appear in single excitons at much lower power levels than those applied in the experiment ($\alpha P \simeq 1$) and should result in different ΔI_{DF} amplitudes and signs of the $2|E|$, $D - |E|$ and $D + |E|$ transitions which does not agree with the experiment.

The lifetime of the triplet pair state ($\sim 10^{-7}$ s) poses an intriguing question as to its nature. From the hopping frequency of the excitons in A/TCNB, it is known that the average time that two excitons can spend on one complex is on the order of 10^{-9} s. It appears that the correlation time between the excitons in a triplet pair is at least two orders of magnitude longer than that. A logical consequence of this fact is that the two excitons forming a pair are space-correlated on a scale much larger than one lattice constant, that is, the excitons do not have to undergo a direct encounter to form the triplet pair.

The conclusion of von Schütz et al. regarding the origin of the ODMR signals agrees with the one drawn by Lesin et al. [173] and Frankevich et al. [174] from similar DF-ODMR experiments conducted on the A/TCNB crystal in high field (hf-ODMR, called by the authors RYDMR (reaction yield detected magnetic resonance). The high field DF-ODMR differs from the previously presented zero field method in only a few aspects. The magnetic field mixes the singlet and quintet states of the transient T-T pair in such a way that there are two states of mixed singlet-quintet character:

$$SQ_1 = \left(\tfrac{1}{3}\right)^{1/2}|S\rangle + \left(\tfrac{2}{3}\right)^{1/2}|Q_0\rangle \qquad (4.18a)$$

$$SQ_2 = \left(\tfrac{1}{3}\right)^{1/2}|S\rangle - \left(\tfrac{2}{3}\right)^{1/2}|Q_0\rangle \qquad (4.18b)$$

and four pure quintet states (the three triplet states do not mix with the singlet and quintet ones). The SQ_1 and SQ_2 states are the only ones which have a nonzero singlet amplitude and therefore exclusively contribute to the delayed fluorescence. The microwave transitions in high field take place between the pure quintet states and these two mixed states. The SQ_1 and SQ_2 levels are underpopulated under stationary conditions by the effective annihilation processes, so the microwave transitions increase their populations thus increasing the intensity of delayed fluorescence. The two observed hf-ODMR signals are therefore always positive (see also the hf-ODMR spectrum of triplet excitons in the Ph/TCNQ crystal in Fig. 2.8). Lesin et al. [173] noticed certain peculiarities in their DF-ODMR spectra, notably a broadening and disappearance of signals in a narrow angular range close to the crossing point of the two signals during rotation of the crystal in the magnetic field, which could only be explained by considering transitions within the transient T-T pair and not in single excitons (the resonance fields of DF-ODMR transitions within the T-T pair are exactly the same as those within single excitons and thus do not give a clue to the mechanism of reaction). Also, a dip at high rf power was discovered in the spectra, exactly as was the case in the zero field spectra of von Schütz et al. [72], although this effect was attributed wrongly to double-quantum transitions.

Frankevich et al. [174] also observed in their DF-ODMR spectra of triplet excitons in A/TCNB, that the signals are not always positive, but become negative at certain orientations of the crystal relative to the magnetic field. This is a phenomenon similar to that observed in the ESR spectra of triplet excitons, although in the case of ODMR, both signals simultaneously change their phases from positive to negative and vice versa at the same angular position of the crystal. The obvious reason of this behavior is optical spin polarization (OSP) discussed in Section IV.B.1.

The problem of OSP influence on the DF-ODMR spectra of triplet excitons in CT crystals was treated by Agostini et al. [104] together with the question of the nature of the transient species within which the microwave-induced transitions take place. The authors developed and discussed three models:

    i. the microwave transitions take place within the triplet pair;

    ii. the microwave transitions take place within single excitons before they collide to form the T-T pair;

    iii. the combined model of cases (i) and (ii) following von Schütz et al. [72], taking into consideration OSP and SLR.

After deriving the corresponding formulae for the DF-ODMR signals of triplet excitons, Agostini et al. [104] compared the results of their calculations with the experimentally observed changes in signal intensity. For A/TCNB in certain planes of rotation, the signals reverse their phases, as was also reported by Frankevich et al. [174]. The important observation is that the phase changes take place exactly at the same angles where the excitonic ESR signals also change phase. A simulation procedure proves that this cannot be accurately explained by model (i), while model (ii) works very well. Thus, according to Agostini et al. [104], the microwave transitions in A/TCNB take place within single excitons rather than within T-T pairs, at least at the low power levels applied (<200 mW), contrary to the conclusions of Frankevich et al. [174] and von Schütz et al. [72]. In the B/TCNB crystal, the situation is different [104]; even at intermediate (100 mW) rf power, the signals are positive and do not change phases, unlike the ESR signals. This can only be explained on the ground of models (i) and (iii), but not model (ii). It seems, therefore, that the microwave transitions in B/TCNB take place within T-T pairs rather than within single excitons. This conclusion, however, was withdrawn in a later paper on B/TCNB by the same group [47], which claimed that in this crystal, the DF-ODMR transitions also occur within single excitons.

The same models presented above allow one to correlate the phase change behavior of DF-ODMR signals in different CT crystals with the mechanism of triplet-triplet annihilation. Thus the appearance of exclusively negative DF-ODMR signals in heavily doped tS/TCNB [151], DPA/TCNB [175] and B/TCNB [176] crystals was explained by the heterofusion mechanism of T-T annihilation. A similar explanation was given to temperature changes of ODMR signal phases in undoped B/TCNB [47]. In the latter case, the increased trapping of triplet excitons at lower temperatures leads to a change of phases, attributed to an increasing domination of the heterofusion process over homofusion. The reader is referred to the original papers for more details on the subject.

To conclude this section covering the mechanisms of T-T annihilation, it should be said that the number of 1:1 CT crystals investigated by DF-ODMR is limited (A/TCNB and B/TCNB only, to our best knowledge). The results of studies conducted on these systems do not allow their properties to be extrapolated to other CT crystals. The experimental method, however, has a significant and largely untapped potential for further examinations as it is one of very few allowing direct observation of mobile triplet excitations as well as other short-lived transient triplet species.

5. Triplet-Triplet Annihilation and Photoelectric Properties

The transient T-T pair, whose nature and characteristics were discussed in the preceding section, has other ways of relaxing besides P-type delayed fluorescence. Frankevich et al. [177] postulated yet another channel of decay, which may be characteristic for CT crystals only. The CT crystal A/DMPM (A/dimethylpyromellitimide) was examined from the point of view of its photoconductive properties. The dependence of the photoconductivity on the exciting light intensity was found to be nonlinear ($n > 1$) at strong light intensities. The photoconductivity also depends on the external magnetic field in a complex way, stemming apparently from two different processes.

The nonlinear dependence of the photoconductivity on the excitation light intensity was explained by the contribution of bimolecular processes, that is triplet-triplet annihilation. According to the model, the transient T-T pair relaxes to a highly excited singlet CT state $^1(D^+A^-)^*$, which may either undergo ISC to a highly excited triplet state $^3(D^+A^-)^*$ or relax further to the lowest excited singlet state (also CT). The complex dependence of the photoconductivity on the external magnetic field was explained by Frankevich et al. as the influence of the field on the ISC rate constant from the highly excited singlet CT state $^1(D^+A^-)^*$ to the triplet $^3(D^+A^-)^*$ state. If, as it is usually assumed, the dissociation of the CT complex into a hole and a charge carrier takes place from the S_1 state, which is populated from the highly excited CT state $^1(D^+A^-)^*$, the change of the ISC rates caused by the magnetic field leads to a change of the population of the S_1 state, thus also affecting the photoconductivity.

The possibility of T-T annihilation playing a role in the creation of charge carriers may be specific to CT crystals because their conduction band is closely correlated with the lowest singlet band [24]. In one-component aromatic crystals on the other hand there is practically no correlation between the two. In A/DMPM the sum of the energies of two triplet excitons $2E_T = 3.66$ eV is larger than the energy of the conduction band (2.2 eV) thus allowing creation of free carriers from the T-T pair. In anthracene itself, this energy would still be lower than that of the conduction band (approx. 4.0 eV).

Very similar effects were found by Frankevich et al. [178] also in A/TCNB crystal, confirming the earlier results on A/DMPM. Additionally, a resonant effect of microwaves on photoconductivity was also observed (Fig. 4.12). This was attributed to the magnetic dipole transitions within the highly excited triplet CT state $^3(D^+A^-)^*$, discussed above, and which can best be described as a loosely bound ion radical pair of very short lifetime (approx. 10^{-9} s as judged from the line

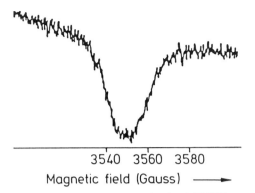

3540 3560 3580
Magnetic field (Gauss) ⟶

Figure 4.12. The change of photoconductivity in A/TCNB as a function of the magnetic field (a RYDMR spectrum) (from [178]).

broadening). The RYDMR signal can be resolved into two Lorentzian lines split by about 8 G. If this splitting is indeed caused by dipolar spin–spin interaction of the two $S = 1/2$ spins residing on two ion radicals, as the authors argue, this highly excited state represents a case of an almost complete separation of the charges. It should be mentioned that the radicals forming the radical pairs in photosynthetic systems, that is, photosynthetic bacteria, exhibit a splitting of the same magnitude (17 MHz) with comparable lifetimes (10–20 ns) [179], so the argument of Frankevich et al. appears plausible.

The RYDMR experiment described above is completely different from that observed by Ziegler and Karl [93] on the photocurrent and discussed in Section III.F. In the latter case, a transient triplet-doublet pair state was observed, and the dipolar interaction within this pair was responsible for the observed large splittings of the resonant signals.

Fünfschilling et al. [180] came to the same conclusions as Frankevich et al. regarding the role and nature of the highly excited CT state, while studying the effect of delayed fluorescence quenching by charge carriers in the A/TCNB crystal. The role of the charge carriers was studied by applying an external electric field to the crystal while measuring the delayed fluorescence. As neither the transient triplet-triplet pair nor the lowest singlet state can be responsible for the interaction with the carriers required to explain the experimental effects reported in the paper, the authors postulated the presence of a long-lived charge carrier pair, which is an exact equivalent of the radical ion pair or highly excited CT state suggested by Frankevich et al.

D. Kinetic Schemes in Various CT Crystals: Summary

The preceding sections have discussed the different kinetic processes present in CT crystals, involving both population and depopulation of the T_1 state. As for populating mechanisms, there appears to be no significant difference between the localized triplet species and mobile excitons. There is, however, a significant difference between the CT crystals and one-component aromatic molecular systems regarding the nature of ISC processes: despite the fact that in numerous CT crystals the T_1 state is limited to one component (as in anthracene-containing systems), the role of the other component is often decisive in the ISC processes and under no circumstances should be neglected (see Section IV.B.1). It appears also that the CT character of the T_1 state is not an important factor in the ISC processes.

The processes leading to a depopulation of the T_1 state are in contrast very different for isolated and mobile species, with the exception of E-type delayed fluorescence and possibly non-radiative decay. For trapped triplet states, the most obvious decay path is radiative (phosphorescence) and nonradiative decay to the ground state with E-type delayed fluorescence appearing in certain cases. For mobile excitons, the direct decay is unimportant and the depopulation of the T_1 band is dominated by the three phenomena: trapping, P-type and E-type delayed fluorescence. The relative role of these processes apparently depends on the temperature and the CT character of triplet excitons, as can be shown by comparing CT crystals differing in their CT character.

In crystals with a practically neutral triplet state, such as A/TCNB, the dominant depopulation mechanism of the triplet band at low temperatures (below 25 K) is trapping, although it is not efficient enough to depopulate the band completely. Some fraction of the mobile excitons annihilate with triplet traps at low temperatures, yielding heterogeneous T-T annihilation. With increasing temperature, trapping becomes less effective and the T-T annihilation changes, gradually acquiring a homogeneous character. The temperature effect on the delayed fluorescence intensity in A/TCNB is very small.

Other CT crystals with a largely neutral CT character of the T_1 state share generally the kinetic properties of A/TCNB with some minor differences. This applies to N/TCPA, P/TCPA, Pyr/PMDA, Pyr/TNB. The two exceptions are Ph/TCNQ and F/TCNQ with the T_1 state localized on TCNQ, which do not show delayed emission at all, the entire triplet excitation relaxing by nonradiative processes (the T_1 states in both systems lie very low on the energy scale) [38,39].

The CT crystals with a moderate to high CT character of the T_1 state also exhibit P-type delayed fluorescence as an important decay pathway over a certain range of temperatures, differing from crystal to crystal and depending on its CT character. Examples are B/TCNB, F/TCNB, HMB/TCNB and P/PMDA. At low temperatures, trapping is so effective in these crystals that mobile triplet excitons, if present, are extremely difficult to detect. At higher temperatures E-type delayed fluorescence begins to play a prominent role in the triplet exciton decay. In B/TCNB, which has a CT character of about 50% and a singlet-triplet energy gap of 1750 cm^{-1}, E-type DF shows up at $T > 250$ K. In P/PMDA ($c_1^2 = 76\%$ and a S_1-T_1 gap of 830 cm^{-1}), this effect is visible above 60 K. The dominant role of this emission in the depopulation scheme is best seen in the drastic shortening of the excitonic lifetime down to 12 μs in P/PMDA and to less than 10 μs in HMB/TCNB at room temperature. Because of the temperature-dependent changes of the dominant depopulation mechanisms, delayed fluorescence intensity in these systems is no longer temperature-independent, as it is for neutral triplet states.

In the case of a nearly ionic triplet state (HMB/PMDA and HMB/TCPA), no P-type delayed fluorescence has ever been observed. The excitons released from defects apparently immediately become activated to the S_1 band due to the very small singlet-triplet energy gap (less than 400 cm^{-1} in both cases). This is accompanied by a drastic shortening of their lifetime to 10 μs at 150 K.

V.   TRIPLET EXCITONS AS PROBES OF THE MOLECULAR DYNAMICS

A.   General

As was pointed out in Section I, many CT crystals are characterized by a relatively large orientational degree of freedom of the component molecules. Due to their sandwich-like structure, this mostly affects the motion about an axis perpendicular to the molecular plane, about which the molecules can often quite freely librate. In Section V.B we discuss how such librations affect the ESR spectra of mobile triplet excitons, in particular the lineshape and linewidth. Special attention is devoted to the two CT crystals in which freezing in of the librations leads to a phase transition of the order-disorder type (A/TCNB and N/TCNB). Although the number of CT crystals in which phase transitions appear is larger, these are the only two systems to the best of our knowledge, in which triplet exciton properties can be correlated with the dynamics of molecules.

Section V.C concerns the effect of molecular motion on the position of

ESR signals, that is, on the zfs tensor. The analysis is done on the same two examples (A/TCNB and N/TCNB), as it turns out that they represent quite different cases.

The conclusions that emerge from these two sections are that triplet excitons can be a useful probe of molecular dynamics, albeit with certain limitations. Consequently, corroboration by experimental methods other than ESR is usually necessary. These are presented in Section V.D, which gives a short discussion of the results of X-ray and neutron diffraction, nuclear magnetic resonance, Raman and calorimetry investigations conducted on CT crystals. As these methods remain in principle out of the scope of this work, only the results pertinent to the two systems discussed in this chapter, i.e. A/ and N/TCNB are discussed. The interested reader can find references to other CT crystals studied by these methods in the original papers.

B.   Effects of Molecular Dynamics on Excitonic ESR Lineshape and Linewidth

1.   A/TCNB crystal

In one of the first papers on the triplet ESR spectroscopy of CT crystals, Möhwald and Sackmann [102] reported an abrupt change of the angular pattern of excitonic signals in A/TCNB at about 205 K, accompanied by an equally sudden change of the linewidth (Fig. 3.2). In view of the structural information available for this crystal which suggests pronounced librations of the anthracene molecules around their out-of-planes axes [181], this effect was attributed to a phase transition. The transition was interpreted as freezing in of the anthracene librations with the molecules assuming one of the extreme positions of the high temperature phase.

This conclusion was confirmed by Ponte Goncalves [35] who reported not only the sudden changes of the ESR signal positions and widths, but also a splitting of the pair of high temperature signals into two pairs below T_c (Fig. 3.3), which was apparently overlooked in [102]. The angular difference of the low temperature patterns of the ESR signals amounts to about 16°, which coincides well with the amplitude of anthracene librations in the high temperature phase [181], and confirms the nature of the phase transition. Also, a slight variation of the orientation of the z (out-of-plane) axis of anthracene, which can not be correlated with the high temperature structure, was found upon passing the phase transition temperature.

Another observation of Ponte Goncalves concerned the temperature dependence of the ESR linewidth. This has already been discussed in

Section III.B in terms of spin-jumping between the sublattices 1 and 2. At the time the quoted article was published the nature of the low temperature phase of A/TCNB was not known, so the author could not decide whether this spin exchange process represented an actual jumping of the excitons between two sublattices or was rather an effect of a static disorder in the crystal below T_c. In the second case, the activation energy of $700 \, cm^{-1}$ associated with the exchange process could be attributed to the thermal activation of this disorder. However, the subsequent X-ray analysis of the low temperature phase of A/TCNB carried out at different temperatures by Stezowski [10], determined this phase to be ordered and consisting of uniformly oriented stacks. It also supplied information about continuous changes of molecular orientations with varying temperature, which may be responsible for the thermal activation of excitonic jumping rates between the differently oriented stacks.

The changes in the ESR of triplet excitons in A/TCNB in the vicinity of the phase transition were studied in more detail by Möhwald et al. [117]. As discussed earlier in Section III, these authors were able to separate effects attributable to molecular dynamics (i.e., an increase in the exciton hopping rate between the sublattices) from those originating from a thermally activated reorientation of molecules. Figure 5.1 illustrates the reorientation of the anthracene molecules in the A/TCNB crystal in terms of the displacement of their long in-plane x-axis relative to the mean position of this axis in the high temperature phase. Those results were deduced from the changes in the ESR spectra by the method discussed in Section III.B. The most interesting detail is a continuous decrease in the displacement upon approaching the phase transition from

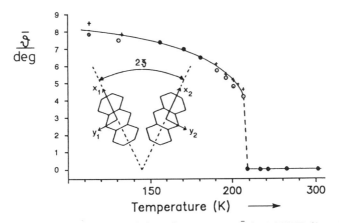

Figure 5.1.   Temperature changes of the order parameter $\bar{\vartheta}_x$ in A/TCNB (from [117]).

below from 8 to 5°, followed by a sudden drop to zero at the phase transition temperature.

Möhwald et al. [117] chose the value ϑ_x as the order parameter characterizing the phase transition; they pointed out that the transition may be first order, as indicated by the sudden drop of ϑ_x to zero, and also by a noticeable temperature region over which the low and high temperature phases coexist.

A small reorientation of the out-of-plane anthracene axes observed initially by Ponte Goncalves was also found by Möhwald et al. [117]. Möhwald et al. suggest that the corresponding value of ϑ_z may be taken as a second order parameter and the coupling of the two parameters could be responsible for a first order phase transition. This was the subject of a theoretical treatment by Erdle and Möhwald in their next paper [182].

The Landau theory of phase transitions [183] was used for calculations on the deuterated A/TCNB crystal (Ad$_{10}$/TCNB). The expression for the free energy as a function of the order parameter ϑ_x is given by

$$F = F_0 + A\,\frac{T - T_c}{T_c}\,\vartheta_x^2 + B\vartheta_x^4 + C\vartheta_x^6 \tag{5.1}$$

F_0, A, B and C are constants, T_c is 194 K for the deuterated crystal, and is taken from the low limit of the phase coexistence as shown in Fig. 5.2.

The minima of the free energy yield the order parameter as a function of temperature:

$$(\vartheta_x)_{\min} = \sqrt{-\frac{B}{3C} + \frac{1}{3C}\sqrt{B^2 - 3AC[(T - T_c)/T_c]}} \tag{5.2}$$

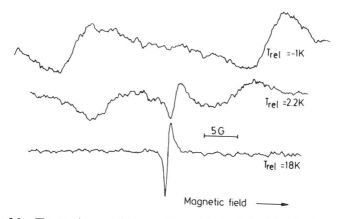

Figure 5.2. The coexistence of phases observed by ESR in Ad$_{10}$/TCNB. T_{rel} is the temperature relative to the transition temperature (from [182]).

From a least squares fit to the experimental values of ϑ_x, the following values of the constants are obtained:

$$A = 2.37 , \qquad B = -0.315 , \qquad C = 1$$

In analogy to the ferroelectric–paraelectric phase transition in solids [184] the negative value of B points to a first order transition in A/TCNB, which is further confirmed by the coexistence of phases (although no hysteresis of T_c was observed in Ad_{10}/TCNB, as required by a first order phase transition). A detailed analysis of the behavior of the second order parameter, ϑ_z which will be omitted here, strongly suggests that the coupling between the two parameters may cause an increase of the phase transition temperature as well as giving it first order character.

The phase transition in the Ad_{10}/TCNB crystal was studied independently by Ponte Goncalves [185] using ESR. His results confirm the lowering of T_c in this crystal, but he found a much larger region of phase coexistence, at least 10 K. His reported T_c value of 174 K (chosen as the temperature at which the concentrations of the low and high temperature phases are equal) is also much lower than that measured by Erdle and Möhwald. To add to this confusion, Dalal et al. [186] reported a T_c value for Ad_{10}/TCNB of 199 K (essentially in agreement with Erdle and Möhwald), but for the protonated A/TCNB crystal they gave 212 K, much higher than the values reported by both Ponte Goncalves [35] and Möhwald et al. [117]. As will be further discussed, the T_c values reported by Reddoch and co-workers for N/TCNB are consistently higher than those established by other authors. One reason may be that they extrapolate their values to zero excitation light intensity, thus eliminating the heating effects (a crystal absorbing light may have a higher actual temperature than the thermocouple placed even very close to it on the holder). This argument fails, however, in the case of other experimental techniques discussed in Section V.D, which do not depend on exciting the crystal with light (NMR, DSC), and where a similar controversy appears.

The molecular dynamics and phase transition in A/TCNB was selected as a subject of investigations by Park and Reddoch [118]. Their insight into the exciton dynamics has already been discussed in Section III.B. Briefly, the authors attributed the temperature dependence of the ESR spectra below T_c to a thermally activated static disorder (domains of "wrongly" oriented anthracene molecules within each stack), which increases when approaching the phase transition from below. Park and Reddoch chose another order parameter defined as $\xi = x_A - x_B = 1 - 2x_B$ to study the nature of the phase transition (x_B is the mole fraction of the "wrongly" oriented molecules in each stack). An analysis based on the Landau theory, analogous to that of Erdle and Möhwald [182] brings inconsistent results. In particular the coefficient B, whose negative sign prompted Erdle and Möhwald to declare the transition first order, did not converge to a negative value in Park and Reddoch's work. Rather, instead of the Landau theory, an analysis of the phase transition by renormalization group theory [187] yielded results agreeing well with the experiments. We do not give a detailed line of thought of Park and Reddoch here because, as it turned out (Section V.D.1), the X-ray diffraction analysis of the low temperature phase of A/TCNB [10] definitely ruled out a possibility of static disorder in the low temperature phase. Thus their model, while certainly appealing, is based on nonphysical assumptions.

The same comments apply to the work of Pasimeni and Corvaja [188] regarding a hf-ODMR study of the A/TCNB crystal in the phase transition temperature region. Delayed fluorescence was monitored during experiments, and the crystal was studied in two orientations relative to the magnetic field: B parallel to one of the low temperature donor x axes (x_D) and B parallel to the acceptor long axis (x_A). Besides a significant broadening of the ODMR signals close to T_c for $B \parallel x_D$, which is an exact equivalent of the ESR broadening for this orientation, the authors also found a smaller but distinct broadening for $B \parallel x_A$. This is impossible to explain by exciton dynamics because whatever the nature of the exciton exchange between the two sublattices, it should not affect the ODMR linewidth in this particular orientation where both sublattices are magnetically equivalent. Pasimeni and Corvaja went to great lengths to explain this phenomenon on the basis of slowing down critical fluctuations in the order parameter defined the same way as Park and Reddoch [118], that is, assuming a static molecular disorder below T_c. The reader is referred to the original article for details.

To conclude the discussion of the numerous ESR results regarding the phase transition in the A/TCNB crystal, we should mention the work of Erdle and Möhwald [120] concerning mixed anthracene/phenanthrene/ TCNB crystals. A/TCNB and P/TCNB are structurally so similar that apparently substituting one donor by the other does not produce a large distortion of the lattice. Hence it is possible to grow crystals with a formula $A_x P_{1-x}/TCNB$ where x varies from 0.001 to 0.75. In the case of $x = 0.001$ these are practically P/TCNB crystals doped with anthracene. No disorder or other molecular dynamic phenomena could be found in this crystal, which is consistent with the crystallographic results of Wright et al. [189]. However, for $x = 0.75$, which represents the case of the mixed crystal $A_{0.75}P_{0.25}/TCNB$, a phase transition was found when observing triplet excitons by ESR at 155 K, much as in the pure A/TCNB crystal although at a much lower temperature.

Triplet excitons prove to be a rather sensitive probe for molecular reorientation phenomena in A/TCNB, as seen from results presented earlier. This is due to the high anisotropy of their ESR signals, making it possible to study orientational changes as small as 0.5°. They also represent the bulk properties of a crystal much better than the localized (trap) states which are discussed in Section VI. The drawback of the exciton ESR studies is the difficulty of separating the effects due to molecular dynamics from those resulting from exciton dynamics. This must always be done with great care, and the results corroborated by other experimental techniques.

2. N/TCNB Crystal

In contrast to A/TCNB, the molecular dynamics in N/TCNB was quite well known by experimental methods other than ESR at the time the first ESR studies were attempted. In particular, the librational motion of naphthalene molecules had been studied by NMR [190–192], which is discussed in Section V.D.2. These studies discovered that donor molecules librate in a double-well potential with the potential minima separated by 36° (see Fig. 5.9). The NMR studies [190] also predicted a phase transition from freezing in the librations at a temperature of 50–60 K, unavailable to that experiment.

In fact, the phase transition was found at 63 K by Macfarlane and Ushioda using Raman scattering spectroscopy [193]. Accordingly, the crystal became the subject of ESR studies conducted on triplet excitons, similar to A/TCNB. Unfortunately, exciton dynamics does not cooperate in the case of N/TCNB in the way it does in A/TCNB, and mobile triplet species are observable in this system only at temperatures above 130 K [40,41,192,194]. The phase transition takes place far below this temperature and thus mobile excitons cannot be used for studying this interesting phenomenon in the manner it was done in A/TCNB.

Erdle and Möhwald [194] tried to overcome this obstacle by observing ESR spectra of localized (trap) states in N/TCNB. A shallow defect characteristic for the 1:1 crystal was observed only at $T < 30$ K, so could not supply any information on the phase transition. However, it allowed one to deduce useful information about the structure of the low temperature phase: the long in-plane naphthalene axes differ by a rotation of 36.8°, which means that the phase transition depends on freezing in naphthalene librations with an amplitude of ±18°, also observed indirectly by X-ray diffraction [195] and NMR [191,192]. Also, the z out-of-plane axes differ by almost 20° in the low temperature phase, information that is unavailable from other methods.

In an attempt to observe the effects of the phase transition in

N/TCNB, Erdle and Möhwald [194] doped the crystal with a small concentration of anthracene creating in this way deep ($3000 \, cm^{-1}$) A/ TCNB traps. This experiment, however, gave results rather inconsistent with the information available from other methods on the molecular dynamics in N/TCNB, and is only briefly discussed in Section VI. In general, it turns out that doping the crystals does not yield information pertinent to 1:1 systems. This was further confirmed by the results of Ponte Goncalves and Vyas [196] who doped N/TCNB with phenazine and observed the molecular dynamics of the dopant by ESR, and by Ripmeester et al. [192] who in principle repeated the ESR experiments of Erdle and Möhwald [194]. The latter authors observed the N/TCNB trap in the ESR spectrum at temperatures as high as 90 K, thus passing through the phase transition region, but they could not notice any effect of the transition on the spectra. It follows that the libration rates of naphthalene molecules both below and above the phase transition are smaller than the ESR time scale. Thus ESR of trap triplet states always "sees" both naphthalene molecules in their extreme positions even above T_c.

This is quite a different situation than the one encountered with mobile triplet excitons in the same crystal, which at temperatures above 130 K always emerge at the average position of the trap signals [41,192]. Ripmeester et al. [192] attempted to evaluate the rate of the process leading to this averaging and obtained a value of $8 \times 10^9 \, s^{-1}$. They were unable, however, to separate the exciton dynamics from the molecular dynamics.

Grupp et al. succeeded in solving exactly this problem in their work [41] by observing the excitonic ESR linewidth at different temperatures as a function of crystal orientation relative to the magnetic field (Fig. 5.3). The linewidth shows a pronounced dependence on the crystal orientation at low temperatures, being the smallest at an orientation corresponding to the magnetic field lying parallel to the averaged long in-plane naphthalene molecular axis. Increasing the temperature decreases the linewidth as well as its dependence on the crystal orientation. Thus the obvious conclusion is that the temperature variations of linewidth for B parallel to x must be attributed to exciton dynamics (Section III.B), while those for other orientations are influenced by both exciton and molecular dynamics. After deducting the contribution from the excitonic motion, the naphthalene libration rates were best-fitted to the experimental curves using the formula

$$\nu_{lib} = (\pi g \beta / 2h\sqrt{3}) B_{12}^2 / (\Delta B_{obs} - \Delta B_0) \tag{5.3}$$

ΔB_{obs} is the observed linewidth for the angle φ between B_0 and x, ΔB_0 is the linewidth for $B_0 \parallel x$.

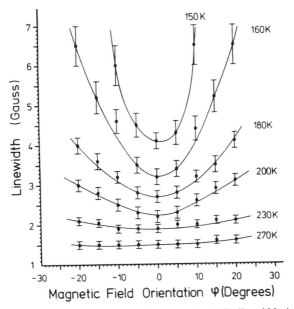

Figure 5.3. Angular dependence of the excitonic ESR linewidth in N/TCNB at different temperatures. The magnetic field is rotated in the xy plane of the zfs tensor, $\varphi = 0$ corresponds to $B \parallel x$ (from [41]).

The libration rates obtained in this way are 7×10^9, 1×10^{10}, 4×10^{10} and 5×10^{10} s^{-1} at 150, 160, 180 and 200 K, respectively. At temperatures higher than 230 K, the possibility of the naphthalene molecules making large (144°) jumps between the extreme positions [191] leads to additional line narrowing.

C.   Influence of Molecular Dynamics on zfs Tensors

In Section III.C we showed using the example of the A/TCPA crystal how triplet exciton dynamics can induce changes in the zfs tensors affecting ESR spectra. Here we will show that molecular librations can cause similar changes in at least two different ways.

Vyas and Ponte Goncalves were the first to report pronounced changes of the fine structure tensor of mobile triplet excitons in the A/TCNB crystal [98]. As was pointed out in Section III.C, the model they developed to account for the observed effect, which assumed excitons jump from the lowest triplet band to a higher one, was not based on a real physical situation. The model of von Schütz et al. [97] which takes into account the librations of anthracene molecules in the high temperature phase explains the temperature changes of zfs parameters equally well, and is based on the physical situation encountered in the A/TCNB crystal.

The way molecular librations might affect the zfs parameters can be deduced from Eqs. (2.6) and (2.7). Libration about an axis lying in the molecular plane leads in principle to a decrease of the parameter D while E remains unchanged. A libration about the out-of-plane axis should not alter D but may reduce E. In any case, librations lead to an averaging of the individual zfs tensors representing the extreme positions of the librating molecule, and thus the measured values are always smaller than those existing in the absence of librations. In the following analysis, the librations of anthracene about its three symmetry axes are assumed to be decoupled, since the electron density in the molecule does not indicate any coupling between the individual libration modes [10]. The basis of the calculations is formed by the zfs tensor measured at the lowest temperature (30 K) at which mobile excitons could be detected by ESR [91]: $|D| = 0.06782$, $|E| = 0.00768 \text{ cm}^{-1}$. At this temperature, it is assumed that no librations take place in the crystal, thus the measured zfs parameters are not influenced by them.

From these values a zfs tensor is calculated, which is averaged by the libration about the x (long in-plane) molecular axis, influencing the tensor in a very profound way:

$$\mathbf{F}_x = 1/2\mathbf{F}(\delta_x + 1/2\omega_x) + 1/2\mathbf{F}(\delta_x - 1/2\omega_x) \qquad (5.4)$$

(δ_i are the mean orientations of anthracene, ω_i are the amplitudes of librations around the given axis.)

This tensor is then used as the basis for a calculation of the zfs tensor averaged as a result of librations about the other in-plane molecular axis (y):

$$\mathbf{F}_{xy} = 1/2\mathbf{F}_x(\delta_y + 1/2\omega_y) + 1/2\mathbf{F}_x(\delta_y - 1/2\omega_y) \qquad (5.5)$$

Analogously, the zfs tensor resulting from librations about the z axis (out-of-plane) is given by

$$\mathbf{F}_{xyz} = 1/2\mathbf{F}_{xy}(\delta_z + 1/2\omega_z) + 1/2\mathbf{F}_{xy}(\delta_z - 1/2\omega_z) \qquad (5.6)$$

The libration amplitudes ω_i were taken from the structural data in [10], extrapolated as necessary to higher temperatures (up to 470 K). The most important value ω_x was also independently fitted to the experimental zfs parameters obtained by Vyas and Ponte-Goncalves [98] and by Sauter [91].

The results of the calculations are presented in Fig. 5.4.a. It can be

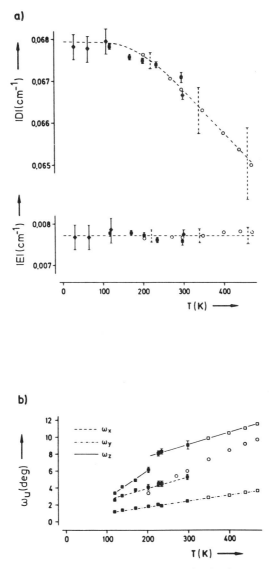

Figure 5.4. (a) Temperature dependence of the excitonic zfs parameters in A/TCNB. Full diamonds represent the experimental results in [91], full squares are the results of the calculations based on the model using the libration amplitudes ω_i known from the structural analysis [10], empty circles are the results of the same calculations using the parameter ω_x best-fitted to the experimental data in [98], represented by the dashed lines. (b) Libration angular amplitudes: full symbols are structural data from [10], empty squares are the results of extrapolation of ω_x and ω_y, empty circles are the results of the fit of the amplitude ω_x to the ESR data in [98] (from [97]).

seen that this method of averaging the zfs tensors by molecular librations works at least qualitatively. As a result of librations (mostly about the x axis) the D parameter decreases above 100 K, while the E parameter remains unchanged within experimental error. Figure 5.4b illustrates the way the libration angular values ω_y and ω_z are extrapolated from the structural data. The results of the fitting of ω_x to the ESR data are also compared to the structural data. Again, at least a qualitative agreement is reached.

This model of how molecular librations can affect the zfs parameters works well in A/TCNB, but fails in the other CT crystal discussed earlier in this section, namely N/TCNB [41]. The temperature changes of the parameters in N/TCNB are shown in Fig. 5.5. There is a distinct increase in the D parameter from 130 K (where the mobile excitons emerge in ESR spectra) up to 300 K, where D appears to approach a steady value. The E parameter increases similarly, but it reaches a maximum at about 250 K, above which it starts to decrease slightly. In order to explain this phenomenon, Grupp et al. [41] quoted the article by Mayoh and Prout [197], according to which the CT interaction is the largest when the donor and acceptor molecules are oriented with their long in-plane axes parallel to each other. The librations taking place in the crystal bring the donor away from this position, and thus decrease the CT character. From Eq.

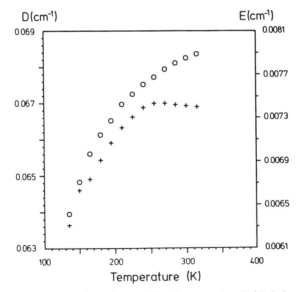

Figure 5.5. Temperature dependence of the zfs parameters D (circles) and E (crosses) of the triplet excitons in N/TCNB (from [41]).

(2.15) the decrease in CT character as a result of naphthalene librations in N/TCNB can be estimated as from 32% at 130 K down to 27% at 300 K.

The temperature effect on the E parameter can be also explained on the same basis. In particular, the small decrease in E with increasing temperature above 250 K can be well understood as the naphthalene molecules overcoming the higher energy barrier (43 kJ/mol) and spinning freely about their z axes (see Fig. 5.9). This increases the spatial symmetry of the triplet state of the complex and hence leads to a decrease in the absolute value of E.

D. Corroboration by Other Experimental Techniques

1. X-Ray and Neutron Diffraction Analysis

The original X-ray diffraction study of the room temperature structure of the A/TCNB crystal by Tsuchiya et al. [181] revealed a structural disorder of the anthracene molecules and suggested the dynamic character of this phenomenon, that is, librations of anthracene around its z axis by ±8.6°. Ponte Goncalves [35] and Möhwald et al. [117] postulated freezing in of this libration as the main cause of the phase transition.

Stezowski [10] confirmed these ideas by an X-ray diffraction study of A/TCNB at different temperatures. In particular, he found that while anthracene (and to a smaller extent TCNB) librate in the high temperature phase over a broad single potential well, in the low temperature phase they assume two distinct positions, that is, they librate within a double well potential. The amplitudes of librations were also determined (Fig. 5.4), along with the ESR results, which they nicely confirm. The results of Stezowski confirm the earlier assumptions of Möhwald et al. [117] of a temperature-dependent reorientation of anthracene molecules in the low temperature phase while simultaneously ruling out those of Park and Reddoch [118] of a disorder in this phase.

The work of Stezowski was criticized by Boeyens and Levendis [198], who noticed that the impossibility of resolving the positions of the α carbon atoms in anthracene in the high temperature phase of A/TCNB could be explained better by a static disorder in the crystal. Whatever the real situation, it should be pointed out that this controversy does not influence the results of the ESR studies, for the simple reason that ESR does not distinguish between a static disorder and a dynamic one, provided the latter is on a time scale larger than the ESR time scale. Thus a thermally activated static disorder of the high temperature phase could explain the temperature changes of the zfs parameters in A/TCNB as equally well as the libration model discussed in Section V.C.

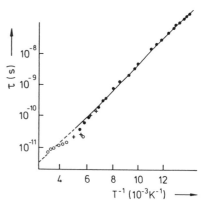

Figure 5.6. Naphthalene libration correlation time versus reciprocal temperature in N/TCNB obtained by QNS (empty circles) compared with NMR results from [192] (full circles) and ESR [41] (crosses) (from [199]).

The room temperature X-ray diffraction study of the N/TCNB crystal [195] also discovered a disorder in the high temperature phase of this crystal, but could not determine its character, that is, whether it is static or dynamic. NMR techniques (see the next section) showed it is dynamic [190–192], while Raman studies [193] first reported the phase transition as an order-disorder type at about 63 K. The neutron diffraction study by Czarniecka et al. [199] supplies the result that in the low temperature phase there is no change of the crystal symmetry (space group remains $C2/m$), however, there is a small discontinuity in the lattice constant b, and small variations of other constants. It turns out that the phase transition is less dramatic than that observed in A/TCNB. Also, the incoherent quasi-elastic neutron scattering (QNS) study reported in [199] gave a very interesting result concerning numerical values of the libration frequencies at different temperatures. These results (Fig. 5.6) nicely corroborate the values extracted from the ESR experiments performed by Grupp et al. [41]. At room temperature, the residence time of a naphthalene molecule in one of its extreme positions is about 8 ps, which yields a libration frequency of 1.2×10^{11} s^{-1}. The QNS study also suggests the possibility of a second phase transition taking place at about 170 K, as seen in Fig. 5.6.

2. Nuclear Magnetic Resonance

NMR studies provide useful information about the molecular dynamics in crystals through the sensitivity of the relaxation rates to molecular reorientation processes. Before we engage in the specific discussion of CT crystals, however, some general remarks are necessary, because too often the processes of molecular reorientation, librations and phonon-assisted relaxation via electron paramagnetic centers are misinterpreted, leading to widespread confusion.

Both the amplitude and time scale of these processes enter in the spin relaxation rates of the nuclei, of which predominantly protons have been studied. Librations in quadratic potentials (to a first approximation) are usually excited even at the lowest temperatures up to frequency ranges far exceeding the Larmor precession frequencies of the nuclei, even in the highest fields available. This also holds for the modulation of the mutual distances between the molecules by phonons. Correspondingly, the relaxation rates of the nuclei are low, on the order of $10^{-2}\,\mathrm{s}^{-1}$ or less [200]. These processes therefore contribute to the measured values only minimally, and are exceeded considerably even by the relaxation of the proton spins by electron spins localized on defects.

Extreme reduction of the elastic forces at a phase transition (softening of the modes) reduces the vibronic frequencies and therefore enhances the relaxation rates considerably. This is demonstrated on the proton relaxation rate T_{1H}^{-1} of A/TCNB at $T_c = 206\,\mathrm{K}$ (see below). It is worth noting the very low relaxation rates of $10^{-3}\,\mathrm{s}^{-1}$ even at T_c.

Molecular reorientation is much more effective for T_1^{-1}. Considering only crystal structures for which just the molecular skeleton is given, researchers very often tend to assume the reorientation to the axis of the smallest moment of inertia, that is, in the case of aromatic molecules, the long in-plane axis. In a space-filling presentation, however, taking into account the electron densities, it becomes obvious that large reorientational jumps (up to 180°) can take place only around the axis perpendicular to the molecular plane, and that the "rounder" the shape of the molecules, for example, benzene or pyrene, the weaker this motion is, as shown experimentally by von Schütz and Weithase [201].

The effectiveness of this kind of motion is due to the commonly large amplitudes and the "matched" time dependence. The latter results from a thermal activation over the barriers given by the intermolecular interaction. The molecules "sit" most of the time only to jump quickly and to "sit" again. When the mean sitting time (correlation time τ) between two jumps is on the order of the inverse Larmor precession frequency ($\omega\tau \sim 1$), the relaxation is the most effective. Applying the Arrhenius law to τ, $\tau = \tau_0 e^{-\Delta E/kT}$, we immediately get a v-shaped temperature dependence for T_1^{-1} on an Arrhenius plot.

Proton spin lattice relaxation rates T_1^{-1} were measured in A/TCNB for two frequencies in the vicinity of T_c by Auch et al. [202]. The results are shown in Fig. 5.7. Two features deserve particular attention: the generally low relaxation rates over the whole temperature range ($T_1^{-1} = 10^{-3}\,\mathrm{s}^{-1}$) and the triangular (v-shaped) maximum near the phase transition temperature. To account for the observed effect, three possible mechanisms were discussed in [202], namely, order–disorder fluctuations in the vicinity of the phase transition, modulation of the proton–proton

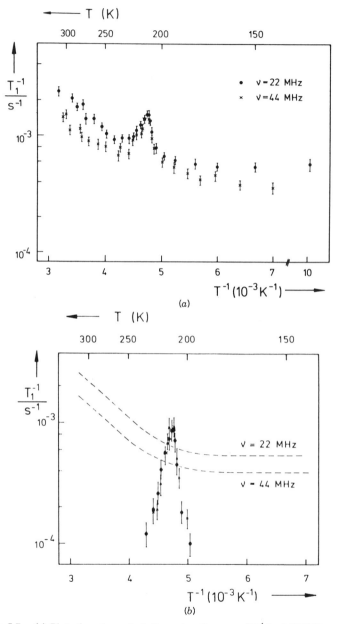

Figure 5.7. (a) Plot of proton spin lattice relaxation rates T_1^{-1} in A/TCNB at 22 MHz (circles) and 44 MHz (crosses) as a function of reciprocal temperature. (b) Attempt to extract the relaxation due to paramagnetic impurities from the order-disorder fluctuations, represented by the dashed lines (from [202]).

and proton–paramagnetic impurity interactions, and thermally activated molecular reorientations.

Of the possible mechanisms, the most reasonable one is a modulation of the proton–proton interactions near T_c, as indicated by Fig. 5.7.b. After the interaction of protons with the paramagnetic impurities is subtracted from the plot, there remains a sharp, triangular maximum, with a distinct asymmetry. This asymmetry is quite similar to those reported for the measurements of excitonic ESR linewidth (Section III.B) and Raman scattering modes (next section).

The thermally activated molecular reorientation mechanism can be ruled out because it needs nonphysical parameters to explain the observed effects, that is, slow libration frequencies below T_c and fast ones above. On the other hand, a model of "rigid molecules" would be in agreement with the results of Boeyens and Levendis [198] who deny the presence of dynamic disorder in A/TCNB, postulating a static one instead.

The usefulness of NMR relaxation studies for molecular dynamics processes can be also shown by the example of the A/TCPA crystal. Proton SLR rates were measured in this crystal in the 190–350 K temperature range [203] (Fig. 5.8). Below 257 K, the relaxation is bi-

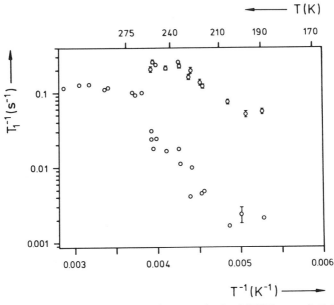

Figure 5.8. Proton spin lattice relaxation rates in the A/TCPA crystal. Below 257 K there are two relaxation components (from [203]).

exponential, that is, there are two T_1 values for a given temperature. Above 257 K the relaxation is mono-exponential. This effect points to a phase transition in A/TCPA, which has in fact been confirmed by X-ray crystallography [11] and DSC [204].

The N/TCNB crystal has been an object of extensive NMR studies regarding its molecular dynamics. Fyfe [190] used broad-line NMR in the 77–400 K range, Fyfe et al. [191] applied pulsed NMR in the same temperature range, while Ripmeester et al. [192] extended these methods below the phase transition temperature. The observables in the experiments were: second moment of the cw-NMR spectra, proton spin lattice relaxation time T_1 and proton spin lattice relaxation time in the rotating frame $T_{1\rho}$. The results can be summarized as follows:

- Naphthalene molecules undergo rapid reorientation between two minima separated by 36° in the molecular plane (low barrier, Fig. 5.9). This motion is responsible for the "V" shape in the SLR time T_1 versus temperature plot, with the minimum at about 100 K. At higher temperatures naphthalene can jump 144° to the other orientation. This additional motion causes a minimum of the SLR time $T_{1\rho}$ at about 250 K. Both motions also cause changes in the second moment of cw-NMR spectra.

- The activation energy values for the two different kinds of motion are obtained from the NMR experiments and calculations [205], and found to be 9.7 and 42.7 kJ/mol, respectively.

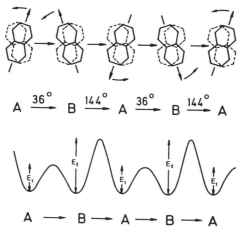

Figure 5.9. N/TCNB crystal, conclusions of NMR studies. Top: schematic representation of the occupancy of the two possible molecular sites by a single naphthalene molecule during a complete rotation in its molecular plane. Bottom: the potential energy profile corresponding to the motion of naphthalene as described above (from [190]).

- The phase transition at about 72 K does not influence the T_1 time, indicating a different mechanism for spin relaxation than in A/TCNB. It causes, however, an anomaly in the rotating-frame SLR time $T_{1\rho}$. The detailed considerations lead to a conclusion that in the low temperature phase, the two minima are no longer energetically equivalent, differing by 7.2 kJ/mol at 62 K. Thus the probabilities of the naphthalene molecule sitting in them are no longer equal and the structure becomes partially ordered. A full order, however, is not reached until $T = 0$ K.

- The T_c, as detected by NMR, is significantly higher than that reported by Macfarlane and Ushioda [193] (63 K), which means that a controversy exists similar to the one surrounding the T_c of the phase transition in A/TCNB.

3. Raman Scattering

Raman scattering spectroscopy is a particularly useful tool for studying changes in the crystal symmetry which appear at phase transitions. It also allows direct insight into the modes responsible for these transitions. Accordingly, it was employed by Möhwald and Thaer [206] to examine the phenomena accompanying the phase transition in A/TCNB occurring at about 206 K.

The Raman spectrum of A/TCNB at 10 K consists of five strong lines at the following energies: 23, 37.5, 64.5, 86.5 and 110.2 cm^{-1} accompanied by some weaker signals. The two high energy signals appear as doublets. The temperature dependence of the intensity of the five prominent signals is shown in Fig. 5.10. It follows that the modes of energies 23 and 64.5 cm^{-1} disappear above T_c, while the intensity of the mode at 37.5 cm^{-1} undergoes a pronounced change at this temperature. This same mode also shows a triangular maximum of its linewidth very similar to that observed in the NMR proton spin relaxation time [202]. Möhwald and Thaer interpreted these changes as resulting from the phase transition at 206 K. Additionally, after performing Raman experiments on the perdeuterated A/TCNB crystal, they attempted to attribute the observed modes to molecular motions in the crystal. Thus the most intense mode at 86.5 cm^{-1} was attributed to "intra-CT", also called "breathing" movements of the CT pair forming the complex. The modes of the lowest frequency (23 and 37.5 cm^{-1}) were attributed to the librations of anthracene about its z (out-of-plane) axis.

Raman experiments were also performed on protonated and deuterated A/TCNB crystals by Dalal et al. [207]. The results of Möhwald and Thaer [206] were essentially confirmed with the exception of the phase transition temperature, which was placed by Dalal et al. at 212 K, that is,

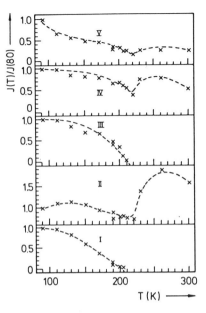

Figure 5.10. Temperature dependence of the intensity of the five most prominent Raman modes in A/TCNB (from [206]).

at least 6 K higher. Thus the controversy stemming from the ESR experiments has not been solved by Raman spectroscopy. Dalal et al. [207] also reported T_c values for partially deuterated crystals, with either α or β positions of anthracene deuterated. Partial deuteration lowers T_c by a comparable amount, while perdeuteration causes a greater decrease, suggesting an additivity of the process (the 9,10 deuterons were reported to increase T_c). Deuteration increases the moment of inertia of a molecule and thus should in principle increase the T_c. Dalal et al. [207] offered an explanation, that the potential wells in which the molecules librate are determined by intermolecular repulsions rather than by charge transfer interactions. Consequently, if the α or β protons, which are most probably involved in the repulsive interactions, are replaced by deuterons, the shortening of the C–D bond may lead to a decrease of T_c as observed experimentally.

Raman investigations were also performed on the N/TCNB crystal by Macfarlane and Ushioda [193]. They obtained the first results showing the existence of a phase transition at about 63 K. A set of low frequency modes was observed in the energy range 23–225 cm^{-1}. The low frequency mode at 23 cm^{-1} undergoes dramatic changes when approaching the phase transition from below: its linewidth increases sharply, while the intensity decreases; the mode disappears completely above T_c. Similarly the next low frequency mode at 44 cm^{-1} shows a significant increase of linewidth at $T > 63$ K from $0.1\,\text{cm}^{-1}$, to as much as $6\,\text{cm}^{-1}$. The rapid broadening of the Raman modes in the vicinity of a phase transition is explained by the onset of disorder which relaxes the $\Delta k = 0$ rule for Raman scattering. The mode at $23\,\text{cm}^{-1}$ is tentatively identified as the "breathing mode" of the CT complex, while the one at $44\,\text{cm}^{-1}$ is assigned to in-plane libration of naphthalene.

The N/TCNB crystal, and also Nd_8/TCNB were also studied by Bernstein et al. [208] by Raman spectroscopy. Their work confirmed the existence of a phase transition in both crystals. However, the transition temperature was found to be significantly higher than that reported by Macfarlane and Ushioda [193]: 69 K for the protonated complex and 62 K for the deuterated one. These values agree well with those obtained from the NMR studies discussed before (72 ± 2 K for Nh_8/TCNB). The reason for the difference remains unsolved, as is the case with the A/TCNB crystal.

4. Calorimetry

The onset of disorder in a crystal is usually accompanied by an anomaly in its heat capacity. Accordingly, phase transitions such as those taking place in CT crystals can in principle be observed by calorimetric studies. Such a study, using differential scanning calorimetry (DSC) measurements of both protonated and perdeuterated A/TCNB crystals was reported by Dalal et al. [186]. Maxima in the thermograms were observed at temperatures 212 and 199 K for the protonated and deuterated crystals, respectively. These values of T_c agree well with those consistently reported by this group in their ESR and Raman studies.

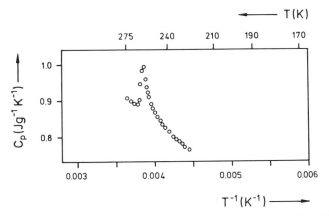

Figure 5.11. Temperature dependence of the specific heat capacity of A/TCPA. The anomaly at about 257 K corresponds to a phase transition (from [204]).

The N/TCNB crystal was also investigated by the DSC technique by Boerio-Goates et al. [209]. An anomaly in the heat capacity of the crystal was found at about 75 K. This value of T_c is slightly higher than that found in NMR studies [192] and Raman [207], and much higher than that reported originally in [193].

A similar anomaly in the heat capacity was found in the A/TCPA crystal at about 257 K [204] (Fig. 5.11). This corresponds to the phase transition detected by both X-ray crystallography [11] and NMR [203].

VI. TRIPLET TRAP STATES IN CT CRYSTALS

A. General

CT crystals, like all other crystals, contain a number of defects. Some of them, having a lower triplet energy than that of the T_1 band, act as traps for triplet excitons propagating through the crystal. Triplet traps are abundant in CT crystals and can be detected by the same experimental methods as mobile excitons. In fact, under certain conditions (particularly at low temperature), they are much easier to detect than the mobile species.

The information obtained about a trap state is relevant to the defect responsible for its creation, but does not necessarily represent the bulk properties of the crystal. Nevertheless, such information can sometimes be useful. For example, it was stated in Section II that it is always important to compare the zfs parameters of mobile excitons with those of a shallow trap in order to eliminate the possible effects of the exciton mobility on these parameters.

In Section VI we do not discuss all the trap states detected in 1:1 CT crystals, but rather concentrate on those properties which make them different from similar defects in one-component molecular systems. This is done in Section VI.B. One such property is the very number of defects. Instead of having to do with only one (or seldom more) well-defined traps as in one-component crystals, the high resolution experiments on CT crystals usually reveal a much greater number of defects, even in crystals grown most carefully from thoroughly purified components. This is shown in Section VI.B.1, which is devoted to optical emission studies, and in Section V.B.2 which deals with ODMR and ESR experiments. Section VI.B.3 discusses the value of ENDOR studies of defects in CT crystals. Section VI.B.4 represents an attempt to summarize these results and to draw conclusions about the possible structure of certain typical defects.

A number of experiments reported in the literature concern defects created purposefully in CT crystals by doping them with small quantities of donor (or more seldom acceptor) molecules, which replace the host donor (or acceptor), and form a CT complex with the other partner. If the triplet energy of such a complex is lower than that of the host triplet band, the defect acts as a trap. Such defects are discussed in Section VI.C together with the two applications of such experiments. As will be shown, the actual value of these experiments turns out to be rather limited.

B.   Defects in 1:1 Crystals

1.   Optical Emission Studies

Contrary to mobile triplet excitons in CT crystals, in most cases trapped species emit strong phosphorescence, which may be studied at sufficiently low temperatures. This is so because the only important triplet depopulation process competing with the $T_1 \rightarrow S_0$ transition is the thermal detrapping. The shape of phosphorescence spectra in CT crystals can vary significantly depending on the CT character of the emitting defect. The reason for this is exactly the same as for the excitation spectra of mobile excitons, which was discussed in detail in Section II.B.2, that is, exciton-phonon coupling. Thus phosphorescence in CT crystals characterized by a low CT character of the T_1 state looks very much like that in one-component crystals: sharp zero-phonon lines in both 0–0 transitions and their vibronic progressions, characteristic for the donor or acceptor, depending on which component the triplet excitation is localized. An example of such a phosphorescence spectrum is presented in Fig. 6.1.

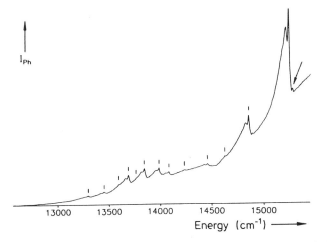

Figure. 6.1.   Phosphorescence spectrum of A/TCPA at 4.2 K. The identified vibronic transitions of the dominant trap are marked by dashes. The resonance band phosphorescence is shown by the arrow (from [11]).

Practically the only difference between the phosphorescence of the A/TCPA crystal and that of anthracene is the blue shift of the former amounting to 536 cm^{-1}. This phenomenon was discussed in Section II.B.1 in terms of stabilizing the ground state of the complex.

A more careful look at the 0–0 region under higher resolution conditions reveals a discrete number of lines corresponding to 0–0 transitions of multiple traps differing in depth (Fig. 6.2). This is a situation commonly encountered in CT crystals. Instead of the very limited number of well-defined traps found in one-component crystals, the number of traps in CT crystals is rarely restricted to one or two. Moreover, their number and depth critically depend on the technique of purifying the components before growing the crystal, as well as on the method employed in growing the crystal. This was pointed out by Haarer in the case of the N/TCPA crystal [69] as well as by Bos and Schmidt in the case of A/TCNB [210]. In the latter crystal, the authors report the observation of phosphorescence from a group of deep defects previously undetected in A/TCNB spectra and possessing very unusual properties. These particular defects are discussed further in following sections.

Increasing the CT character of the T_1 state in a CT crystal changes the shape of the low temperature phosphorescence, just as it does to the triplet excitation spectra. The zero-phonon transitions become accompanied by phonon wings. With a large CT character the zero-phonon lines disappear under the dominant phonon wings, and the spectra become

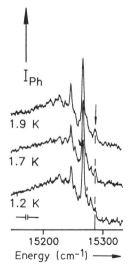

Figure 6.2. Phosphorescence of A/TCPA under a better resolution (1.8 Å). The arrow marks the resonance band phosphorescence (from [11]).

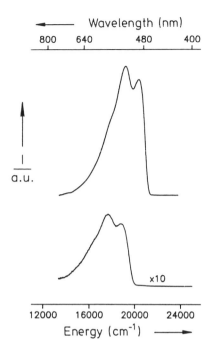

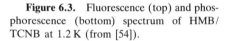

Figure 6.3. Fluorescence (top) and phosphorescence (bottom) spectrum of HMB/TCNB at 1.2 K (from [54]).

almost unstructured, as shown in Fig. 6.3 for HMB/TCNB. The phosphorescence spectra in such cases look very much like fluorescence.

Due to the lack of structure in the phosphorescence spectra of crystals characterized by a high CT character of the T_1 state, it is usually difficult to evaluate the number of emitting species. However, it is evident by studying the temperature dependence of phosphorescence that this number is no less than in other CT crystals. Upon elevating the temperature, the phosphorescence usually shifts gradually towards lower energies, which is also accompanied by a decrease of its intensity. This is a result of shallow traps being thermally depopulated and deeper defects becoming relatively more populated [159].

The presence of multiple defects in practically all CT systems studied is confirmed by magnetic resonance experiments and its possible reasons are discussed in Section VI.B.4.

2.   ODMR and ESR Studies

The existence of multiple triplet trap states differing in depth as seen in Fig. 6.2 suggests immediately an ODMR experiment monitoring phosphorescence from individual defects. One expects that the zfs parameters of these defects should also be different. In fact, both ODMR and ESR experiments often reveal a surprising variety of defects characterized by a wide range of zfs parameters. This has led in the past to considerable disagreements between authors studying one system under different conditions. The N/TCNB crystal may serve as a good example:

In an early ESR study Krebs et al. [107] reported the zfs parameters of a presumed defect in N/TCNB, $D = \pm 0.0358$, $E = \mp 0.0224\,\mathrm{cm}^{-1}$, which, when related to the values of naphthalene itself, gave a CT character of the order of as much as 50%. These values agree very well with those of the isolated CT complex in a liquid crystal ($D^m = 0.0514\,\mathrm{cm}^{-1}$); they are at odds, however, with the results of the optical work of Iwata et al. [211] who describe the T_1 state of a defect in N/TCNB as a predominantly local naphthalene state. The studies of Yagi et al. [108] and Hayashi et al. [109] reporting ODMR studies on the same crystal identify three distinct sites (defects) with very different zfs parameters:

$$\text{Site 1: } |D| = 0.0863\,, \qquad |E| = 0.0215\,\mathrm{cm}^{-1}$$
$$\text{Site 2: } |D| = 0.0629\,, \qquad |E| = 0.0182\,\mathrm{cm}^{-1}$$
$$\text{Site 3: } |D| = 0.048\,, \qquad |E| = 0.015\,\mathrm{cm}^{-1}$$

Site 3 can be readily identified with the defect observed by Krebs et al. [107] after recalculating the zfs parameters of the latter according to the ODMR convention, in which D must be greater than $3|E|$. Site 1 can be identified as a slightly perturbed naphthalene state (zfs parameters of naphthalene are $|D| = 0.0994$, $|E| = 0.0154\,\mathrm{cm}^{-1}$ [212]), while site 2 represents a triplet trap with a moderate CT character of approximately 30%. These results are in agreement with the phosphorescence-microwave double resonance (PMDR) spectra also reported by Hayashi et al. [109]. Site 1 gives a PMDR spectrum almost identical to naphthalene, site 2 shows phonon wings accompanying the sharp zero-phonon lines, while site 3 gives a PMDR spectrum characteristic for triplet states of high CT character (no zero-phonon line discernible).

Erdle and Möhwald [194] examined by ESR a defect in N/TCNB, which has the zfs parameters $D = \pm 0.068$, $E = \mp 0.009\,\mathrm{cm}^{-1}$, and CT character of about 30%, comparable to site 2 of Hayashi et al.

Ponte Goncalves and Hunton [213] re-examined the N/TCNB crystal and also the isolated complex in low temperature glass, detecting in the latter case two sites representing two different configurations of the complex with different CT character. In a subsequent study, Blackadar and Ponte Goncalves [40] repeated in principle the experiments of Hayashi et al. [109]. While the actual ODMR transition frequencies reported by them are at variance with those of Hayashi et al., they also detected defects 1 and 2, but failed to observe defect 3 of high CT character.

The controversy surrounding the defects in N/TCNB was eventually resolved by Grupp [110] and Grupp et al. [41] through detailed ODMR and ESR studies. They found at least three dominant triplet traps in the crystal, characterized by the following zfs parameters:

$$\text{Defect 1:} \quad D = \pm 0.0936, \quad E = \mp 0.0140 \text{ cm}^{-1} \quad (c_1^2 = 5\%)$$
$$\text{Defect 2:} \quad D = \pm 0.0680, \quad E = \mp 0.0097 \text{ cm}^{-1} \quad (c_1^2 = 28\%)$$
$$\text{Defect 3:} \quad D = \pm 0.0422, \quad E = \mp 0.0052 \text{ cm}^{-1} \quad (c_1^2 = 50\%)$$

The nature of these defects is discussed further in Section VI.B.4, but it should be said that it is the defect 2 which most adequately represents the bulk properties of the crystal, having zfs parameters almost identical to those of the mobile excitons appearing in the crystal at higher temperatures [40,41,192].

A similar situation to that in N/TCNB is encountered in the P/TCPA crystal. Although early optical and ESR experiments conducted on an isolated complex [214] suggested a significant CT character of the T_1 state (50%), the results of Krzystek et al. [44] revealed a different situation. Figure 6.4 shows the ODMR spectrum of the crystal at 1.2 K. There are clearly three defects discernible: two shallow traps B and C whose zfs parameters are close to each other and suggest a CT character of only

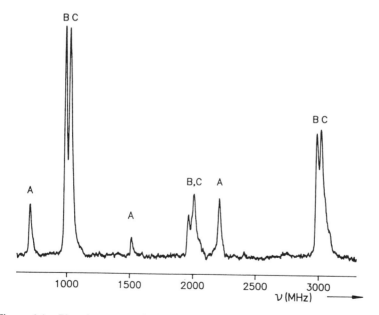

Figure 6.4. Phosphorescence-detected ODMR spectrum of P/TCPA at 1.2 K. Three defects (A, B and C) contribute to the spectrum (from [44]).

about 30%, and a deeper one A, whose CT character is in agreement with the result of Yu [214]. The comparison with the zfs parameters of mobile excitons appearing in the crystal above 30 K suggests that it is trap C which most closely resembles the properties of the exciton, and perhaps could even be tentatively identified as a "self-trap" [121].

A less clear situation exists in the A/TCNB crystal. It was mentioned earlier in Section VI.B.1 that in this system Bos and Schmidt [85,210] identified a group of defects whose presence had not been detected in earlier examinations. These are deep traps, more than $400\,\text{cm}^{-1}$ deep, and characterized by zfs tensors very different from those of the earlier identified defects, which always have zfs characteristic for anthracene with negligible CT character. In particular, the zfs parameters of the deep defects are much larger than those of anthracene and comparable to those of TCNB ($D = -0.1101$, $E = +0.0054\,\text{cm}^{-1}$). The orientation of the zfs tensor also does not coincide with that of anthracene. One of these defects was extensively examined by ENDOR with the aim of establishing its nature and structure. This is discussed in Section VI.B.3.

A characteristic feature of the magnetic resonance signals of triplet traps is their spectral width. In zero field experiments the linewidth is determined principally by the distribution of traps in the crystal with slightly different zfs parameters, that exceed the line broadening due to the second order hfs. In certain crystals, characterized by a static molecular disorder, such as F/TCNB or F/TCNQ, this distribution causes ODMR linewidths to reach as much as 60 MHz [51,38]. In high field experiments (ESR and hf-ODMR), the typical linewidth is on the order of 10 G. The main factor responsible for this is the presence of inhomogeneous broadening due to hyperfine interactions in the absence of motional effects. In many cases a well-resolved hyperfine structure can be observed in the ESR spectra (Fig. 6.5).

In order to interpret the proton hyperfine structure of ESR spectra, the spin Hamiltonian (2.10) has to be extended by the addition of three extra terms and takes the following form:

$$\mathbf{H} = \mathbf{H}_{\text{EZ}} + \mathbf{H}_{\text{ZFS}} + \mathbf{H}_{\text{NZ}} + \mathbf{H}_{\text{HFS}} + \mathbf{H}_{\text{Q}} \tag{6.1}$$

where $\mathbf{H}_{\text{EZ}}$ is the electron Zeeman term, $\mathbf{H}_{\text{ZFS}}$ is the zfs term (both were discussed in Section II.C.1), $\mathbf{H}_{\text{NZ}}$ is the nuclear Zeeman term, $\mathbf{H}_{\text{HFS}} = \bar{\mathbf{S}} \cdot \mathbf{A} \cdot \bar{\mathbf{I}}$ is the hyperfine term ($\bar{\mathbf{S}}$ is the electron spin, $\bar{\mathbf{I}}$ is the nuclear spin and $\mathbf{A}$ is the hfs tensor), $\mathbf{H}_{\text{Q}} = \bar{\mathbf{I}} \cdot \mathbf{Q} \cdot \bar{\mathbf{I}}$ is the quadrupole term ($\mathbf{Q}$ is the quadrupole tensor), which is nonzero for $I > 1/2$.

Each nonequivalent nucleus is characterized by its own hyperfine tensor, which is a 3×3 diagonal tensor in its unique coordinate system.

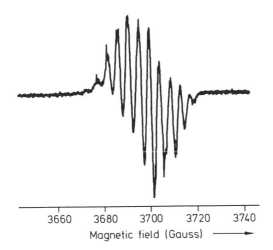

Figure 6.5. The upper part of the ESR spectrum of the A/TCPA crystal with magnetic field parallel to the x axis at 3.8 K. The hfs structure originates from two anthracene molecules in two slightly nonequivalent orientations as determined by the low temperature crystal structure (from [37]).

The diagonal elements are also called the hfs constants. Each of them is a sum of the isotropic (Fermi or contact) and anisotropic (magnetic dipole-dipole) interactions. The isotropic constant a can be extracted from the experimental hfs tensor **A** according to

$$a = 1/3 \, \mathrm{Tr} \, \mathbf{A} \qquad (6.2a)$$

The anisotropic elements of the tensor **B** can be then obtained from

$$B_{ii} = A_{ii} - 1/3a \qquad (6.2b)$$

It is well known that the values of the hyperfine constants in aromatic molecules depend on the electron density on the carbon atom adjacent to the nucleus interacting with the electron spin. Thus a charge transfer taking place in the excited state of the complex should in principle lower the hfs constants of the donor molecule due to the reduced spin density. The hfs constants of the acceptor should be also influenced by the charge transfer, but most of the acceptors forming the weak CT crystals discussed in this work have very few protons and a small spin density on carbon atoms adjacent to them, thus their contribution to the hfs of a complex is barely observable (exception: TCNQ complexes, where the T_1 state is localized on TCNQ).

Dalal et al. [83] described the observed hfs constants of the CT complex just like the fine structure, that is, by a linear combination of the hfs constants of the donor and the CT zero order states:

$$A_{ii}(\text{compl}) = c_1^2 A_{ii}(\text{CT}) + (1 - c_1^2) A_{ii}(\text{D}) \qquad (6.3)$$

where $A_{ii}(\text{compl})$ is the iith element of the hfs tensor of the CT complex, $A_{ii}(\text{CT})$ is the corresponding element of the CT zero order state, and $A_{ii}(\text{D})$ is the corresponding element of the donor molecule.

A problem arises immediately, similar to that with the zfs parameters of the zero order CT state, namely the evaluation of the hfs constants of the CT state. Dalal et al. assumed that these constants can be approximated by those of a radical ion pair, which are approximately half of those of the donor due to the split spin density:

$$A_{ii}(\text{CT}) \simeq 1/2 A_{ii}(\text{D})$$

Equation (6.3) is therefore reduced to

$$A_{ii}(\text{compl}) = A_{ii}(D) - 1/2 c_1^2 A_{ii}(D) \qquad (6.4)$$

It is then possible to calculate the CT character of the given defect from the formula

$$c_1^2 = 2 \frac{A_{ii}(\text{D}) - A_{ii}(\text{compl})}{A_{ii}(\text{D})} \qquad (6.5)$$

The CT character of a number of triplet traps in different CT crystals, calculated from the hfs splittings in their ESR spectra, is shown in Table IX together with the values estimated from fine structure data.

TABLE IX
CT Character of Triplet Traps in CT Crystals Calculated
from hfs and zfs Splittings

Crystal	c_1^2 from hfs	c_1^2 from zfs	Ref.
A/TCNB	<0.05	0.05	[91]
A/TCPA	<0.04	0.03	[37]
N/TCPA	0.10	<0.05	[43]
N/TCNB	0.28	0.3	[188]
P/PMDA	0.56	0.76	[46]
Ph/TCNQ	<0.09	–	[79]

The results obtained from the hfs splittings generally agree well with those derived from zfs parameters, with the exception of the only system of high CT character that was studied by this method (P/PMDA). In this particular case, the agreement is rather poor. Keijzers and Haarer [46] suggest that the reason for this discrepancy is the assumption made before formulating Eq. (6.4). In fact, triplet states and radical-ion pairs may have slightly different spin molecular orbitals and hence also spin densities. Thus triplet states resembling to a significant degree ionic states, as in P/PMDA, are influenced by this difference.

3.   ENDOR Studies

The ENDOR technique is particularly useful in the determination of the hfs constants of various nuclei in molecules. Its main advantage over ESR is the greatly reduced number of signals: for an electron spin $S = 1/2$ and n nuclei of spin $I = 1/2$, the number of ENDOR transitions is $2n$, while the number of ESR lines is 2^n, making the interpretation of ESR spectra sometimes very difficult, if not impossible.

The fundamentals of ENDOR may be found in [215]. We only draw the basic outlines of this technique here.

The result of Hamiltonian (6.1) acting on the triplet spin wave function is shown in terms of sublevel splitting in Fig. 6.6, which illustrates the case of one $I = 1$ nucleus, for example, deuterium or nitrogen.

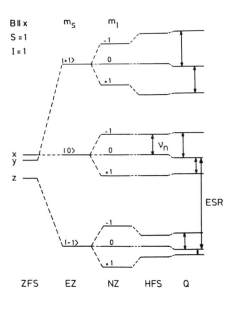

Figure 6.6. Schematic nuclear sublevel splitting in the magnetic field for $S = 1$, $I = 1$, and the field parallel to the x zfs axis. ν_n is the nuclear Zeeman frequency at the given magnetic field. ESR marks the (not to scale) ESR transition being saturated. Small arrows represent the ENDOR transitions. ZFS is the zero field splitting, EZ is the electron Zeeman splitting, NZ is the nuclear Zeeman splitting, HFS is the hyperfine splitting, and Q is the quadrupole splitting.

The ENDOR experiment involves partial saturation of an ESR transition in a fixed magnetic field corresponding to the ESR resonance, then scanning the radio frequency applied to the sample. The absorption of radio waves is detected indirectly by a change of the ESR amplitude. Because both the hfs and quadrupole interactions are anisotropic, the full characterization of all the parameters of the spin Hamiltonian requires repetition of this experiment for a number of orientations of the crystal in the magnetic field rotating it in at least two planes, similar to the way the zfs parameters of a triplet state are obtained in a single crystal.

The first ENDOR experiment on a CT crystal was performed by Grupp [110] and Grupp et al. [41] on Nd_8/TCNB. Of the three defects present in this system, the one with characteristics similar to the bulk properties of the crystal (CT character of about 28%) was chosen, being also the most abundant one. A typical ENDOR spectrum for a fixed orientation of the crystal relative to the magnetic field is shown in Fig. 6.7. It contains lines originating from the deuterons in the naphthalene molecule as well as the protons and nitrogen nuclei in TCNB (the latter not shown in the figure).

The information obtained from the experiments of Grupp is twofold. First, it was possible to extract the numerical values of the hfs tensors of all deuterons, and to separate the isotropic constants a from the anisotropic tensors. The isotropic hfs constant for each proton (or deuteron) is related to the spin density on the adjacent carbon atom by the empirical McConnell formula [216] $a = Q_{CH}\rho$ (Q_{CH} is a coefficient, ρ is the spin density). Grupp used the coefficient $Q = -72.4\,MHz$, obtaining in this way a complete map of spin density on the naphthalene molecule in the

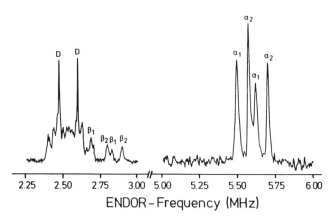

Figure 6.7. ENDOR spectrum of Nd_8/TCNB for the magnetic field parallel to the x axis detected by saturating the $|0\rangle - |1\rangle$ ESR line (from [41]).

Nd_8/TCNB complex. The sum of the individual densities obtained is 0.84, which gives a CT character of the triplet state equal to 0.16, somewhat smaller than the one evaluated from the zfs parameters. The spin density distribution was also found to have a center of symmetry on the naphthalene molecule. From this, it was concluded that the spin density of 16% is distributed symmetrically over two TCNB molecules located above and below naphthalene in the stack. Thus the excited N/TCNB complex is a trimer

$$TCNB^{-0.08}N^{+0.84}TCNB^{-0.08}$$

This is in agreement with the results of Keijzers and Haarer [46] reported for the P/PMDA system and discussed in Section III.C.4.

Second, the important information obtained from the ENDOR experiment regards the orientation of the molecules in the excited defect (the low temperature phase structure of N/TCNB is unknown). This information was derived from the orientation of the quadrupole tensors of the deuterons. For the defect in N/TCNB, the result is not only a deviation of the molecular x axis from the a' direction by $\pm16°$ (which was to be expected from freezing in the librations about the z axis), but also a tilting of the molecular z axis from the stack c axis by $\pm23°$. Additionally it was found that the zfs principal axes differ from the molecular symmetry axes of naphthalene despite the relatively low CT character of the complex. This deviation is $(x, x') = 4°$, $(y, y') = 11°$, $(z, z') = 11°$. Experiments on isotopically mixed crystals were used to check that the above numbers were characteristic for the N/TCNB complex and did not originate from a misorientation of the defect relative to the bulk of crystal.

It should in principle be possible to determine the position of the TCNB molecule in the defect from the quadrupole tensors of the nitrogen nuclei. Unfortunately, the amount of data was not sufficient to extract meaningful information.

The ENDOR technique in its optically detected version (analogous to high-field ODMR) was employed also in the investigations of a deep defect in the A/TCNB crystal by Bos and Schmidt [85]. As discussed before, optical and ESR experiments detected the presence of a group of very unusual defects in this crystal. The most peculiar properties are the very large zfs parameters, and the orientation of the zfs tensors not corresponding to any molecular axis system.

The ENDOR experiments conducted in the manner similar to Grupp et al. on one of these defects in A/TCNB gave the following results:

- The sum of spin densities on individual carbon atoms of anthracene obtained from the isotropic hyperfine constants is only 0.5. This means that the CT character of the defect is practically 100%.
- The anthracene molecule is rotated around its x (long in-plane) symmetry axis by 13°.
- The spin density is distributed solely over one complex pair, and not as in N/TCNB over a trimer.

These are rather surprising results, particularly the discrepancy between the large zfs parameters, which initially prompted Bos and Schmidt to interpret this trap state as a $n\pi^*$ state localized on TCNB, and the very small spin density on anthracene, which points at a 100% CT state. Similarly surprising is the large angular value of 13° between the z axis of anthracene in the ground state of the complex, which is known from crystallographic studies, and in the excited defect as derived from hfs tensors of protons in the T_1 state.

ENDOR experiments prove their value particularly in obtaining precise orientation of the defect under study relative to the crystal axes.

4. The Nature of Defects

Since defects represent either foreign molecules in a small concentration, or local distortions of the lattice, all information regarding their nature must be obtained by indirect methods. In particular, this information may be obtained from:

- the depth of the trap (in the singlet and triplet manifold);
- its zfs parameters and tensor orientation;
- the orientation of the molecules forming the defect as derived from hyperfine and quadrupole tensor orientation.

In many cases these results may lead to conclusions with a high degree of probability. In the following we attempt to systematize the information available on the structure of defects typically found in CT crystals:

- Defects having zfs parameters characteristic of one component of the complex. Such defects have been found in many CT crystals such as A/TCNB, N/TCNB or Ph/TCNQ. It was postulated by Grupp [110] that the nature of these defects may lie in a stoichiometric error in the crystal, for example, in N/TCNB, one TCNB molecule gets replaced by naphthalene. A stack containing the defect can be symbolically represented by the following picture:

$$\cdots -A-D-A-D-A-D-D^*-D-A-D-A-D-A-\cdots$$

It is easily understood that the zfs parameters of such a defect should have zfs parameters close to those of the donor. Also, since the CT character of such a defect is close to zero, its energy should not be changed much by CT interactions and the depth of it is usually small. A similar interpretation was invoked for shallow traps in A/TCNB [85]. In the case of systems with very small CT character of the T_1 state such as A/TCNB and A/TCPA, however, it is impossible to distinguish between stoichiometric defects and those representing a slightly distorted complex pair. For example, it was found in Ph/TCNQ that a defect having zfs parameters characteristic for TCNQ had a depth in the triplet manifold of only 4 cm^{-1}, but in the singlet state its depth was on the order of several hundred cm^{-1} [38]. Since the S_1 state in Ph/TCNQ, as in all other weak CT crystals, is largely a CT state in which small distortions of the complex geometry may have profound influence on the energy, the triplet trap despite its small depth must also involve both the donor and acceptor. A related kind of substitutional defect was proposed in the A/TCPA crystal [37]. Due to the specific low temperature phase structure of this crystal (four complexes in the unit cell [11]), a wrongly built-in TCPA molecule (turned by 180°) may induce no less than four different defects. As the T_1 state in this crystal is localized on anthracene, the T_1 energy of such defects should be only slightly altered relative to the band. In fact, several traps only a few cm^{-1} deep were found in the DF-ODMR spectrum of A/TCPA [37] (the phosphorescence revealed also deeper defects of the kind described below):

- Defects whose zfs parameters are similar to those of mobile excitons. Such defects can be detected in most if not all CT crystals. They are equivalent to the well-known X-traps in one-component molecular crystals. The difference lies in the depth of such traps. Depending on the CT character, it may reach values of up to 200 cm^{-1} since the energy of such a trap is sensitive to the mutual orientation of both components of the complex. Such defects, however, must be distinguished from the "self traps" described in Section III.B, which have been observed by ESR in certain CT crystals, such as P/TCPA [121], F/TCNB [51] or B/TCNB [217].

- Defects whose zfs parameters are smaller than those of mobile excitons. These defects are usually induced by significant distortions of the lattice and have therefore usually a depth of a few hundred cm^{-1}. Traps of this kind were identified in N/TCNB [110] and P/TCPA [44]. In both systems, the CT character of these traps was estimated at about 50%, while that of mobile excitons was found to be about 30%. As a possible cause of the distortion, Grupp [110] suggests the stoichiometric defect in a stack. While the donor molecule whose "mistaken" insertion constitutes a defect of zero CT character and small depth, the neighboring complex pair may be disturbed to a much larger degree and thus may form a deep trap. It is worth noticing that quite often the zfs parameters of such a deep defect agree very well with those of the CT complex isolated in a glass, as is the case with the N/TCNB and P/TCPA complexes. It was therefore suggested by Krzystek et al. [44] that since complexes in glasses represent well-defined pairs of molecules, the same may be true of the aforementioned deep defects. In other words, while in defects representing only slightly distorted complexes the triplet wave function may be distributed over three molecules, as suggested by Keijzers and Haarer [46] and Grupp [110], in the deep defects this distribution may be limited strictly to one complex pair, with the consequence of changing the zfs parameters.

• Defects impossible to be categorized into any class, detected in A/TCNB by Bos and Schmidt [85,210]. Their presence merely underscores the difficulties in obtaining CT crystals of very good quality. In fact, there is no ideal method of purifying the components and growing the crystals. Bos and Schmidt extensively purified both components by column chromatography but apparently did not submit anthracene to zone-refining, the procedure otherwise universally applied to purifying this donor. This single fact may account for the difference between the number and nature of defects observed by them and by other authors.

C. Defects Created by Doping

1. CT Character of Dopant Complexes

As in any other molecular crystal, doping a CT crystal with small quantities of an additional component creates defects in the host lattice. If the dopant is structurally similar to the host donor it may replace the host in the lattice. Unlike in one-component crystals, the inclusion of a guest donor in a 1:1 crystal creates a CT complex between the guest and the host acceptor. Depending on the triplet energy of the guest complex, it may act as a triplet trap in the system. Similarly, doping a CT crystal with an electron acceptor creates a CT complex between it and the host donor, which depending on its triplet energy, may act as a trap. The most important factors in a successful creation of such traps are: the structural similarity of the dopant to one of the host molecules; and the energy of the dopant, which must be lower than that of the molecule it is supposed to replace in the lattice.

A significant number of articles devoted to the optical and particularly magnetic properties of chemical triplet traps in CT crystals have been published. Möhwald and Sackmann [48,218] reported ESR measurements on a number of chemical traps created in the N/TCNB, D/TCNB, P/TCNB and B/TCNB crystals by doping them with pyrene, phenanthrene, durene, and anthracene in small quantities (0.1–1%). In each of these cases, the guest molecule replaces the host donor. The authors checked the orientation of one guest complex, A/TCNB in N/TCNB by measuring the zfs tensor orientation, and came to the conclusion that anthracene replaces naphthalene preserving its orientation (a substitutional crystallization). It was assumed that in other complexes, such a crystallization also takes place. It turns out that in some cases, the zfs parameters of the guest complex are very similar to those of the same complex in a 1:1 crystal. For example, the parameters of the A/TCNB trap in four different host systems (N/TCNB, D/TCNB, B/TCNB and P/TCNB) are practically the same as those in the 1:1 A/TCNB crystal, pointing to the almost zero CT character of its T_1 state. However, in other cases there are serious discrepancies: the Pyr/TCNB trap in the N/TCNB crystal has a CT character estimated at 36%, in the B/TCNB crystal, 15%, and an isolated complex Pyr/TCNB has a CT character of only 9%.

The paper by Dalal et al. [83] reported the first successful measurements of the hyperfine structure of the chemical trap A/TCNB, P/TCNB and Pyr/TCNB in the N/TCNB crystal, as well as a calculation of the CT character of these traps from the hfs splittings along the lines discussed in Section VI.B.2.

The two articles by Möhwald and Böhm [82,219] described traps created by doping the N/TCNB crystal with unconventional acceptors like hexacyanobenzene, pentacyanotoluene, TNB, chloranil, riboflavin and FAD (flavin-adenosine dinucleotide). Also, N/TCNB was also doped with a rarely used donor, coronene. Two results are worth mentioning:

- The hfs splittings of the complex N/hexacyanobenzene allows one to interpret the symmetry of the excited complex as that of a dimer and not a trimer as found in 1:1 crystals P/PMDA [46] and N/TCNB [41]. This was explained by a breakdown of the inversion symmetry of the host crystal by the creation of the defect.

- The CT complex N/chloranil had distinctly small zfs parameters: $|D| = 0.0081$, $|E| = 0.0011 \, \text{cm}^{-1}$. These values are one order of magnitude smaller than those of both naphthalene and chloranil, and point out the very high CT character of the excited state of this complex.

The method of Möhwald et al. was followed by Yu and Lin [220] who investigated the complexes A/TCPA, P/TCPA and Ph/TCPA created in the N/TCPA crystal by doping it with respective donors. A striking result is that both A/TCPA and P/TCPA traps have a CT character much larger than that of corresponding 1:1 crystals: A/TCPA, 18% (1:1 crystal 3–4% [37]); P/TCPA, 45% (1:1 crystal 30% [44]). The authors discovered that the traps are not populated via the host exciton band, but rather directly via ISC from singlet traps, similar to the defects in many 1:1 systems such as Ph/TCNO [38] and A/TCPA [37]. Also, Yu and Lin checked the validity of the assumption of Möhwald et al. regarding the substitutional crystallization of the dopants. It turns out that in many cases the guest donor is oriented in the host lattice at a certain angle in respect to the host donor.

A number of traps created by doping have been investigated by Corvaja, Pasimeni et al. In [103] defects created in N/TCNB by doping with TCNQ and p-chloranil were examined. The complex B/TCNQ created as a trap in the B/TCNB crystal was also studied in [176]. In effect, the triplet states of the B/ and N/TCNQ complexes were found to be largely local states of TCNQ, in agreement with the study of the 1:1 crystals Ph/TCNQ and F/TCNQ [38]. The B/TCNB crystal served as a host to another CT complex embedded as a trap, N,N'-dimethyldihydrophenazine/TCNB [221]. This triplet trap is characterized by very small zfs parameters $D = \mp 0.0165$, $E = \pm 0.00095 \text{ cm}^{-1}$, which point to the predominantly ionic character of this triplet state.

The defect created by a *trans*-stilbene (tS) impurity present as an impurity in the DPA/TCNB crystal was investigated by Corvaja et al. [49] using optical spectroscopy, by Agostini et al. [152,222] using ESR, and Maniero and Corvaja [223] using ENDOR. The triplet state of the tS/TCNB complex was found to be a largely local tS state, which allows one to draw conclusions about the T_1 state of *trans*-stilbene, itself difficult to access experimentally by direct methods. From the ENDOR experiments, the detailed orientation of tS regarding the host crystal structure was derived, and more importantly, the spin densities in the triplet state were obtained from the isotropic hfs constants.

The two papers by Corvaja et al. [224,225] reported on ESR and ENDOR measurements of the isolated complex Pyr/TCNB in N/TCNB, while the one by Maniero et al. [226] presented ENDOR of the isolated complex P/TCNB in the same host. Both papers concern the problem of molecular dynamics in the host crystal and are discussed in Section VI.C.2.

Summarizing the results of the investigations conducted on the defects created by doping CT crystals, it should be said that despite the considerable effort and a corresponding number of papers, the results regarding the properties of the created traps, like the CT character and orientation in the host crystal, are of limited value. This was acknowledged by Möhwald and Erdle in the last paper of their series [227]. In particular, the substitutional crystallization of the dopants seems to be the exception rather than the rule. In general, the dopant complex assumes an orientation which is forced by the steric interactions in the host crystal rather than the one it assumes in a 1:1 crystal. This may lead to such a mutual orientation of the guest complex molecules that the electron spin density in the triplet state is quite distorted in comparison to a 1:1 crystal, with the result of changing the zfs and hfs parameters of this state, sometimes radically. Hence the CT character values of guest CT complexes say very little about the corresponding values of the same complexes in a relaxed configuration in 1:1 crystals or even in complexes isolated in low temperature glasses. Additionally, one has to take under consideration the possibility of the dopant molecules undergoing librations, which may influence the measured zfs parameters [227].

2. *Dopant Complexes as Probes of Molecular Dynamics*

Section V of this work was devoted to the problem of molecular dynamics in CT crystals, concentrating on two systems, A/TCNB and N/TCNB. It was shown that in the case of A/TCNB, mobile triplet excitons may serve as a useful probe of the molecular reorientation processes taking place in A/TCNB, while in N/TCNB their usefulness is limited due to the fact that they appear only at high temperature (>130 K). Consequently, some information on the molecular structure of N/TCNB was derived from the defect in the 1:1 crystal [194], but this defect in turn gets depopulated at temperatures too low to study the effect of the phase transition taking place at about 63 K.

N/TCNB is a system particularly suitable for doping by other donors due to the high energy of its T_1 state ($21,352 \text{ cm}^{-1}$, see Table I). This was recognized by Erdle and Möhwald [194] who doped it with anthracene and by Möhwald and Erdle [227] who doped it also with pyrene and phenanthrene, creating very deep triplet traps. The behavior of ESR signals originating from these traps has been studied in detail, particularly in the vicinity of the phase transition taking place in the host crystal at about 63 K. It turns out that each of the guest donors behaves differently in the host lattice. Phenanthrene assumes only a single orientation in both the low temperature and high temperature phases of the N/TCNB host. This is in strong contrast to naphthalene in the same crystal, which assumes two orientations differing by a $\pm18.5°$ rotation about the z axis and $\pm23°$ about the x axis. The phase transition can be observed by a

broadening of the hfs of P/TCNB in a narrow temperature range about
63 K (Fig. 6.8). Since there is only one potential well for phenanthrene in
the N/TCNB crystal, this molecule does not undergo librations.

Anthracene assumes two orientations in the N/TCNB crystal, similar
to the host donor. These orientations differ by ±12.5° in the xy plane and
±5° in the zy plane, both values smaller than those of naphthalene.
Above the phase transition, anthracene librates between the two orientations with a frequency of about $5 \times 10^9 \text{ s}^{-1}$, that is, somewhat faster than
the host donor. Also, the activation energy of libration (450 cm^{-1}) is
smaller than for naphthalene (700 cm^{-1}).

Pyrene also assumes two orientations in the low temperature phase of
the crystal. The angular difference of these orientations is, however,
much larger than for naphthalene, ±40°. Above the phase transition,
pyrene librates like naphthalene and anthracene, but the potential well
for the librations is radically changed; the extrema of librations differ
pronouncedly and the libration frequency is one order of magnitude
larger than for naphthalene.

The above results of Möhwald and Erdle bring the following conclusions regarding the usefulness of deep traps created by doping CT crystals
for studying molecular dynamics:

• It is possible to draw certain qualitative conclusions regarding the
  temperature of the phase transition in the host crystal.

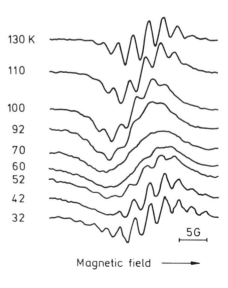

Figure 6.8. Hyperfine structure in the triplet ESR spectrum of the P/TCNB complex in the N/TCNB crystal at different temperatures (from [226]).

- The frequency, amplitude and activation energy of guest librations do not correspond to those of the host molecules, as for each of them the potential well determining these parameters is different.

- The librations of the molecules may strongly influence the measured zfs parameters, thus special care is necessary before evaluating the CT character of defects of this type.

These conclusions were confirmed by Ripmeester et al. [192] who also doped N/TCNB with anthracene, and by Ponte Goncalves and Vyas [196], who used phenazine as a dopant in the same system.

The two deep triplet traps examined by Möhwald and Erdle, namely Pyr/TCNB and P/TCNB were subsequently investigated by Corvaja et al. [224,225] and Maniero et al. [226] using ENDOR. The measurements of the isotropic hfs constants gave the spin densities on the pyrene carbon atoms in Pyr/TCNB, the sum of these densities then gave the CT character of the triplet trap Pyr/TCNB of 14–16% [224]. The anisotropic elements of the hfs tensor were also evaluated and agreement between the experiment and calculations could be obtained assuming librations of pyrene about its out-of-plane axis by $\pm 5°$ and about its long in-plane axis by as much as $\pm 18°$. The first value stands in poor agreement with the one obtained from ESR data by Möhwald and Erdle ($\pm 28°$) [227].

Subsequent ENDOR work on the same system done at lower temperatures (20 K) [225] centered on another defect, whose orientation in the crystal lattice corresponds to that of the defect observed also by Erdle and Möhwald [227].

The paper by Maniero et al. [226] concerned the P/TCNB trap in the N/TCNB host crystal. Their results confirmed the presence of a single orientation of the guest donor in the host lattice as opposed to that of the host donor and a lack of librations of phenanthrene. The CT character of the defect (40%) was obtained from the isotropic hfs constants by the method of Dalal et al. [83] although it was found that the spin densities on the phenanthrene molecule were not uniformly reduced by CT interactions in the triplet state, thus the estimation of the CT character from hfs splittings gave only a rough value of this parameter.

ACKNOWLEDGMENTS

We are deeply indebted to Professor H.C. Wolf for his continuous interest in this work. The cooperation of all the staff, graduate students and students of the 3. Physikalisches Institut and Kristallabor, Universität Stuttgart is also appreciated. Special thanks are due to Dr. M. Krebs. We also thank Dr. James A. van Zee who critically read the manuscript, contributing valuable remarks. The following institutions supported the research which has led to this publication: Universität Stuttgart; Stiftung Volkswagenwerk; Polish Academy of Sciences (program CPBP 01.12); Grimmke Stiftung; Max-Planck-Gesellschaft; Deutsche Forschungsgemeinschaft; Heraeus Stiftung. We also wish to thank the authors and/or publishers of the reproduced figures, for granting permission.

REFERENCES

1. R. S. Mulliken and W. B. Person, *Molecular Complexes*, Wiley, New York, 1969.

2. H. Eyring, J. Walter, and G. E. Kimball, *Quantum Chemistry*, Wiley, New York, 1944, p. 213.

3. S. Iwata, J. Tanaka, and S. Nagakura, *J. Chem. Phys.* **47**, 2203 (1967).

4. J. Czekalla, G. Briegleb, W. Herre, and H. J. Vahlensieck, *Z. Elektrochem.* **63**, 715 (1959).

5. G. Briegleb, *Elektronen-Donator-Acceptor-Komplexe*, Springer, Berlin, 1961.

6. S. Nagakura, in *Excited States*, Vol. 2, E. C. Lim, Ed., Academic Press, New York, 1975.

7. F. H. Herbstein, in *Perspectives in Structural Chemistry*, Vol. 4, J. D. Dunitz and J. A. Ibers, Eds., Wiley, New York, 1971.

8. C. Dean, M. Pollak, B. M. Craven, and G. A. Jeffren, *Acta Crystallogr.* **11**, 710 (1958); F. H. Herbstein, *Acta Crystallogr.* **18**, 997 (1965).

9. J. Trotter, *Can. J. Chem.* **39**, 1574 (1961).

10. J. J. Stezowski, *J. Chem. Phys.* **73**, 538 (1980).

11. W. Mühle, J. U. von Schütz, H. C. Wolf, R.-D. Stigler, and J. J. Stezowski, unpublished.

12. J. B. Torrance, in *Low-Dimensional Conductors and Superconductors*, J. Jerome and L. G. Caron, Eds., NATO ASI Series B, Vol. 155, Plenum Press, New York, 1987.

13. J. B. Torrance, J. E. Vazquez, J. J. Mayerle, and V. Y. Lee, *Phys. Rev. Lett.* **46**, 253 (1981).

14. P. Erk, S. Hünig, G. Klebe, M. Krebs, and J. U. von Schütz, *Chem. Ber.* **124**, 2005 (1991).

15. M. D. Fayer, in *Spectroscopy and Excitation Dynamics of Condensed Molecular Systems*, V. M. Agranovich and R. M. Hochstrasser, Eds., North-Holland, Amsterdam, 1983.

16. R. Silbey, in *Spectroscopy and Excitation Dynamics of Condensed Molecular Systems*, V. M. Agranovich and R. M. Hochstrasser, Eds., North-Holland, Amsterdam, 1983.

17. H. Haken and G. Strobl, *Z. Phys.* **262**, 185 (1973).

18. J. Frenkel, *Phys. Rev.* **37**, 17, 1276 (1931).

19. A. S. Davydov, *Theory of Molecular Excitons*, Plenum Press, New York, 1971.

20. G. H. Wannier, *Phys. Rev.* **52**, 191 (1937).

21. N. Geacintov and M. Pope, *J. Chem. Phys.* **47**, 1194 (1967).

22. P. Reineker, B. J. Schmid and P. Petelenz, *Chem. Phys.* **91**, 59 (1984).

23. V. M. Agranovich and A. A. Zakhidov, *Chem. Phys. Lett.* **50**, 278 (1977).

24. D. Haarer and M. R. Philpott, in *Spectroscopy and Excitation Dynamics of Condensed Molecular Systems*, V. M. Agranovich and R. M. Hochstrasser, Eds., North-Holland, Amsterdam, 1983.

25. D. Haarer, *Chem. Phys. Lett.* **27**, 91 (1974).

26. E. Betz, H. Port, W. Schrof and H. C. Wolf, *Chem. Phys.* **128**, 73 (1988).

27. S. Maier, Diplomarbeit, Universität Stuttgart, 1985.

28. S. Kaesdorf, Diplomarbeit, Universität Stuttgart, 1982.

29. D. Burger, Diplomarbeit, Universität Stuttgart, 1981.

30. J. U. von Schütz, D. Burger, R. Krauss, W. Mühle, and H. C. Wolf, *J. Luminesc.* **24/25**, 467 (1981).

31. S. Maier and H. Port, *Z. Naturforsch. A* **42**, 1261 (1987).

32. H. Nissler, Diplomarbeit, Universität Stuttgart, 1980.

33. D. Haarer, *J. Chem. Phys.* **67**, 4076 (1977).

34. W. Steudle, J. U. von Schütz and H. Möhwald, *Chem. Phys. Lett.* **54**, 461 (1978).

35. A. M. Ponte Goncalves, *Chem. Phys.* **19**, 397 (1977).

36. W. Mühle, Ph.D. thesis, Universität Stuttgart, 1987.

37. W. Mühle, J. Krzystek, J. U. von Schütz and H. C. Wolf, *Z. Naturforsch. A* **44**, 610 (1989).

38. D. Gundel, J. Frick, J. Krzystek, H. Sixl, J. U. von Schütz, and H. C. Wolf, *Chem. Phys.* **132**, 363 (1989).

39. E. L. Frankevich, A. N. Chaban, M. M. Triebel, J. U. von Schütz, and H. C. Wolf, *Chem. Phys. Lett.* **177**, 283 (1991).

40. R. L. Blackadar and A. M. Ponte Goncalves, *Chem. Phys. Lett.* **85**, 483 (1982).

41. A. Grupp, H. C. Wolf, and D. Schmid, *Chem. Phys. Lett.* **85**, 330 (1982).

42. C.-T. Yu and T.-S. Lin, *Chem. Phys. Lett.* **60**, 122 (1978).

43. C.-T. Yu and T.-S. Lin, *Chem. Phys.* **39**, 293 (1978).

44. J. Krzystek, J. U. von Schütz, H. C. Wolf, R.-D. Stigler, and J. J. Stezowski, *Z. Naturforsch. A* **42**, 622 (1987).

45. B. Kozankiewicz, *Chem. Phys. Lett.* **134**, 323 (1987).

46. C. P. Keijzers and D. Haarer, *J. Chem. Phys.* **67**, 925 (1977).

47. C. Corvaja, B. Kozankiewicz, L. Pasimeni, and J. Prochorow, *Chem. Phys.* **119**, 399 (1988).

48. H. Möhwald and E. Sackmann, *Z. Naturforsch. A* **29**, 1216 (1974).

49. C. Corvaja, B. Kozankiewicz, L. Montanari, and J. Prochorow, *Chem. Phys.* **142**, 83 (1990).

50. L. Pasimeni, C. Corvaja, and D. A. Clemente, *Mol. Cryst. Liq. Cryst.* **104**, 231 (1984).

51. W. Mühle, J. Krzystek, J. U. von Schütz, H. C. Wolf, R.-D. Stigler, and J. J. Stezowski, *Chem. Phys.* **108**, 1 (1986).

52. M. Krebs, Diplomarbeit, Universität Stuttgart, 1987.

53. B. Kozankiewicz and J. Prochorow, *Chem. Phys.* **135**, 307 (1989).

54. J. Krzystek, M. Krebs, J. U. von Schütz, H. C. Wolf, and J. Prochorow, *Mol. Cryst. Liq. Cryst.* **148**, 267 (1987).

55. B. Kozankiewicz and J. Prochorow, *J. Mol. Struct.* **114**, 203 (1984).

56. J. Krzystek, unpublished.

57. J. Krzystek, Ph.D. thesis, Institute of Physics, Polish Academy of Sciences, Warsaw, 1983.

58. D. Haarer, C. P. Keijzers, and R. Silbey, *J. Chem. Phys.* **66**, 563 (1977).

59. J. Czekalla, G. Briegleb, and W. Herre, *Z. Elektrochem.* **63**, 712 (1959).

60. J. Czekella, G. Briegleb, W. Herre, and R. Glier, *Z. Elektrochem.* **61**, 537 (1957).

61. K. Huang and A. Rhys, *Proc. Roy. Soc. London Ser. A* **208**, 351 (1951).

62. D. B. Fitchen, R. H. Silsbee, T. A. Fulton, and E. L. Wolf, *Phys. Rev. Lett.* **11**, 275 (1963).

63. D. Haarer, *Chem. Phys. Lett.* **31**, 192 (1975).

64. D. Haarer, M. R. Philpott, and H. Morawitz, *J. Chem. Phys.* **63**, 5238 (1975).

65. A. Brillante, M. R. Philpott, and D. Haarer, *Chem. Phys. Lett.* **56**, 218 (1978).

66. M. R. Philpott and A. Brillante, *Mol. Cryst. Liq. Cryst.* **50**, 163 (1979).

67. A. Brillante and M. R. Philpott, *J. Chem. Phys.* **72**, 4019 (1980).

68. R. L. Beckmann and G. J. Small, *Chem. Phys.* **30**, 19 (1978).

69. D. Haarer, *J. Luminesc.* **18/19**, 453 (1979).

70. M. A. El-Sayed, in *Excited States*, Vol. 1, E. C. Lim, Ed., Academic Press, New York, 1974.

71. R. H. Clarke and J. M. Hayes, *J. Chem. Phys.* **57**, 679 (1972).

72. J. U. von Schütz, W. Steudle, H. C. Wolf, and V. Yakhot, *Chem. Phys.* **46**, 53 (1980).

73. W. G. Dorp, T. Schaafsma, M. Soma, and J. H. van der Waals, *Chem. Phys. Lett.* **21**, 221 (1973).

74. R. H. Clarke and R. E. Connors, *Chem. Phys. Lett.* **33**, 365 (1975).

75. R. H. Clarke, Ed., *Triplet State ODMR Spectroscopy*, Wiley, New York, 1982.

76. C. A. Hutchison and B. W. Mangum, *J. Chem. Phys.* **34**, 908 (1961).

77. H. Hayashi, S. Nagakura, and S. Iwata, *Mol. Phys.* **13**, 489 (1967).

78. D. Gundel, Ph.D. thesis, Universität Stuttgart, 1985.

79. H. Dörner and D. Schmid, *Chem. Phys.* **13**, 469 (1976).

80. H. Hayashi, S. Iwata, and S. Nagakura, *J. Chem. Phys.* **50**, 993 (1969).

81. H. Beens, J. De Jong, and A. Weller, Colloque Ampere **15**, 289 (1969).

82. H. Möhwald and A. Böhm, *Z. Naturforsch. A* **31**, 1324 (1976).

83. N. S. Dalal, D. Haarer, J. Bargon, and H. Möhwald, *Chem. Phys. Lett.* **40**, 359 (1976).

84. K. Minamata and M. Iwasaki, *Mol. Phys.* **23**, 1115 (1972).

85. F. C. Bos and J. Schmidt, *J. Chem. Phys.* **84**, 584 (1986).

86. C. B. Chesnut and W. D. Phillips, *J. Chem. Phys.* **35**, 1002 (1961).

87. R. Mason, *Acta Crystallogr.* **17**, 547 (1964).

88. D. Haarer and H. C. Wolf, *Mol. Cryst. Liq. Cryst.* **10**, 359 (1970).

89. J. U. von Schütz, F. Gückel, W. Steudle, and H. C. Wolf, *Chem. Phys.* **53**, 365 (1980).

90. L. Pasimeni, G. Guella, C. Corvaja, D. A. Clemente, and M. Vicentini, *Mol. Cryst. Liq. Cryst.* **91**, 25 (1983).

91. R. Sauter, Ph.D. thesis, Universität Stuttgart, 1983.

92. D. Haarer and N. Karl, *Chem. Phys. Lett.* **21**, 49 (1973).

93. J. Ziegler and N. Karl, *Chem. Phys.* **40**, 207 (1979).

94. J. Frick, Diplomarbeit, Universität Stuttgart, 1987.

95. W. Mühle, J. U. von Schütz, and H. C. Wolf, *Chem. Phys. Lett.* **131**, 175 (1986).

96. R. Krauss, W. Schrof, H. C. Wolf, and D. Schmid, *Chem. Phys.* **73**, 55 (1982).

97. J. U. von Schütz, J. Krzystek, W. Mühle, R. Sauter, and H. C. Wolf, *Proc. XIth Molecular Crystal Symposium*, Lugano, Switzerland, 1985, p. 310.

98. H. M. Vyas and A. M. Ponte Goncalves, *Chem. Phys. Lett.* **51**, 556 (1977).

99. E. L. Frankevich, A. I. Pristupa, and V. I. Lesin, *Chem. Phys. Lett.* **47**, 304 (1977).

100. H. Möhwald and E. Sackmann, *Chem. Phys. Lett.* **21**, 43 (1973).

101. G. Agostini, C. Corvaja, G. Giacometti, L. Pasimeni, D. A. Clemente, and G. Bandoli, *Mol. Cryst. Liq. Cryst.* **141**, 165 (1986).

102. H. Möhwald and E. Sackmann, *Solid State Commun.* **15**, 445 (1974).

103. L. Pasimeni, G. Guella, and C. Corvaja, *Chem. Phys. Lett.* **84**, 466 (1981).

104. G. Agostini, C. Corvaja, G. Giacometti, and L. Pasimeni, *Chem. Phys.* **77**, 233 (1983).

105. L. Pasimeni and C. Corvaja, *Mol. Cryst. Liq. Cryst.* **100**, 17 (1983).

106. D. A. Lowitz, *J. Chem. Phys.* **46**, 4698 (1967).

107. P. Krebs, E. Sackmann and J. Schwarz, *Chem. Phys. Lett.* **8**, 417 (1971).

108. M. Yagi, N. Nishi, M. Kinoshita, S. Nagakura, and H. Hayashi, *Mol. Phys.* **30**, 147 (1975).

109. H. Hayashi, M. Yagi, and N. Nishi, *J. Luminesc.* **12/13**, 169 (1976).

110. A. Grupp, Ph.D. thesis, Universität Stuttgart, 1982.

111. D. L. Evans and W. T. Robinson, *Acta Crystallogr.* **B 33**, 2891 (1977).

112. C. P. Keijzers and D. Haarer, *Chem. Phys. Lett.* **49**, 24 (1977).

113. P. W. Anderson and P. R. Weiss, *Rev. Mod. Phys.* **25**, 269 (1953).

114. A. Carrington and A. D. McLachlan, in *Introduction to Magnetic Resonance*, Harper and Row, London, 1969.

115. R. H. Hochstrasser, S. K. Lower, and C. Reid, *J. Chem. Phys.* **41**, 1073 (1964).

116. P. Petelenz and V. H. Smith, *Chem. Phys. Lett.* **82**, 430 (1981).

117. H. Möhwald, E. Erdle, and A. Thaer, *Chem. Phys.* **27**, 79 (1978).

118. J. M. Park and A. H. Reddoch, *J. Chem. Phys.* **74**, 1519 (1981).

119. E. Erdle and H. Möhwald, *Phys. Stat. Sol.* (*b*) **98**, 617 (1980).

120. E. Erdle and H. Möhwald, *Z. Naturforsch.* A **35**, 236 (1980).

121. J. Krzystek and J. U. von Schütz, *Mater. Sci.* (*Wroclaw*) **13**, 133 (1987).

122. J. Allgeier, V. Macho, D. Stehlik, H. M. Vieth, W. Auch, and J. U. von Schütz, *Chem. Phys. Lett.* **86**, 522 (1982).

123. A. Salm, Diplomarbeit, Universität Stuttgart, 1985.

124. M. S. de Groot, I. A. M. Hesselman, and J. H. van der Waals, *Mol. Phys.* **16**, 61 (1969).

125. W. Steudle and J. U. von Schütz, *J. Luminesc.* **18/19**, 191 (1979).

126. J. U. Schütz, W. Steudle, H. C. Wolf, P. Reineker, and U. Schmid, *Chem. Phys. Lett.* **79**, 1 (1981).

127. U. Schmid and P. Reineker, *Phys. Stat. Sol.* (*b*) **109**, 265 (1982).

128. A. H. Francis and C. B. Harris, *Chem. Phys. Lett.* **9**, 181 (1971).
129. J. M. Park and A. H. Reddoch, *Chem. Phys. Lett.* **91**, 117 (1982).
130. F. C. Bos and J. Schmidt, *Mol. Phys.* **58**, 561 (1986).
131. W. Steudle, Ph.D. thesis, Univeristät Stuttgart, 1978.
132. W. Auch and J. U. von Schütz, *Phys. Stat. Sol.* (*b*) **101**, 287 (1980).
133. J. Allgeier, A. Gutsze, D. Stehlik, and H. M. Vieth, *J. Luminesc.* **24/25**, 449 (1981).
134. D. Stehlik, in *Excited States*, Vol. 3, E. C. Lim, Ed., Academic Press, New York, 1977.
135. D. Stehlik and K. H. Hausser, in *Advances in Magnetic Resonance*, Vol. 3, J. Waugh, Ed., Academic Press, New York, 1968.
136. H. Haken and E. Schwarzer, *Chem. Phys. Lett.* **27**, 41 (1974).
137. H. M. Vieth, V. Macho and D. Stehlik, *Chem. Phys. Lett.* **60**, 368 (1979).
138. D. Stehlik, A. Doehring, J. P. Colpa, E. Callaghan and S. Kesmarky, *Chem. Phys.* **7**, 165 (1975).
139. D. A. Antheunis, J. Schmidt, and J. H. van der Waals, *Mol. Phys.* **35**, 1521 (1974).
140. W. Auch, Ph.D. thesis, Universität Stuttgart, 1980.
141. J. U. von Schütz and H. C. Wolf, *Z. Naturforsch. A* **27**, 42 (1972).
142. F. Noack, *Prog. Nuclear Magn. Res. Spectrosc.* **18**, 171 (1986).
143. E. Schwarzer and H. Haken, *Phys. Stat. Sol.* (*b*) **84**, 25 (1977).
144. J. U. von Schütz, W. Auch, W. Güttler, and H. C. Wolf, in *Magnetic Resonance and Related Phenomena*, E. Kundla, E. Lippmaa and T. Saluvere, Eds., Springer, Berlin, 1979.
145. K. H. Hausser and H. C. Wolf, *Adv. Magn. Res.* **8**, 85 (1976).
146. C. P. Keijzers, J. Duran, and D. Haarer, *J. Chem. Phys.* **69**, 3563 (1978).
147. J. H. van der Waals, and M. S. de Groot, in *The Triplet State*, A. B. Zahlan, Ed., Cambridge University Press, Cambridge, 1967.
148. C. P. Keijzers, P. S. Bagus, and J. P. Worth, *J. Chem. Phys.* **69**, 4032 (1978).
149. L. Montanari, L. Pasimeni, and C. Corvaja, *Mol. Cryst. Liq. Cryst. Lett.* **1**, 191 (1985).
150. M. Yagi, N. Nishi, M. Kinoshita, and S. Nagakura, *Mol. Phys.* **35**, 1369 (1978).
151. C. Corvaja and L. Pasimeni, *Chem. Phys. Lett.* **88**, 347 (1982).
152. G. Agostini, C. Corvaja, G. Giacometti, and L. Pasimeni, *Chem. Phys.* **85**, 421 (1984).
153. R. A. Schadee, J. Schmidt, and J. H. van der Waals, *Chem. Phys. Lett.* **41**, 435 (1976); R. A. Schadee, J. Schmidt, and J. H. van der Waals, *Mol. Phys.* **36**, 177 (1978).
154. B. E. Robertson and J. J. Stezowski, *Acta Crystallogr. B* **34**, 3005 (1978).
155. B. T. Lim, S. Okajima, A. K. Chandra, and E. C. Lim, *Chem. Phys. Lett.* **79**, 22; B. T. Lim, S. Okajima, and E. C. Lim, *J. Chem. Phys.* **77**, 3902 (1982).
156. C. A. Parker, in *Photoluminescence of solutions*, Elsevier, Amsterdam, 1968.
157. H. Beens and A. Weller, in *Molecular Photoluminescence*, E. C. Lim, Ed., Benjamin, New York, 1969.
158. B. Kozankiewicz, *Mat. Sci.* (*Wroclaw*) **10**, 137 (1984).
159. B. Kozankiewicz and J. Prochorow, *Mol. Cryst. Liq. Cryst.* **148**, 93 (1987).

160. B. Kozankiewicz and J. Prochorow, *Chem. Phys. Lett.* **61**, 347 (1979).

161. A. A. Avdeenko, V. A. Karachevtsev, and Yu. V. Naboikin, *Zurn. Prikl. Spektr.* **37**, 666 (1982).

162. A. A. Avdeenko, V. A. Karachevtsev, and Yu. V. Naboikin, *Ukr. Fiz. Zurnal* **29**, 1873 (1984).

163. V. A. Karachevtsev, *Fiz. Tverd. Tela* **28**, 1400 (1986) (*Sov. Phys. Solid State* **28**, 788 (1986)).

164. D. D. Dlott, M. D. Fayer, and R. D. Wieting, *J. Chem. Phys.* **67**, 3803 (1977); R. D. Wieting, M. D. Fayer, and D. D. Dlott, *J. Chem. Phys.* **69**, 1996 (1978); D. D. Dlott, M. D. Fayer, and R. D. Wieting, *J. Chem. Phys.* **69**, 2752 (1978).

165. H. Port, D. Rund, and H. C. Wolf, *Chem. Phys.* **60**, 81 (1981).

166. R. M. Shelby, A. H. Zewail, and C. B. Harris, *J. Chem. Phys.* **64**, 3192 (1976).

167. W. Güttler, J. U. von Schütz, and H. C. Wolf, *Chem. Phys.* **24**, 159 (1977).

168. Ch. E. Swenberg and N. E. Geacintov, in *Organic Molecular Photophysics*, Vol. 1, J. B. Birks, Ed., Wiley, New York, 1973.

169. V. Ern, *Mol. Cryst. Liq. Cryst.* **18**, 1 (1972).

170. B. Kozankiewicz, Ph.D. thesis, Institute of Physics, Polish Academy of Sciences, Warsaw, 1978.

171. B. Kozankiewicz, J. Prochorow, and J. Krzystek, *Mol. Cryst. Liq. Cryst.* **75**, 17 (1981).

172. R. C. Johnson and R. E. Merrifield, *Phys. Rev. B* **1**, 896 (1970).

173. V. I. Lesin, V. P. Sakun, A. I. Pristupa, and E. L. Frankevich, *Phys. Stat. Sol.* (*b*) **84**, 513 (1977).

174. E. L. Frankevich, A. I. Pristupa, and V. I. Lesin, *Chem. Phys. Lett.* **54**, 285 (1978).

175. L. Pasimeni, C. Corvaja, G. Agostini, and G. Giacometti, *Z. Naturforsch. A* **39**, 427 (1984).

176. L. Pasimeni, C. Corvaja, G. Agostini, and G. Giacometti, *Chem. Phys.* **97**, 357 (1985).

177. E. L. Frankevich, M. M. Tribel, I. A. Sokolik, and B. V. Kotov, *Phys. Stat. Sol.* (*a*) **40**, 655 (1977).

178. E. L. Frankevich, M. M. Tribel, I. A. Sokolik, and A. I. Pristupa, *Phys. Stat. Sol.* (*b*) **87**, 373 (1978).

179. J. Greis, Diplomarbeit, Universität Stuttgart, 1988.

180. J. Fünfschilling, M. Samoc, and D. F. Williams, *Chem. Phys. Lett.* **96**, 157 (1983).

181. H. Tsuchiya, F. Marumo, and Y. Saito, *Acta Crystallogr. B* **28**, 1935 (1972).

182. E. Erdle and H. Möhwald, *Solid State Commun.* **26**, 327 (1978).

183. L. D. Landau and E. M. Lifschitz, *Statistical Physics*, Pergamon Press, New York, 1980.

184. C. Kittel, *Introduction to Solid State Physics*, Wiley, New York, 1986.

185. A. M. Ponte Goncalves, *Mol. Cryst. Liq. Cryst.* (*Lett.*) **56**, 163 (1980).

186. N. S. Dalal, J. A. Ripmeester, A. H. Reddoch, and D. F. Williams, *Mol. Cryst. Liq. Cryst.* (*Lett.*) **49**, 55 (1978).

187. S.-K. Ma, *Modern Theory of Critical Phenomena*, Benjamin, Reading, MA, 1976.

188. L. Pasimeni and C. Corvaja, *Solid State Commun.* **53**, 213 (1985).

189. J. D. Wright, T. Ohta, and H. Kuroda, *Bull. Chem. Soc. Jpn.* **49**, 2961 (1976).

190. C. A. Fyfe, *J. Chem. Soc. Faraday Trans.* 2 **70**, 1642 (1974).

191. C. A. Fyfe, D. Harold-Smith, and J. Ripmeester, *J. Chem. Soc. Faraday Trans.* 2 **72**, 2269 (1976).

192. J. A. Ripmeester, A. H. Reddoch and N. S. Dalal, *J. Chem. Phys.* **74**, 1526 (1981).

193. R. M. Macfarlane and S. Ushioda, *J. Chem. Phys.* **67**, 3214 (1977).

194. E. Erdle and H. Möhwald, *Chem. Phys.* **36**, 283 (1979).

195. S. Kumakura, F. Iwasaki and Y. Saito, *Bull. Chem. Soc. Jpn.* **40**, 1826 (1967).

196. A. M. Ponte Gonvalves and H. M. Vyas, *J. Chem. Phys.* **70**, 1560 (1979).

197. B. Mayoh and C. K. Prout, *J. Chem. Soc. Faraday Trans.* 2 **68**, 1072 (1972).

198. J. C. A. Boeyens and D. C. L. Levendis, *J. Chem. Phys.* **80**, 2681 (1984).

199. K. Czarniecka, J. M. Janik, J. A. Janik, J. Krawczyk, I. Natkaniec, J. Wasicki, R. Kowal, K. Pigon, and K. Otnes, *J. Chem. Phys.* **85**, 7289 (1986).

200. G. E. Pake and T. L. Estle, *The Physical Principles of Electron Paramagnetic Resonance*, Benjamin, Reading, MA, 1973.

201. J. U. von Schütz and M. Weithase, *Z. Naturforsch. A* **30**, 666 (1976).

202. W. Auch, W. Steudle, J. U. von Schütz, and H. C. Wolf, *Phys. Stat. Sol.* (a) **49**, 563 (1978).

203. W. Höptner, Diplomarbeit, Universität Stuttgart, 1978.

204. Kristallabor Stuttgart, measurements performed by N. Karl.

205. U. Shmueli and I. Goldberg, *Acta Crystallogr. B* **29**, 2466 (1973).

206. H. Möhwald and A. Thaer, *Phys. Stat. Sol.* (a) **50**, 131 (1978).

207. N. S. Dalal, L. V. Haley, D. J. Northcott, J. M. Park, A. H. Reddoch, J. A. Ripmeester, D. F. Williams, and J. L. Charlton, *J. Chem. Phys.* **73**, 2515 (1980).

208. H. J. Bernstein, N. S. Dalal, W. F. Murphy, A. H. Reddoch, S. Sunder, and D. F. Williams, *Chem. Phys. Lett.* **57**, 159 (1978).

209. J. Boerio-Goates, E. F. Westrum, and C. A. Fyfe, *Mol. Cryst. Liq. Cryst.* **48**, 209 (1978).

210. F. Bos and J. Schmidt, *Chem. Phys. Lett.* **108**, 565 (1984).

211. S. Iwata, J. Tanaka, and S. Nagakura, *J. Am. Chem. Soc.* **89**, 2813 (1967).

212. M. Schwoerer and H. C. Wolf, *Mol. Cryst.* **3**, 177 (1967).

213. A. M. Ponte Goncalves and R. J. Hunton, *J. Phys. Chem.* **79**, 71 (1975).

214. C.-T. Yu, Ph.D. thesis, Washington University, St. Louis, 1979.

215. L. Kevin and L. D. Kispert, *Electron Spin Double Resonance Spectroscopy*, Wiley, New York, 1976.

216. H. M. McConnell, *J. Chem. Phys.* **24**, 764 (1956).

217. G. Agostini, D. Carbonera, G. Giacometti, A. L. Maniero, and L. Pasimeni, *Chem. Phys. Lett.* **167**, 78 (1990).

218. H. Möhwald and E. Sackmann, *Chem. Phys. Lett.* **26**, 509 (1974).

219. H. Möhwald and A. Böhm, *Chem. Phys. Lett.* **43**, 49 (1976).

220. C.-T. Yu and T.-S. Lin, *J. Phys. Chem.* **83**, 3397 (1979).

221. G. Agostini, C. Corvaja, G. Giacometti, C. Mancabelli, and L. Pasimeni, *Mol. Cryst. Liq. Cryst. Lett.* **3**, 155 (1986).

222. G. Agostini, C. Corvaja, G. Giacometti, L. Pasimeni and D. A. Clemente, *J. Phys. Chem.* **92**, 997 (1988).
223. A. L. Maniero and C. Corvaja, *Chem. Phys.* **135**, 277 (1989).
224. C. Corvaja, A. L. Maniero, and L. Pasimeni, *Chem. Phys.* **100**, 265 (1985).
225. C. Corvaja, A. Lodolo, and A. L. Maniero, *Z. Naturforsch. A* **45**, 1317 (1990).
226. A. L. Maniero, C. Corvaja, and L. Pasimeni, *Chem. Phys.* **122**, 327 (1988).
227. H. Möhwald and E. Erdle, *Phys. Stat. Sol.* (*b*) **103**, 757 (1981).

FLOW BEHAVIOR OF LIQUID CRYSTALLINE POLYMERS

GIUSEPPE MARRUCCI AND FRANCESCO GRECO

Department of Chemical Engineering, University of Naples, P. Tecchio, 80125, Italy

CONTENTS

Advances in Chemical Physics, Volume LXXXVI, Edited by I. Prigogine and Stuart A. Rice.
ISBN 0-471-59845-3 © 1993 John Wiley & Sons, Inc.

Liquid crystalline polymers (LCPs) have recently become a very active research topic in polymer science, both in view of possible applications and because they have represented a scientific challenge. Monographs on LCPs are now available [1,2], to which we refer for a general introduction to LCP chemistry and phase stability. The physics of nonpolymeric liquid crystals can also be found in various textbooks (compare, e.g., [3]); the concepts required here will nevertheless be reintroduced and thoroughly discussed where needed. In these introductory remarks, therefore, we recall a few general definitions, and indicate the limitations and the scope of the present review.

Two main categories of LCPs can be distinguished, depending on the location in the molecular structure of the units, or moieties, that generate the liquid crystalline phase. These units can either form an integral part of the chain backbone (main-chain LCPs) or be attached to it (side-group LCPs). In this review, only main-chain LCPs are considered, their rheology being more important for applications and more interesting conceptually. On the other hand, from the available information on the flow behavior of side-group LCPs, it appears that they behave essentially, although possibly not entirely, like ordinary polymers [4].

Most main-chain LCPs form a nematic phase, which is the liquid crystalline state characterized by the fact that the chains are roughly parallel to one another, while their positions remain randomly distributed. In LCP processing, whenever flow is involved, advantage is typically taken of the very low viscosity of the nematic state. This chapter therefore concentrates on nematic LCPs. Cholesterics, which are the nematic phases formed by chiral molecules, are also automatically included, because in fact cholesteric LCPs are undistinguishable from nematics during flow [5]. Smectic phases formed by main-chain LCPs, when they exist, are much less mobile than nematic ones; their rheology remains outside the scope of this work. In conclusion, among the various categories of LCPs that now exist, only polymeric nematics are considered here, due to the special importance of their flow behavior.

Apart from their use in applications, polymeric nematics indeed show several peculiar flow phenomena that are not observed either in other polymers or in nematics of low molecular weight. In this regard, it is worth noting that the flow behavior of small molecule nematics is well understood in its fundamentals, as described by the Leslie–Ericksen theory [3]. Similarly, the flow dynamics of ordinary polymers, studied from many viewpoints and over many decades, has also reached a reasonable level of interpretation [6]. Now, although the flow of LCPs resembles that of small molecule nematics in some respects, and that of ordinary polymers in others, the peculiar phenomena alluded to previously cannot be explained by any of the respective theories, nor by any trivial combination of them.

Various authors have recently contributed a number of works from which a consistent interpretation of LCP flow behavior is beginning to appear. The main objective of this paper is that of presenting, as much as possible in a systematic way, this emerging interpretation of LCP rheology. With respect to the theories of small molecule nematics on the one hand, and of isotropic polymeric liquids on the other, the LCP dynamic theory can be seen as an autonomous topic in itself. The emphasis of this review being on theory, all references to experiments are only made for comparison purposes. In other words, no attempt is made here to review the numerous experimental works on flowing LCPs. It should also be admitted that not all experimental results are as yet fully understood; the contents of this chapter are, in part at least, work in progress.

Another point that must be made clear from the start has to do with the distinction between thermotropic and lyotropic LCPs, that is, between polymers for which the nematic state is obtained in a range of temperatures (thermotropics) and those for which the isotropic–nematic transition is induced by changing the concentration in a suitable solvent (lyotropics). Although this distinction is very important in several respects, primarily because these two categories of LCPs are suitable for different applications [2], in this chapter we refer to thermotropics and lyotropics almost interchangeably. Indeed, as shown by some experiments (to be discussed in the following), the flow behavior of both categories is much less dissimilar than it might superficially appear, and is rooted in the essential characteristic of main-chain LCPs, namely in their molecular rigidity. At the same time, however, some distinction in flow behavior remains, just because the molecular rigidity is not alike for all polymers. Roughly speaking, lyotropic LCPs can afford to be more rigid than thermotropic ones since the liquid state is obtained with the help of a solvent in the lyotropic case. The more rigid the chain, the more its behavior in flow is expected to conform to that predicted by the rod-like molecular theory to be described below.

The paper is organized as follows. First, we briefly recall the classical theory of nematodynamics of Frank, Leslie, and Ericksen, and discuss the limitations that are expected when trying to apply this theory to a nematic phase made up of polymeric molecules. In Section II, the basic molecular model of LCPs, namely the rigid rod, is presented, together with a rich variety of predictions of flow behavior, in excellent qualitative agreement with the experiments at high shear rates. One of the basic predictions of the molecular theory, as well as many experimental data at low shear rates, demonstrate the necessity of accounting for spatial inhomogeneities of the nematic phase. These are examined in Section III. The concluding section summarizes the key points, and indicates possible further progress in LCP rheological theory.

I. THE STATUS OF THE LESLIE–ERICKSEN THEORY

The first systematic presentation of the statics of nematics is due to Frank [7], who clearly recognized that the continuum mechanical description of a nematic phase required the use of a new variable, indicative of the local anisotropy. Such an indicator is a unit vector **n** known as the "director" (actually a pseudo-vector, since **n** and −**n** are physically undistinguishable), which specifies the anisotropy axis. The theory of Frank deals with orientational "distortions", that is, with spatial dependencies of the director. By assuming that the energetic cost of these distortions is quadratic in the gradients of the **n**-field, he demonstrated than an arbitrary distortion can be decomposed into three basic ones (splay, twist, bend). Correspondingly, three material constants are needed to measure the resistance offered by the nematic to these orientational distortions. It is worth noting for future reference that the quadratic energy assumption made by Frank implies that the distortions are small (in some sense to be made precise). Indeed, since the undistorted nematic represents an energy minimum, small deviations from such a minimum can be described with a quadratic law in absolute generality.

In their original works, Ericksen [8] and Leslie [9] extended the continuum mechanical approach used by Frank in statics to encompass the dynamical behavior of nematics. Their theory (henceforth referred to as LE theory) couples the director and the velocity fields via the appropriate choice of a constitutive equation for the stress tensor. Such a theory, and a few consequences that are relevant for LCPs, are briefly described in Sections I.A, I.B and I.C. Sections I.D and I.E are meant to illustrate the "status" of LE theory in the context of polymeric nematics; LE theory was in fact developed to describe nematics of low molecular weight.

A. The Equations of LE Theory

A dynamic theory must of course account for viscous dissipation. As is well known, the viscous stress in ordinary liquids is simply proportional to the rate of strain, which is the symmetric part of the velocity gradient. In the case of anisotropic liquids, however, dissipation is expected to depend on the orientation of the director with respect to the velocity gradient. Thus, although the linearity between stress and velocity gradient is maintained in LE theory, one should not be surprised that the expression of viscous stresses in nematics is quite complex compared to that for isotropic liquids. In fact, the viscous (or "Leslie–Ericksen") stress $\boldsymbol{\sigma}_{LE}$ comes out as the sum of six terms, as follows:

$$\boldsymbol{\sigma}_{LE} = \alpha_1 \mathbf{A}:\mathbf{nnnn} + \alpha_2 \mathbf{nN} + \alpha_3 \mathbf{Nn} + \alpha_4 \mathbf{A} + \alpha_5 \mathbf{nn} \cdot \mathbf{A} + \alpha_6 \mathbf{A} \cdot \mathbf{nn} \qquad (1)$$

where the vector $\mathbf{N}$ is given by

$$\mathbf{N} = \dot{\mathbf{n}} - \mathbf{W} \cdot \mathbf{n} \qquad (2)$$

and $\mathbf{A}$ and $\mathbf{W}$ are the symmetric and antisymmetric part of the velocity gradient, respectively; the constitutive parameters $\alpha_1 \ldots \alpha_6$ in Eq. (1) (dimensions of viscosity) are known as Leslie coefficients. Notice that the director $\mathbf{n}$ appears in these equations also through its rate of change $\dot{\mathbf{n}}$ (see Eq. (2)). The time derivative $\dot{\mathbf{n}}$ is a "convected" or "substantial" derivative, that is, it describes the rate of change of $\mathbf{n}$ as it is observed while "following the particle" in its motion.

The expression for the viscous stress in nematics, as given in Eqs. (1) and (2), is the most general constitutive equation that can be written while preserving linearity between stress and velocity gradient. Needless to say, Eqs. (1) and (2) obey frame indifference, that is, material "objectivity". As much as six "viscosities" appear here, as opposed to a single one in ordinary isotropic liquids. To be precise, the independent Leslie coefficients are only five, because thermodynamics dictates the equality $\alpha_2 + \alpha_3 = \alpha_6 - \alpha_5$, known as the Parodi relationship [10]. Notice further that not all Leslie coefficients are required to be positive, and in fact some of them are often negative. However, thermodynamics poses various restrictions on combinations of these coefficients, in the form of inequalities, so that dissipation is always positive.

To the viscous stress so far considered, an elastic stress arising from orientational distortions must be added. The starting point to obtain this elastic contribution to the stress tensor is the expression of Frank's energy density E_F, that is, the most general expression quadratic in $\nabla \mathbf{n}$ [7]:

$$E_F = \frac{1}{2} K_1 (\nabla \cdot \mathbf{n})^2 + \frac{1}{2} K_2 (\mathbf{n} \cdot \nabla \times \mathbf{n})^2 + \frac{1}{2} K_3 (\mathbf{n} \times \nabla \times \mathbf{n})^2 \qquad (3)$$

where K_1, K_2, and K_3 are the elastic constants (dimensions of force) for splay, twist, and bend distortions, respectively. The elastic (or "Frank") stress $\boldsymbol{\sigma}_F$ is then derived from E_F by performing the following operation:

$$\boldsymbol{\sigma}_F = \frac{\partial E_F}{\partial \nabla \mathbf{n}} \cdot (\nabla \mathbf{n})^T \qquad (4)$$

The sum of the viscous and elastic contributions gives the total stress σ. As is usual in liquids, however, σ remains defined only to within an arbitrary pressure p because of the incompressibility condition. The total stress σ then enters the usual momentum balance equation

$$\rho_0 \dot{v} = \nabla \cdot \sigma - \nabla p \tag{5}$$

where ρ_0 is the mass density.

In ordinary liquids, Eq. (5) and the continuity equation $\nabla \cdot v = 0$ (together with a constitutive equation for the stress tensor) are sufficient to solve for the pressure and the velocity fields. Conversely, in the case of nematic liquids, where the $\mathbf{n}$-field must also be determined, an additional balance equation is explicitly required, namely the angular momentum balance. In ordinary liquids, the latter equation merely dictates the symmetry of the stress tensor, thus posing a restriction on the stress constitutive equation. Conversely, in nematics, because body couples can in fact be exerted, the stress symmetry is generally not preserved, and the angular momentum must be balanced in a nontrivial way.

The angular momentum equation is written as

$$\left(-\gamma_1 \mathbf{N} - \gamma_2 \mathbf{A} \cdot \mathbf{n} + \nabla \cdot \frac{\partial E_F}{\partial \nabla \mathbf{n}} - \frac{\partial E_F}{\partial \mathbf{n}} \right) \times \mathbf{n} = \mathbf{0} \tag{6}$$

where the first two terms describe the viscous contribution to the torque, and the second two the elastic contribution. Notice that no angular inertia appears in Eq. (6) (differently from the original works) because it is commonly accepted nowadays that the inertia to be associated with director reorientation is negligible. Notice further that we have omitted in Eq. (6) torques generated by external fields (e.g., magnetic) because they are not required here. Notice finally that the material constants γ_1 and γ_2 are not additional constitutive parameters because they are linked to the α's of Eq. (1) through the relationships: $\gamma_1 = \alpha_3 - \alpha_2$; $\gamma_2 = \alpha_2 + \alpha_3$. Thus, among the six Leslie coefficients, α_2 and α_3 play a special role since the viscous torque is determined by them.

The balance equations, Eqs. (5) and (6), together with the constant density condition, $\nabla \cdot v = 0$, and the constitutive equations, Eqs. (1) and (3), form a complete set, that is, once the appropriate initial and boundary conditions are specified, the pressure, velocity, and director fields remain determined.

B. Shear Flow Predictions: Flow-Aligning and Tumbling

The general equations of the previous section can of course be special-ized, and simplified, for classes of flows. The important case of a simple shear flow is recalled here, and thoroughly discussed, because it allows one to discriminate between two quite distinct dynamic behaviors that are rooted in the nature of the nematic, that is, in its constitutive parameters.

A nematic liquid will in fact belong to one of two categories, those of "flow-aligning" and "tumbling" nematics, respectively. In flow-aligning nematics, the director wants to orient at a particular angle θ_0 with the shear direction, as shown in Fig. 1.1. This preferred direction lies in the "plane of shear", that is, in the plane formed by the shear and the gradient directions (the plane represented in Fig. 1.1). The angle θ_0, known as the Leslie angle, obeys the relationship

$$tg^2\theta_0 = \frac{\alpha_3}{\alpha_2} \tag{7}$$

which obviously requires α_3/α_2 to be a positive quantity. Also, since nematics made of rod-like molecules (as opposed to disc-like ones) always have $|\alpha_3/\alpha_2| < 1$, the angle θ_0 is usually small, smaller than 45° in any event.

Equation (7) is better understood if one considers the torque balance, Eq. (6), which is rewritten here as

$$\dot{\theta} = -\dot{\gamma}\; \frac{\sin^2\theta - \dfrac{\alpha_3}{\alpha_2}\cos^2\theta}{1 - \dfrac{\alpha_3}{\alpha_2}} \tag{8}$$

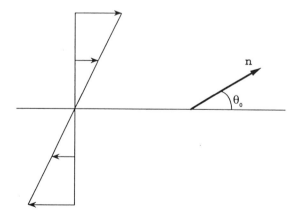

Figure 1.1 A flow-aligning nematic in shear flow. The director **n** lies in the plane of shear, at an angle θ_0 from the shear direction; θ_0 is the Leslie angle, and is determined by the Leslie coefficients α_2 and α_3 (see Eq. (7)).

This equation is obtained from Eq. (6) by assuming that the elastic torque is nil, that is, that there are no distortions. Equation (8) further assumes that the director belongs to the plane of shear, so that a single angle θ (measured from the shear direction) is sufficient to determine the director $\mathbf{n}$. We discuss the out-of-plane situation later. In Eq. (8), $\dot{\gamma}$ is the shear rate and $\dot{\theta}$ is the rate of change of angle θ.

The dynamics of the director, as described by Eq. (8), is readily understood. If α_3/α_2 is positive, $\dot{\theta}$ changes sign across $\theta = \pm\theta_0$. More precisely, since $\alpha_3/\alpha_2 < 1$, $\dot{\theta}$ is positive in the interval $-\theta_0 < \theta < \theta_0$ and negative outside. (In view of the $\mathbf{n}$ and $-\mathbf{n}$ equivalence, we can limit our analysis to the θ-interval from $-\pi/2$ to $\pi/2$.) Thus, for all initial orientations of the director different from $\theta = \pm\theta_0$, the shear flow will make the angle θ converge towards the value $+\theta_0$, which corresponds to a stable equilibrium; the angle $-\theta_0$, although also an equilibrium state, is unstable. As shown by Eq. (7), the angle θ_0 which the director of a flow-aligning nematic wants to make with the shear direction is fully "constitutive", that is, it is determined by the ratio of the Leslie coefficients α_3/α_2 and by nothing else.

Nematics which have a negative value of α_3/α_2 are called "tumbling" nematics. Indeed, as shown by Eq. (8), $\dot{\theta}$ is negative for all values of θ in such a case. Therefore, if no other effect enters the picture (see below), the director keeps rotating indefinitely in the plane of shear. In fact, since Eq. (8) is integrable in closed form, all details of the director dynamics can be made explicit. For the particular case of a constant $\dot{\gamma}$, the tumbling period, that is, the time required for the director to go through the full θ-interval from $-\pi/2$ to $\pi/2$, is obtained as

$$T = \frac{\pi}{\dot{\gamma}} \left(\lambda^{1/2} + \lambda^{-1/2} \right), \qquad \lambda = -\frac{\alpha_3}{\alpha_2} > 0 \qquad (9)$$

Again considering Eq. (8), it is worth noting that the time variable is in fact fictitious. Since $\dot{\gamma}$ can always be written as $d\gamma/dt$, where γ measures the shear deformation, it follows that Eq. (8) can be rephrased in terms of the derivative $d\theta/d\gamma$. In other words, the dynamics described by Eq. (8) is indifferent to the rate of shear; how much has the director changed its orientation only depends upon the "units of shear" γ by which the sample has been deformed. Thus, for example, a full turn of the director in the tumbling case requires a number of shear units which, from Eq. (9), is given by $\pi(\lambda^{1/2} + \lambda^{-1/2})$, again a fully constitutive quantity.

We have so far assumed that the director belonged to the shear plane. However, the general picture outlined above does not change too much when that restriction is removed, that is, when the director is allowed to span all possible orientations in space (all points on a sphere of unit radius). In the flow-aligning case ($\alpha_3/\alpha_2 > 0$), the single stable orientation of the director remains that given by Eq. (7). In other words, whatever the initial orientation (an arbitrary point on the unit sphere), the shear flow will always carry the director towards the stable alignment in the plane of shear, as depicted in Fig. 1.1. Needless to say, the approach "trajectory" on the unit sphere is described by a (slightly) more complex equation than Eq. (8).

Similarly, in the tumbling case ($\alpha_3/\alpha_2 < 0$), it remains true that the director continuously moves in a periodic fashion, describing trajectories that are closed curves on the unit sphere. One of these trajectories is the previously considered circle in the plane of shear, while the others are more complex curves. All of these trajectories, however, are travelled with the same rate, in the sense that the angle θ made by the projection of the director onto the shear plane still obeys Eq. (8). Thus, in particular, the period given by Eq. (9) is always the same.

The constitutive distinction between flow-aligning and tumbling nematics also remains important when elastic effects due to orientational distortions (so far ignored) enter the picture. Again restricting ourselves, as a temporary simplification, to in-plane directors, inclusion of orientational distortions modifies Eq. (8) in the following way:

$$\dot{\theta} = -\dot{\gamma} \, \frac{\sin^2\theta - \dfrac{\alpha_3}{\alpha_2}\cos^2\theta}{1 - \dfrac{\alpha_3}{\alpha_2}} + \frac{K}{\alpha_3 - \alpha_2} \frac{\partial^2\theta}{\partial y^2} \tag{10}$$

Two further simplifications have been used in writing Eq. (10). We have adopted the widespread assumption known as "one-constant approximation", whereby the three constants of Frank elasticity are taken to be equal ($K_1 = K_2 = K_3 = K$). We have also assumed that everything repeats itself along the shear direction, and that the distortions only develop along the shear gradient direction (the vertical direction in Fig. 1.1). Equation (10) can be seen as describing the competition between the viscous and elastic torques. If the latter is absent (no distortions), then also the viscous torque must be zero; such a particular condition was expressed by Eq. (8).

In the absence of distortions ($\partial^2\theta/\partial y^2 = 0$ in Eq. (10)), we have seen that flow-aligning nematics in shear achieve a steady state ($\dot{\theta} = 0$) with the director at $\theta = \theta_0$. In any practical case, however, there will exist physical boundaries of the sheared sample that may impose some particular director orientation θ_w at the walls. Typical examples of this "strong anchoring" effects are the "planar" or the "homeotropic" conditions, imposing a specific direction along the wall or the orthogonal direction, respectively. If such is the case, the θ_0 angle cannot be maintained near the boundaries of the sample where, therefore, distortions must be present.

However, to solve this boundary value problem Eq. (10) alone is not sufficient, because two unknown functions, namely $\theta(y, t)$ and $v(y, t)$, appear in that equation ($v(y, t)$ is hidden within $\dot{\gamma}$). On the other hand, in a shear flow, the momentum equation (Eq. 5) can be solved independently, thus completely specifying the total shear stress σ as a function of y and t (possibly a constant). The problem then closes by expressing σ as a function of θ and $\dot{\gamma}$, that is, by specializing the constitutive equations, Eqs. (1) and (4), for the shear case:

$$\sigma = \frac{1}{2}\,\alpha_1\dot{\gamma}\sin 2\theta \cos 2\theta - \alpha_2\left(\dot{\theta} + \frac{1}{2}\,\dot{\gamma}\right)\sin 2\theta - \alpha_3\left(\dot{\theta} + \frac{1}{2}\,\dot{\gamma}\right)\cos 2\theta$$

$$+ \frac{1}{2}\,\alpha_4\dot{\gamma} + \frac{1}{2}\,\alpha_5\dot{\gamma}\sin 2\theta + \frac{1}{2}\,\alpha_6\dot{\gamma}\cos 2\theta \qquad (11)$$

Notice that in this case the total shear stress coincides with the viscous contribution; the elastic contribution to σ vanishes because the **n**-field does not vary along the shear direction.

For a flow-aligning nematic, the steady state solution of the set of Eqs. (10) and (11) shows that, unless the sample is very thin, the Leslie angle θ_0 will prevail throughout the bulk, while distortions are confined to boundary layers. The characteristic length over which the transition occurs, from the wall orientation θ_w to the bulk value θ_0, can in fact be readily derived from order-of-magnitude arguments on Eq. (10) alone [3]. By calling ξ the thickness of the boundary layer, the elastic torque (per unit volume) is expressed as K/ξ^2. Equating this quantity to the viscous torque, ξ is of order

$$\xi = \sqrt{\frac{K}{\eta\dot{\gamma}}} \qquad (12)$$

where η stands for a characteristic viscosity of the nematic, and $\eta\dot{\gamma}$ is the applied shear stress σ.

The situation for a tumbling nematic is very different. Indeed, in discussing Eq. (8) for such a case, we have already seen that no steady state solution exists in the absence of distortions. The more general Eq. (10) shows, however, that, due to the elastic term, $\dot{\theta}$ can in fact become zero. Thus, orientational elasticity can in principle stop actual director tumbling of an intrinsically tumbling nematic, that is, of a material for which α_3/α_2 is negative.

Time independent solutions of the set of Eqs. (10) and (11) in the tumbling case have been known for some time [11,12]. The most important feature of such solutions is the "winding up" of the director field along the direction of the velocity gradient, that is, the occurrence of orientational distortions throughout the entire sample thickness, not just in boundary layers as in the previous case. The physical reason for such a "structure" of the solution is the necessity of compensating the viscous torque (which cannot vanish) with an elastic one at all points. Recent results show that the in-plane solution is in fact unstable, that is, the director "escapes in the third dimension" thus relaxing part of the distortion [13,14]. The fact remains, however, that the tumbling nature of the nematic necessarily generates complex patterns of the director field in a shear flow. Sometimes, these convoluted patterns have been pictorially referred to as "director turbulence" [15], although of course no inertial effects are involved.

C. Peculiar Scaling Properties

We have seen that LE nematics display both viscous and elastic properties. One might therefore be tempted to consider these fluids as viscoelastic. On the other hand, viscoelastic fluids show relaxation times that are intrinsic to the material, that is, constitutive relaxation times. As discussed in this section, LE nematics do not conform to this rule. Thus they should not be considered viscoelastic, at least not in the usual sense.

The constitutive parameters of LE fluids are the five independent Leslie coefficients, collectively represented with the symbol η (dimensions of a viscosity), and the three independent constants of Frank elasticity, collectively indicated with the single constant K (dimensions of a force). From η and K, no intrinsic time constant can be derived. Dimensional analysis shows that a quantity τ with dimensions of time can be obtained from η and K only in conjunction with some length l as

$$\tau = \frac{\eta l^2}{K} \tag{13}$$

Now, since the fluid does not exhibit any constitutive length scale, a characteristic time τ can occur only when some "external" length has been fixed, for example an apparatus dimension. In other words, if and when relaxation effects are found in nematics obeying LE theory, the observed relaxation time is not constitutive, but it must also depend upon the relevant size of the experimental geometry.

The scaling equation, Eq. (13), can well be read in the opposite direction: if the time scale of the experiment has been forced in some way, then a length scale will also "spontaneously" appear, which depends on both the constitutive parameters of the fluid and the external action. We have already shown one such example when discussing the boundary layer thickness ξ arising in sheared flow-aligning nematics. In Eq. (12), which has the same structure as Eq. (13), the externally imposed time scale appears through $\dot{\gamma}$; the system reacts by creating the length ξ.

It should be apparent that the arguments developed above, although important in understanding that LE fluids are not viscoelastic in the usual sense, are but trivial applications of dimensional analysis. If only four dimensional quantities are involved in a problem, then a single non-dimensional group can be defined which remains constant in all possible realizations. By defining the nondimensional group,

$$Er = \dot{\gamma}\,\frac{\eta l^2}{K} \tag{14}$$

known as the Ericksen number, Eq. (13) can be read as $Er = 1$.

There are instances, however, where the nonconstitutive quantities, a time and a length, can both be externally imposed, thus forcing the Ericksen number defined through them to be different from unity. These cases necessarily involve a fifth dimensional quantity, and thus a second nondimensional group. In a typical example, the size of the apparatus (the capillary diameter, say) is chosen as l and, simultaneously, the characteristic time is fixed through the kinematics of the flow (by imposing the flow rate in the capillary, say). A dynamical quantity is necessarily involved, however (the stress σ at the wall, which directly relates to the pressure gradient along the capillary in this example). The following relationship is then dictated by dimensional analysis:

$$\frac{\sigma}{\eta\dot{\gamma}} = f(Er) \tag{15}$$

The scaling predicted by Eq. (15), explicitly considered by Ericksen [16], is peculiar to nematics and has been experimentally verified [17]. Some of its implications are discussed in the following section.

Before concluding with scaling laws based on LE theory, it may be worth reconsidering the case where Frank elasticity is neglected altogether, as in Eq. (8). For a vanishing K, Eq. (15) reduces to

$\sigma/(\eta\dot{\gamma}) = 1$. Such a result, just like Eq. (15) itself, holds for a steady state; for time dependent situations, dimensional analysis predicts

$$\frac{\sigma}{\eta\dot{\gamma}} = f(\dot{\gamma}t, \text{init}) \tag{16}$$

where "init" indicates initial values in nondimensional form. Equation (16) shows that the elapsed time plays a role only through the deformation $\gamma = \dot{\gamma}t$ occurring in that time. This peculiar time dependence of the stress σ is at odds with classical viscoelasticity; indeed, Eq. (16) describes the time dependence of a purely viscous stress. The reason why, for a constant value of $\dot{\gamma}$, the viscous stress of a nematic may vary with time is that, as shown by Eq. (11) (or more generally by Eq. (1)), the viscous stress does depend on director orientation θ and director rate of change $\dot{\theta}$, and these two quantities will vary with time unless a steady state is reached. Thus, also by recalling the discussion following Eqs. (8) and (9), we can conclude by saying that in purely viscous nematics both the director orientation and the nondimensional stress at time t only depend on the initial **n**-value, and on the deformation occurring from $t = 0$ up to current time.

D. Linearity versus Nonlinearity

When introducing viscous stresses and torques, we have emphasized their linearity in $\nabla\mathbf{v}$ (compare Eqs. (1), (2), and (6)). Such a linear character of LE theory makes this theory the equivalent for nematics of Stokes' law of friction for isotropic fluids; in other words, LE theory is expected to apply to nematics much like Stokes' law applies to "Newtonian" fluids, that is, typically, to liquids of low molecular weight. Yet, small molecule nematics are known to display non-Newtonian behavior. For example, Poiseuille's law is not obeyed; the apparent viscosity varies with the flow rate in the capillary, a clear manifestation of nonlinearity.

Viscous nonlinearities in LE nematics are due to the coupling between velocity and director fields. At the end of the previous section, we have explicitly noted that the viscous stress depends on director orientation. Reciprocally, we had seen previously that the thickness of an orientational boundary layer depends, through $\dot{\gamma}$ (see Eq. (12)), on the intensity of the flow field. We discuss here in some detail how these interactions may create a nonlinear overall response in an intrinsically linear material.

The simplest example is offered by the same sheared layer of a flow-orienting nematic which was used to introduce the concept of orientational boundary layers. The steady-state fields of both velocity and director are found by solving the set of Eqs. (10) and (11), with $\dot{\theta} = 0$, and with σ a constant throughout the thickness l of the sheared layer. Without going into the details of the solution, two extreme situations can be readily examined. If the applied shear stress σ is large, then the boundary layer depth ξ is vanishingly small (see Eq. (12)), and the director sets at the Leslie angle θ_0 virtually through the entire thickness l. Conversely, if the shear stress is made to vanish, the value of ξ diverges, that is, the director orientation at the wall θ_w "propagates" to the whole thickness of the sheared layer. The viscosities that would be measured in these two extreme cases are quite different. They can be calculated (as $\sigma/\dot{\gamma}$) from Eq. (11) by setting (together with $\dot{\theta} = 0$) $\theta = \theta_0$ or $\theta = \theta_w$, respectively. In conclusion, the nematic layer will show a viscosity that varies with the applied shear stress (or shear rate), that is, the nematic appears to be non-Newtonian.

We are dealing here with a peculiar variety of non-Newtonian behavior, however. As a rule, non-Newtonian fluids are expected to be "intrinsically" nonlinear, in the sense that their viscosity, although varying with shear rate (or shear stress), does not depend on "external" variables, such as an apparatus dimension for example. Quite to the contrary, the nonlinearity of LE nematics is sensitive to such an external variable. Indeed, the scaling law described by Eq. (15) shows this peculiarity. The apparent viscosity $\sigma/\dot{\gamma}$ is a function of the Ericksen number $\dot{\gamma}\eta l^2/K$, and therefore, for a constant $\dot{\gamma}$, the viscosity will vary with the gap thickness l. On the contrary, by varying the shear rate, but if at the same time l is changed in order to maintain Er a constant, then the apparent viscosity can be kept fixed. The latter situation corresponds to self-similarity of the director profiles across the gap [1].

We conclude this section by renewing the emphasis on the intrinsically linear character of LE theory. The reader should not be misled by nonlinearities due to coupling of the velocity and director fields which are, in a way, secondary. It remains true that, in spite of the complications due to the unavoidable introduction of vector $\mathbf{n}$, LE theory is but the analogue for a nematic continuum of the Newtonian fluid assumption for the isotropic one. Thus, just as ordinary non-Newtonian polymeric liquids approach linearity in the limit of slow flows, one would expect that LCPs obey LE theory in the same limit. Unfortunately, further complications may arise, as discussed in the following section.

E. Frank Elasticity and Defects

It is a fact that nematic samples often contain "defects", lines or points where the director field is singular [3,18], therefore called "disclinations". Figure 1.2 shows some director patterns due to typical defects; right at the singularity, the director is of course undefined.

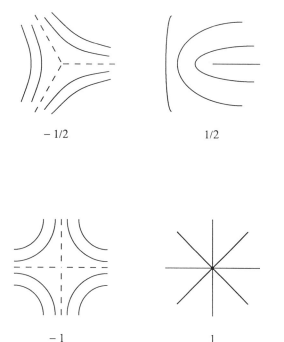

$-1/2$ $1/2$

-1 1

Figure 1.2 Director field around typical "wedge" defect lines in nematics. The defect lines are orthogonal to the plane of the figure; the positive and negative numbers indicate the "defect strength" [3]. In the same reference [3], the more complex "twist" defect lines (often encountered in LCPs) are explained.

It can be readily verified that, while approaching the singularity, $\nabla\mathbf{n}$ diverges and the energy density of Frank's elasticity (see Eq. (3)) becomes infinite. What is perhaps even more disturbing is that such a divergence is often nonintegrable; the overall Frank's energy of a sample volume containing a defect line is itself infinite. Needless to say, these divergencies frustrate the possibility of studying defect configurations by using minimum energy arguments.

On second thoughts, however, the fact that Frank's energy density diverges while approaching a defect is less surprising, and less worrysome, than at first sight. We have noted previously that the quadratic law of Eq. (3) implies that the orientational distortions are small. Now, whatever small may mean for a distortion (a definition to be given later), there is no question that $\nabla\mathbf{n}$ approaching infinity can only mean that distortions become very large. Hence, the infinite energy paradox is eliminated by the simple consideration that Eq. (3) cannot hold true in close proximity of defects; a different expression for the energy density will prevail at large distortions, such that no divergencies in fact arise because of defects.

We have thus found out another important limitation of the LE theory. Not only is the viscous part of the theory limited to the linear case (small molecules or slow flows), the theory is also fundamentally uncapable of dealing with defects, since these inevitably give rise to large orientational distortions. Such a limitation is particularly serious for LCPs, which are known to be defect-laden more often than not [5,19].

What then is the status of LE theory in the context of LCP flow behavior? The answer is at least twofold. On the one hand, whatever theory is developed to deal with the intrinsic nonlinearities arising in fast flows and/or with the problem of large orientational distortions (defects), it must include LE theory as a limiting case. In other words, the work of Frank, Leslie and Ericksen forms a sort of "ground-state" theory of nematics. Secondly, but equally importantly, the theory can be applied as such in some selected situations. For example, in order to decide on the important question of whether or not LCPs are tumbling nematics in their "ground state", defect-free samples of LCPs have been prepared, and a slow shear deformation has been applied to them [20,21]. Even in defect-laden samples, LE theory is expected to apply to the material in between defects, as long as the flow is not too fast. Indeed, we often refer to this basic theory in the rest of our presentation, either to find confirmations or to emphasize differences.

It is anticipated, however, that we certainly need to go beyond LE theory in order to deal with fast flows and/or defects. In this regard, it is perhaps better to make it clear from the start that the best (if not the only) way of dealing with constitutive theories that want to go beyond the "ground state" is that of following a "molecular" approach. Continuum mechanics cannot handle constitutive nonlinearities in any effective, predictive way [22]. Arbitrary assumptions, very hard to control and interpret, become necessary sooner or later. Molecular models are better in that respect; simplifying assumptions, if and when required, usually have a clear physical interpretation. Comparison with experiments of predictions obtained by a molecular theory can be meaningful even in case of disagreement; the discrepancies may serve to indicate which aspect of the model requires to be modified. Molecular theories of LCP flow behavior are now considered.

II. THE ROD-LIKE MOLECULAR MODEL

The only molecular theory of LCP rheology that has reached a satisfactory level of development is based on the rigid rod-like model of the polymer molecule. Polymer chemists will certainly find the rigid rod a very crude model of their sophisticated molecules; at the same time, however, they would readily agree that some degree of axial rigidity is the distinctive property of main-chain LCPs. Only with such a simplified version of the actual molecular configuration can one manage the complex dynamic behavior displayed by these materials.

The great advantage of the rod-like model is that the molecular conformation is fixed once and for all, and therefore, among the "molecular coordinates", only position and orientation in space remain to be specified. Since nonpolar molecules are considered here, orientation in space is described by a unit pseudo-vector **u**. The "molecular" vector **u** should not be confused with the director **n** of the continuum theory. As shown in the following, the director can in fact be related to an averaging over the **u**'s; just because such an averaging over molecules is implied, vector **n** is sometimes referred to as the "molecular director". No confusion should be made, however, between an individual orientation **u** and the collective orientation **n**.

Individual molecular orientations generally show a spread, even in highly ordered nematic phases. The dispersion of orientations is suitably described by a distribution function $f(\mathbf{u})$ which gives the probability density that a molecule be found with its orientation at **u**, that is, if Ω indicates solid angles, $f(\mathbf{u})\, d\Omega$ measures the fraction of molecules whose orientation falls in the neighborhood $d\Omega$ of a given **u**. The orientational distribution function $f(\mathbf{u})$ plays a central role in the theory of rod-like molecules. The evolution equation for such a function is presented in Section II.A; Sections II.B and II.C complete the molecular picture and show how to link the model to macroscopic properties; finally, Sections II.D, II.E and II.F illustrate the predictions and compare them with the observed flow behavior.

A.   The Equation for the Orientational Distribution

Under equilibrium conditions, distribution functions are known in general terms. In the isotropic phase in particular, $f(\mathbf{u})$ is obviously a constant since all directions are equally probable. The value of this constant is $1/4\pi$ in view of the fact that the integral over the full solid angle must be unity (normalization). In a nematic phase, $f(\mathbf{u})$ is no longer a constant, because in fact a preferred direction does exist. Fig. 2.1 shows schematically the dispersion around a preferred orientation, the latter defining the director **n**. Since no other preferred direction exists, vectors **u** are distributed "uniaxially" about **n**, that is, if **n** is used as the polar axis, the orientational dispersion shows no dependence on the azimuthal angle.

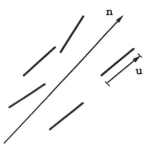

Figure 2.1 The unit vector **u** indicates "individual" molecular orientation. The director **n** gives the "average" orientation.

Because equilibrium distributions are Boltzmann distributions in all cases, we may write

$$f(\mathbf{u}) = C \exp\left(-\frac{V(\mathbf{u})}{kT}\right) \tag{17}$$

where C is a normalization factor, kT is Boltzmann's constant times the absolute temperature, and $V(\mathbf{u})$ is the "nematogenic" potential. Indeed, the molecular alignment characteristic of a nematic phase arises because of some kind of interaction between molecules, which takes the form of a "mean-field" potential; $V(\mathbf{u})$ gives the average action of all other molecules upon the one oriented along **u**. We will see, in a later section, specific forms for this potential. It should be added that Eq. (17) also holds true for the case of an "external" field (e.g., magnetic), or for the "total" potential, the sum of the mean-field contribution and the external contribution.

Moving on to dynamics, a time dependent distribution function $f(\mathbf{u}, t)$ must be introduced. First considering the isotropic phase, the simplest example of rod-like molecule dynamics is the relaxation of some orientational disturbance (Kerr effect). Assume that a magnetic field, say, has altered the uniform orientational distribution of the isotropic phase; when the external field is switched off, the system relaxes towards isotropy as a consequence of thermal agitation. Such an effect is described by the "diffusion" equation

$$\frac{\partial f}{\partial t} = \nabla_{\mathbf{u}} \cdot (D \nabla_{\mathbf{u}} f) \tag{18}$$

The coefficient D in this equation is a "rotational" diffusivity, representing the mean square angle spanned per unit time by a rod-like molecule undergoing Brownian motion. Similarly, the operator $\nabla_{\mathbf{u}}$ is a gradient in **u**-space, that is, over the unit sphere of orientations (see Appendix A).

The right-hand side of Eq. (18) has not been written as $\nabla_u^2 f$ (where ∇_u^2 is the Laplacian in u-space) because, in the general case, D is not constant with u (see below).

In dilute solutions, the rotational diffusivity of the rod-like molecule is u-independent, and attains its maximum value D_0, given by (see, e.g., [6])

$$D_0 = \frac{kT}{\zeta} L^{-3} \ln\left(\frac{L}{d}\right) \qquad (19)$$

where L and d are the rod length and diameter, respectively, and ζ is the friction coefficient per unit rod length, that is, to within a numerical factor, ζ is the viscosity of the solvent. Since the logarithmic term depends weakly on the rod geometry, D_0 (essentially) varies inversely with the third power of the rod length L, and is independent of the transversal dimensions of the rod-like polymer molecule.

In more concentrated solutions, the rotational diffusivity D_c is strongly affected by the "entanglements", that is, by the fact that Brownian rotations are hindered because of collisions among rods. The effect of concentration on the rotational diffusivity in isotropic conditions has been estimated by Doi [23] as

$$D_c = \frac{D_0}{1 + \beta c^2 L^6} \qquad (20)$$

where c is the polymer concentration and β is a numerical factor. Although β has been found to be a small number (see [6]), Eq. (20) shows that, above some critical concentration proportional to $1/L^3$, D_c/D_0 drops considerably.

Doi and Edwards [24] extended the entanglement calculations to the anisotropic case, showing two new effects. On the one hand, the rotational diffusivity becomes u-dependent. Secondly, and in fact more importantly, an u-averaged value of D is larger in the anisotropic situation than in the isotropic one. This increase in diffusivity is brought about by the fact that, the more parallel the rods become, the less probable are the collisions. The increase in D with increasing anisotropy is particularly evident at the transition to the nematic phase.

The u-dependence of D in the ∇_u-operator of the diffusion equation is usually neglected; recent calculations have confirmed that this approximation is permissible [25]. Conversely, the fact that the rotational diffusivity in concentrated systems is a functional of the distribution cannot be neglected. Indeed, the easy processability of LCPs is ultimately

350   GIUSEPPE MARRUCCI AND FRANCESCO GRECO

due to this property; the low viscosity of the nematic phase follows directly from the large molecular diffusivity in the oriented state. A specific form for the functional dependence of D is indicated in a later section (see Eq. (36)).

The relaxation equation, Eq. (18), can be generalized to include some external forcing action. For the specific case of a flow of interest here, Eq. (18) becomes

$$\frac{\partial f}{\partial t} = D\nabla_u^2 f - \nabla_u \cdot (f\dot{u}) \tag{21}$$

where the new term contains the rate of change $\dot{u}$ of a rod orientation u "convected" by the flow

$$\dot{u} = \kappa \cdot u - (u \cdot \kappa \cdot u)u \tag{22}$$

Here, κ is the velocity gradient, and the second term in Eq. (22) arises from the constant length condition (see Fig. 2.2). Equations (21) and (22) were written long ago by Kirkwood and Auer [26] with reference to dilute solutions. The same equations also apply to the concentrated case, however, as long as the effect of rod-rod interactions is limited to the change in rotational diffusivity discussed above.

Finally, when rod–rod interactions also produce thermodynamic effects, the diffusion equation becomes

$$\frac{\partial f}{\partial t} = D\nabla_u \cdot \left(\nabla_u f + \frac{1}{kT} f\nabla_u V \right) - \nabla_u \cdot (f\dot{u}) \tag{23}$$

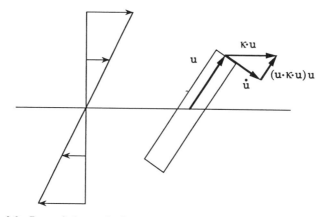

Figure 2.2 Rate of change $\dot{u}$ of a rod u in a velocity gradient κ (see Eq. (22)). The component $(u \cdot \kappa \cdot u)u$ must be subtracted from $\kappa \cdot u$ to obtain $\dot{u}$.

where V is the same mean-field nematogenic potential mentioned previously or, to be precise, the total potential, that is, the sum of a mean field resulting from molecular interactions and of external contributions, if any are present. Equation (23) is the basic equation for the description of the dynamics of rod-like polymers in both the isotropic and the nematic phases. As will be shown in the following sections, it adequately describes the complex nonlinearities encountered in fast flows of LCPs. Of course, Eq. (23) also includes the equilibrium result given by Eq. (17); indeed, if there is no flow, and the time dependence is extinguished, the Boltzmann distribution is recovered from equating the term $(\nabla_u f + f\nabla_u V/kT)$ to zero. Although Eq. (23) belongs to the class of dynamic equations named after Smoluchowski, the merit of first writing this equation with a nematogenic potential, and of recognizing its role in predicting rheological properties of nematic phases, goes to Hess [27] and, even more so, to Doi [28].

B. From the Distribution to the Stress Tensor

The distribution function obtained by solving a dynamical equation is used to calculate averages that correspond to macroscopic observables. Since the only molecular coordinate in our model is the orientation $\mathbf{u}$, an observable quantity Q is obtained from the appropriate molecular expression $q(\mathbf{u})$ weighted with the distribution function, that is,

$$Q(t) = \int q(\mathbf{u})f(\mathbf{u}, t)\, d\Omega \equiv \langle q \rangle \tag{24}$$

where it is understood that the integral extends over the full solid angle, and $f(\mathbf{u}, t)$ is normalized. The symbol $\langle\ \rangle$ typically indicates molecular ensemble averages.

A very important average is the so-called "order parameter tensor" $\mathbf{S}$ given by

$$\mathbf{S} = \langle \mathbf{uu} \rangle \tag{25}$$

which is the second-order moment of the orientational distribution. Tensor $\mathbf{S}$ is obviously symmetric and positive-semidefinite; hence, its geometrical representation is an ellipsoid. Because the trace of $\mathbf{S}$ is always unity, a relationship exists between the lengths of the three principal axes of the ellipsoid.

We have mentioned previously that, in an equilibrium nematic phase, the molecular distribution is uniaxial about the director $\mathbf{n}$; correspondingly, the ellipsoid representing $\mathbf{S}$ is itself axially symmetric, that is, the two principal axes orthogonal to **n** have equal length. Under these conditions, **S** can be written in the form

$$\mathbf{S} = S(\mathbf{nn} - 1/3\boldsymbol{\delta}) + 1/3\boldsymbol{\delta} \tag{26}$$

where the scalar factor S is called the "order parameter" of the nematic phase. Equation (26) satisfies the condition of unit trace for whatever value of S, $\boldsymbol{\delta}$ being the unit tensor. It is readily verified that the condition of positive-definiteness of tensor **S** restricts the scalar S to the range $-1/2 < S < 1$.

To understand the physical meaning of the order parameter S, it is expedient to replace **S** in Eq. (26) with its definition as the molecular average $\langle \mathbf{uu} \rangle$; then, by taking the scalar product with the dyad **nn**, one obtains

$$\langle (\mathbf{u} \cdot \mathbf{n})^2 \rangle = \langle \cos^2\varphi \rangle = \frac{2}{3}S + \frac{1}{3} \tag{27}$$

where φ is the angle that a molecule makes with the director. Therefore, if all molecules are exactly parallel to one another, $\langle \cos^2\varphi \rangle = 1$ and $S = 1$ (a "perfect" nematic). Conversely, should all molecules lie in planes orthogonal to the director, $\langle \cos^2\varphi \rangle = 0$ and $S = -1/2$. Finally, for random molecular orientations, one finds $\langle \cos^2\varphi \rangle = 1/3$, hence $S = 0$. The negative range of S-values in fact corresponds to unstable conditions for an equilibrium nematic phase; thus, the physically significant range of the order parameter of rod-like molecules is $0 < S < 1$. The lower limit $S = 0$ obviously corresponds to isotropy; the ellipsoid of tensor **S** becomes a sphere, and the director is undefined. The larger the value of S, the smaller is the molecular spread around the director, and the ellipsoid becomes more and more elongated.

Under dynamic conditions, tensor **S** generally loses the uniaxial symmetry. Postponing detailed discussions on this important effect, the concept can be illustrated here on an intuitive basis. Consider, for example, a flow-aligning nematic; we know already that the director sets at the Leslie angle with the shear direction in such a case (see Fig. 1.1). One might then imagine that the individual orientations of the molecules are uniaxially distributed around the Leslie angle with the same spread as at equilibrium. This image is not correct, however. The shear flow exerts a torque on molecules, resulting in a modification of the angular spread in the shear plane, but not orthogonally to it; the uniaxial symmetry of the orientational distribution is therefore not maintained.

Tensor **S** is a macroscopically observable quantity. Indeed, to within a constant factor, it coincides with the optical or birefringence tensor. Under equilibrium conditions, in view of the uniaxial symmetry, only **n** and the scalar S are relevant. The latter quantity can be measured in various ways, for example, by NMR techniques [1–3]. In some approximate sense, tensor **S** is often considered as a "substitute" for the whole distribution function, although of course the infinite sequence of moments, not just the second order one, would be required.

Before concluding on **S**, it should be mentioned that the most common choice for the order parameter tensor is $\mathbf{Q} = \mathbf{S} - (1/3)\boldsymbol{\delta}$, which is a traceless tensor. Although **Q** and **S** are mathematically equivalent, we prefer to work with **S**. Indeed, since **S** is positive-definite, it can be easily visualized as an ellipsoid, reverting to a sphere under isotropic conditions.

The next observable of great relevance is the stress tensor **σ**. Two contributions to the stress must be distinguished, one of them of a purely viscous nature, the other intimately related to the free energy of the system. The viscous contribution $\boldsymbol{\sigma}_v$ arises from the relative motion between the rod and the surrounding medium due to molecular inextensibility. Figure 2.2 shows that the longitudinal velocity per unit rod length is given by $(\mathbf{u} \cdot \boldsymbol{\kappa} \cdot \mathbf{u})\mathbf{u}$, from which it follows that

$$\boldsymbol{\sigma}_v = \frac{ckT}{D_0}\, \boldsymbol{\kappa} : \langle \mathbf{uuuu} \rangle \qquad (28)$$

where use has been made of Eq. (19) linking the friction coefficient per unit rod length ζ to the dilute solution diffusivity D_0. The appearance of the diffusivity D_0 does not imply that the validity of Eq. (28) is limited to the dilute case; indeed, because the relative motion generating $\boldsymbol{\sigma}_v$ is "along the rod", the hindrance to rotations that decreases the angular diffusivity in concentrated systems is irrelevant here. The purely viscous character of $\boldsymbol{\sigma}_v$ is shown by the fact that, upon switching off the velocity gradient $\boldsymbol{\kappa}$, $\boldsymbol{\sigma}_v$ instantly vanishes.

The thermodynamic (or "elastic") contribution to the stress tensor is calculated by using the "virtual work" method, that is, by imagining that an infinitesimal instantaneous deformation is applied to the system and by equating the corresponding change in free energy to the work of the stress. Temporarily assuming that the potential V acting on the molecules is external, the expression of the free energy per unit volume is given by

$$A = c(kT\langle \ln f \rangle + \langle V \rangle) \qquad (29)$$

where the first term arises from the orientational entropy of the rods. Since the virtual deformation is instantaneous, the free-energy change can be calculated by assuming that all rods rotate "affinely" with the deformation, which gives for the stress σ_E (here again $\nabla_\mathbf{u}$ has the meaning explained in Appendix A):

$$\sigma_E = c(kT\langle\nabla_\mathbf{u}(\ln f)\mathbf{u}\rangle + \langle\nabla_\mathbf{u}V\mathbf{u}\rangle) \tag{30}$$

Using Eq. (17), it is readily verified that σ_E vanishes at equilibrium (in spite of the anisotropic conditions induced by V), as is appropriate for a liquid. Also, it can be shown [24] that, in absolute generality, the average $\langle\nabla_\mathbf{u}(\ln f)\mathbf{u}\rangle$ can be replaced by $3\langle\mathbf{uu}\rangle - \boldsymbol{\delta}$. Then, dropping the irrelevant isotropic term (in view of incompressibility), Eq. (30) is rewritten in the simpler form

$$\sigma_E = c(3kT\mathbf{S} + \langle\nabla_\mathbf{u}V\mathbf{u}\rangle) \tag{31}$$

Several other comments on the expression of σ_E are in order. First, it should be noted that the second term in Eq. (31) is generally not symmetric. Thus, to avoid possible confusion, we specify that the order of $\nabla_\mathbf{u}V$ and $\mathbf{u}$ in the average above corresponds to the convention for the stress tensor whereby σ_{xy}, say, indicates the x-component of the force on the unit surface orthogonal to the y-axis (not the other way around).

Secondly, it can be shown that Eq. (31) also applies to the case where V is a mean-field potential rather than an external one, or when it is a combination of the two. Indeed, although the contribution to free energy of a mean-field potential is $1/2c\langle V\rangle$ instead of $c\langle V\rangle$ (because the interactions must not be counted twice), the $1/2$ factor cancels out again in applying the virtual work algorithm. This occurs because the mean-field potential depends on $\mathbf{u}$ "twice", so to speak; once directly, and a second time as a functional of the whole distribution, because the mean-field expresses the influence on the given rod of all others. That Eq. (31) must apply for whatever potential can also be understood from the "mechanical" viewpoint; since the value of $\nabla_\mathbf{u}V$ calculated at $\mathbf{u}$ expresses the torque on a molecule oriented at $\mathbf{u}$, the contribution to the stress tensor of such molecules cannot discriminate on the source of the potential acting on them. It is noted, however, that the average $\langle\nabla_\mathbf{u}V\mathbf{u}\rangle$ becomes a symmetric tensor for a mean-field potential. This symmetry is dictated by the fact that no net torque can exist in the absence of external potentials.

The last comment on Eq. (31) has to do with the name of σ_E, usually called the "elastic" stress. To appreciate the elastic (or rather the "viscoelastic") character of σ_E, let us imagine that the system has been brought away from the equilibrium state by, say, a flow process. Upon switching off the flow, the stress σ_E does not vanish instantly. For σ_E to disappear, it is required that the molecular orientations go back to their equilibrium distribution; a "relaxation" process (due to Brownian motion) which will occur over a nonzero time lapse (of the order of $1/D$, cf. Eq. (23)). The name "elastic" used in molecular modeling should not be confused, therefore, with that of Frank elasticity; as previously discussed, the latter is related to "spatial" gradients which are completely ignored in the molecular theory so far presented. The viscoelasticity embodied by Eq. (31) is entirely constitutive; the time constant arising here, $1/D$ depends on the material and on nothing else.

The concept of orientational relaxation regulated by D can be used to estimate the order of magnitude of σ_E in a flow process. Indeed, if κ is the magnitude of the velocity gradient $\boldsymbol{\kappa}$ of the flow, the nondimensional ratio κ/D can be taken to represent the competition between the tendency of the flow to carry the system away from equilibrium and that of Brownian motion to bring it back. Thus, κ/D must also represent the offset from equilibrium when a balance between the two counteracting effects is reached, that is, when a steady state is achieved. It can be concluded that the order of magnitude of the elastic stress must be $\sigma_E = ckT(\kappa/D)$. This estimate is compared with that for the viscous stress, which, from Eq. (28), is given by $\sigma_v = ckT(\kappa/D_0)$. In view of the large difference between D and D_0 in concentrated systems, the viscous contribution $\boldsymbol{\sigma}_v$ is usually negligible with respect to the elastic one $\boldsymbol{\sigma}_E$. In highly ordered nematics, however, D approaches D_0, and the viscous contribution may again become important (as in dilute solutions). Furthermore, when nonlinear effects at high κ-values are considered (see a later section), it can be shown that $\boldsymbol{\sigma}_E$ may approach a saturation, thus "loosing pace" with respect to $\boldsymbol{\sigma}_v$, the latter continuing to increase linearly with κ. In some cases, therefore, $\boldsymbol{\sigma}_v$ must also be accounted for.

C. Nematogenic Mean-Field Potential

Before attempting to solve the dynamical equation, Eq. (23), a form for the nematogenic potential must be chosen. Starting from the pioneering works of Onsager [29] and Flory [30] on the free energy of rod-like systems, several forms of nematogenic potentials have been proposed, the motivation being a better prediction of the equilibrium properties of either small molecule nematics or of LCPs. In studying LCP dynamics, however, due to the intrinsic additional complexities, only the simplest (or else the most fundamental) choices were maintained.

The simplest and most popular form of nematogenic potential is the "ansatz" proposed by Maier and Saupe [31] for the case of small molecule nematics. For the uniaxial situation of the equilibrium state, the original proposal was of the form

$$V(\mathbf{u}) = -kTUS(\mathbf{n} \cdot \mathbf{u})^2 = -kTUS\cos^2\varphi \qquad (32)$$

where U is a temperature-dependent nondimensional intensity factor (a function also of concentration in lyotropic systems). It should be noted that the magnitude of the potential depends on the degree of order, through the order parameter S, vanishing under isotropic conditions. Of course, $V(\mathbf{u})$ is minimum at $\mathbf{u} = \mathbf{n}$, that is, along the director ($\varphi = 0$); the quadratic dependence on $\mathbf{n} \cdot \mathbf{u}$ is respectful of the "quadrupolar" nature of the nematic interaction. Maier and Saupe demonstrated that, by increasing the potential intensity U, a first-order transition from the isotropic phase to the nematic one occurs at $U = 6.8$, the order parameter at the transition jumping from zero to $S = 0.43$.

The Maier–Saupe expression has been extended by Hess [27] and Doi [28] to the biaxial situation, generally encountered in dynamics, in the form

$$V(\mathbf{u}, t) = -kTUS(t) : \mathbf{uu} \qquad (33)$$

which, in view of Eq. (26), properly reduces to Eq. (32) in the uniaxial case ($\mathbf{u}$-independent terms can be dropped from $V(\mathbf{u})$ as long as only $\nabla_{\mathbf{u}}V$ is relevant, as in Eqs. (23) and (31)). The potential given by Eq. (33) will be most frequently encountered in the following.

The alternative form of nematogenic potential used in some dynamic studies goes back to the work of Onsager [29], and is based on the excluded volume interaction between long rigid rods in dilute solutions. Such an excluded volume crucially depends on the relative orientation of the interacting rods, ranging from a value of order d^2L when the two rods are parallel (d and L are rod diameter and length, respectively) to dL^2 in the perpendicular case. Thus, if L/d is a large number, d^2L can be neglected with respect to dL^2, and the excluded volume of arbitrarily oriented rods can be written with a good approximation as

$$v_{\text{excl}} = dL^2 \sin\alpha \qquad (34)$$

where α is the angle between the two interacting rods. From Eq. (34), an interaction free energy term is obtained by averaging over orientations; hence, the orientational potential $V(\mathbf{u})$. Unfortunately, in spite of the simplicity of Eq. (34), the ensuing form for $V(\mathbf{u})$ proves difficult to use. The expansion of $V(\mathbf{u})$ in spherical harmonics is often used in computations (see, e.g., [28]); note that the first term of such an expansion has the form of Eq. (33).

In spite of its greater complexity, the full Onsager potential has been used in some cases to obtain dynamic predictions (see below). Although based on binary interactions only, the Onsager potential is probably the most appropriate choice for the case of lyotropic LCPs having a really rigid rod-like structure. For semi-rigid polymers, especially in the thermotropic case, nematogenic interactions other than the excluded volume one are expected to be important, so that the Onsager form is certainly not expected to apply as such; like small molecule nematics, semi-rigid LCPs are perhaps better described by using the simpler Maier–Saupe form [1,2]. One way or the other, the results to be discussed in the following show that, fortunately, the choice of a particular form for the nematogenic potential does not appear to be crucial, at least insofar as qualitative features of flow behavior are concerned. Thus, also in view of the fact that dynamic calculations are limited from the very start (the choice of the molecular model) to the essential aspects, the simplest form of $V(\mathbf{u})$ will often be preferred.

D. Slow Flow Results

In order to obtain rheological predictions from the molecular model, the procedure is as follows. The velocity gradient tensor $\boldsymbol{\kappa}$ is assigned, either as a constant value or as a function of time (thus fully specifying $\dot{\mathbf{u}}$, see Eq. (22)); the distribution function $f(\mathbf{u}, t)$ is then derived by integrating Eq. (23), so that the averages expressing the stress tensor can be calculated.

The difficult step in the procedure is that of solving the nonlinear differential equation, Eq. (23). The nonlinearity appears through the $f\nabla_{\mathbf{u}}V$ term of this equation, as a result of V being a mean-field potential, itself related to the function f through some integral over orientations. For example, if the Maier–Saupe form of the potential is adopted, as given by Eq. (33), the tensor $\mathbf{S}$ appearing there is in fact the integral (cf. Eqs. (24) and (25)):

$$\mathbf{S}(t) = \int \mathbf{u}\mathbf{u}f(\mathbf{u}, t)\, d\Omega \tag{35}$$

The situation simplifies in slow flows, however, because it is expected that the distribution function will differ only slightly from an equilibrium distribution in such a case; therefore, by expanding about an equilibrium state, a linear problem is recovered from Eq. (23). How "slow" is a flow is determined by the previously mentioned ratio κ/D. If κ/D is much smaller than unity, then the offset from equilibrium is also very small, in fact proportional to κ/D.

The fact that Eq. (23) becomes linearized in the limit of slow flows ultimately implies a linear relationship between the stress tensor and the velocity gradient, that is, the situation that is assumed from the start in the continuum theory of Leslie and Ericksen. In other words, the LE theory is properly recovered as the slow flow limit of the rod-like molecular theory. An important "caveat" is in order, however. The molecular theory so far presented ignores spatial dependencies; thus, Frank elasticity or, more generally, the influence of orientational distortions, is not yet included. From the slow flow limit of the molecular theory considered so far, we therefore recover a "spatially uniform" version of LE theory, that is, only the LE viscous stress of Eq. (1) together with the viscous torque terms of Eq. (6). On the other hand, the six Leslie coefficients are predicted explicitly from the molecular theory; that is, as shown below, they are all expressed in terms of the characteristic viscosity ckT/D, and of the order parameter S of the equilibrium state.

The mathematical problem of LCPs in slow flows, although a linear one, remains nontrivial. One reason is that the equilibrium state (about which to expand) is orientationally degenerate. In other words, while the equilibrium condition dictates a unique form (the Boltzmann expression) to the orientational distribution of the rods "about" the director, the director $\mathbf{n}$ itself remains undetermined. On the other hand, we know from LE theory that the flow, no matter how slow, has a strong effect on the director. If a steady state is reached, $\mathbf{n}$ will assume a particular orientation with respect to the velocity gradient; alternatively, it may keep tumbling. In all cases, in the absence of external potentials, the pair $\mathbf{n}$ and $\dot{\mathbf{n}}$ will be such as to satisfy the condition of zero torque, the latter condition also implying a symmetric stress. On the contrary, in order to derive the nonsymmetric general expression of the stress as given by Eq. (1), it is required that $\mathbf{n}$ be assigned independently of the velocity gradient. A simple way out of all these difficulties is that of superimposing an external field, a magnetic one for example, which will fix $\mathbf{n}$, both at equilibrium and when flow is switched on. A supplementary advantage is that a steady state is automatically guaranteed, also for tumbling nematics.

The problem of rod-like LCPs in slow flows was solved by Semenov [32], and independently by Kuzuu and Doi [33]. Although in both these works, the formal solution for an arbitrary mean-field potential is derived, Semenov eventually shows detailed results obtained by using the Maier–Saupe form, while Kuzuu and Doi opt for the Onsager potential. As anticipated above, the Leslie coefficients are all expressed in terms of a "reference nematic viscosity" $\eta = ckT/D$, where D is the rotational diffusivity in the equilibrium nematic state. For rod-like polymers, the relationship between D and the corresponding isotropic value D_c, given by Eq. (20), had been previously derived by Doi and Edwards [24] in the form

$$D = D_c \left[\frac{4}{\pi} \int\int \alpha(\mathbf{u}, \mathbf{u}') f(\mathbf{u}) f(\mathbf{u}') \, d\Omega \, d\Omega' \right]^{-2} \cong D_c (1 - S^2)^{-2} \quad (36)$$

where $\alpha(\mathbf{u}, \mathbf{u}')$ is the angle between the directions $\mathbf{u}$ and $\mathbf{u}'$, and the latter expression, containing the order parameter S, is an approximation suggested by Doi [28].

Kuzuu and Doi show that three of the Leslie coefficients, namely α_1, α_2, and α_5, are of order η, while α_3, α_4, and α_6 are much smaller. As regards signs, α_1, α_2 and α_6 are found to be positive, while α_3, α_4, and α_5 are negative. All these results apply to the whole concentration range corresponding to the nematic phase. In particular, the ratio α_3/α_2 comes out to be always negative; thus, the rod-like molecular theory would predict that LCPs are invariably tumbling nematics (compare Section I.B). It should be mentioned, however, that the Onsager potential used by Kuzuu and Doi also predicts that the smallest value of the order parameter (i.e., at the isotropic–nematic transition) is as large as ca. 0.8. Perhaps, the tumbling character of the nematic is to be associated essentially with large values of S.

This conjecture is reinforced by the results of Semenov who, using the Maier–Saupe potential, was able to explore a wider range of S-values (the transition value is as small as 0.43 in this case). Indeed, Semenov found that α_3/α_2 is positive below $S = 0.53$ and negative above (see Fig. 2.3). Two important conclusions can then be drawn. First, nematics are expected to be flow-aligning at low values of the order parameter, whereas they tend to be of the tumbling type when more perfectly ordered. The second conclusion is that, since LCPs usually show rather high values of S, they should exhibit the tumbling behavior.

The relevance of these conclusions was not clearly recognized at the time. From the psychological viewpoint, people tended to associate with these new materials, the LCPs, the concept of order and alignment, and thus instinctively refused the tumbling idea, which somehow implies disorder and misalignment. Secondly, a mathematical approximation introduced in the theory [27,28] had previously indicated the opposite result, namely that rod-like polymer nematics are always flow-aligning [34,35]. Notice in this regard that α_3/α_2 is very small in magnitude; thus, although the exact calculations indicated this ratio to be negative, it

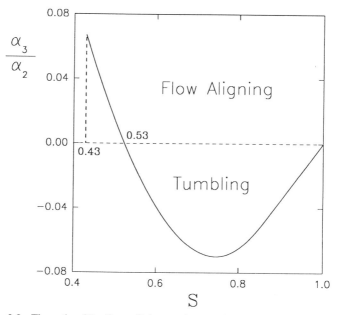

Figure 2.3 The ratio of Leslie coefficients α_3/α_2 as a function of the order parameter S for the case of the Maier–Saupe potential. Tumbling behavior is predicted for $S > 0.53$.

appeared possible that modifications of the physical assumptions of the theory (e.g., to account for polydispersity) could easily determine a crossing of the zero for this quantity, that is, a change of type for the nematic. The ambiguity on whether or not LCPs are in fact tumbling nematics was removed by the results on nonlinear properties (to be presented in the next section), which link a simple rheological observable to the nature of the nematic. Furthermore, recent optical experiments on carefully prepared, defect-free ("monodomain") samples directly indicate the tumbling character of the LCPs [20,21].

E. Nonlinear Behavior of Flowing LCPs

The nonlinear dynamical equation, Eq. (23), cannot be solved in closed form, not even with the simplest choice for the nematogenic potential. Thus, if one wants to avoid the use of complex numerical procedures, mathematical approximations seem necessary. We have mentioned, however, that one such approximation proved not to work properly in the limit of slow flows. A different approach to obtain an approximate solution was later proposed, which is based on a fictitious restriction to two dimensions. With specific reference to a shear flow, Marrucci and

Maffettone [36] assumed that all rods were parallel to the shear plane, so that the molecular orientation **u** is determined by a single angle θ in that plane. Although the mathematics are considerably simplified in this way, the essential physics is hopefully preserved; in analogy with the results of LE theory for the director motion, the off-plane component of the orientation possibly plays a secondary role.

The two-dimensional version of Eq. (23) for a shear flow is [36,37]

$$\frac{\partial f}{\partial t} = D \frac{\partial}{\partial \theta} \left(\frac{\partial f}{\partial \theta} + \frac{f}{kT} \frac{\partial V}{\partial \theta} \right) + \dot{\gamma} \frac{\partial}{\partial \theta} (f \sin^2\theta) \qquad (37)$$

where $\dot{\gamma}$ is the shear rate, θ is the angle made by the rod with the shear direction, and of course $f(\theta, t)$ replaces $f(\mathbf{u}, t)$ here. In two dimensions, the Maier–Saupe potential takes the form

$$V(\theta, t) = -kTU[\langle \cos(2\theta) \rangle \cos(2\theta) + \langle \sin(2\theta) \rangle \sin(2\theta)] \qquad (38)$$

The predictions of the set of Eqs. (37) and (38) are considered first. We then report the results of Larson [38] who developed a numerical code to deal with the complete three dimensional problem.

1. Dynamic Transition From Tumbling to Flow Alignment

For time independent values of the shear rate $\dot{\gamma}$, the nonlinear problem described by Eqs. (37) and (38) contains two, and only two, nondimensional parameters; the intensity U of the potential and the ratio of $\dot{\gamma}$ to some characteristic rotational diffusivity (such a ratio, a nondimensional shear rate, will still be indicated as $\dot{\gamma}$ in the following). The U parameter is entirely constitutive, while the second, through the shear rate, is a "process" parameter. Marrucci and Maffettone [36] first looked for the conditions, if any existed, under which the problem admitted a stationary solution, that is, a time independent orientational distribution $f(\theta)$. They found two quite distinct situations in which a steady state is achieved; these are described separately below.

For U smaller than $U^* = 2.41$, the stationary state exists for all value of $\dot{\gamma}$. Because the interval $0 < U < 2$ corresponds (in two dimensions) to the isotropic phase, and $U > 2$ to the nematic one, a first conclusion is reached: a nematic having U in the interval $2 < U < 2.41$ achieves a steady state in a shear flow, that is, it behaves in a "flow-aligning" way for all values of $\dot{\gamma}$. If this prediction is compared with the results of Semenov [32] for the linear limit (see previous section), we find confirmation that flow-aligning behavior prevails for nematics of low order parameter; indeed, for the two-dimensional case, $U = 2.41$ corresponds to an equilibrium value of the order parameter, $S = 0.55$, very close to $S = 0.53$ found by Semenov in three dimensions for the value separating flow-aligning nematics from tumbling ones (see Fig. 2.3). The nonlinear calculations add some new features to the flow-aligning behavior, however; at high shear rates, the Leslie angle becomes shear rate dependent, progressively decreasing as the shear rate increases. It should be mentioned in this regard that a director (and therefore a Leslie angle) can still be defined in the nonlinear situation, in spite of the fact that the symmetry of the distribution function is now destroyed. For example, the director can be defined as the "main" of the three principal directions (two in two dimensions, of course) of tensor **S**. From now on, it is then understood that the terms "director" and "Leslie angle" can also refer to nonlinear alignments.

For $U > U^*$, no stationary solution is found at low shear rates. The absence of a steady state in this range of parameters suggests, consistently with the linear results in three dimensions of Semenov [32] and Kuzuu and Doi [33], that tumbling behavior must prevail here; subsequent calculations for the time dependent two-dimensional problem explicitly confirmed and completed this result, adding further details (to be reported when discussing Fig. 2.4). In this range of U values, the really important new information was that a stationary solution "can still exist"

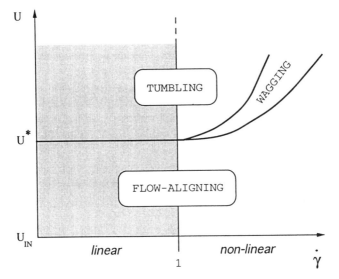

Figure 2.4 Dynamic behavior of rod-like nematics as a function of shear rate $\dot{\gamma}$ and potential intensity U. In the nonlinear range, a transition from tumbling to flow-aligning occurs by increasing $\dot{\gamma}$, for any U larger than U^*.

if and only if $\dot{\gamma}$ is larger than some critical value $\dot{\gamma}_{crit}$. Thus, for any assigned U larger than U^*, a "dynamic transition" takes place from tumbling to flow-aligning with increasing $\dot{\gamma}$. The nonlinear nature of such a transition is particularly evident in this case, because the Leslie angle right after the transition is "negative", that is, the director orients somewhat "below" the shear direction instead of above it [36]. By further increasing the shear rate, the Leslie angle grow to positive values, goes through a maximum, then asymptotically decreases towards zero; this ultimate asymptotic decrease is in fact U-independent, that is, the Leslie angle is exclusively determined by $\dot{\gamma}$ in this region. In general terms, a "competition" between shear flow and nematogenic potential can be envisaged throughout the entire range of shear rate values; at low shear rates, the potential is strong enough (in relative terms) to keep the molecules co-aligned, so that they all tumble together; at high shear rates, the potential becomes comparatively weak, and rods can rotate "individually" so that their distribution becomes stationary [39].

The role of the two parameters, U and $\dot{\gamma}$, in determining flow-aligning or tumbling is summarized, in qualitative terms, in Fig. 2.4, which is a sort of "state diagram" showing dynamic transitions as boundary curves between zones of like behavior. In the left part of the diagram, that is, at shear rates smaller than ca. 1 (we recall that $\dot{\gamma}$ is nondimensional), one must find the linear response; the nematic is flow-aligning below ($U < U^*$) and tumbling above ($U > U^*$), the boundary line being horizontal. In the right part of the diagram, as a consequence of nonlinear effects, the boundary line turns upwards. Thus, the transition to a stationary solution may occur in two ways, either by decreasing U (as in the linear case), or by increasing $\dot{\gamma}$ up to $\dot{\gamma}_{crit}$. In fact, the transition from tumbling to a stationary state (with the director fixed at the negative Leslie angle mentioned above) occurs through an intermediate dynamical state, called "wagging" by Larson [38], where the director oscillates up and down between two limiting orientations, instead of going through all values of θ as in the tumbling regime. The amplitude of this oscillation decreases with increasing $\dot{\gamma}$, up to the point where it vanishes (at $\dot{\gamma} = \dot{\gamma}_{crit}$) and the stationary solution prevails.

Notice finally that in Fig. 2.4 the lowest value of U corresponds to U_{IN}, the potential intensity at the isotropic–nematic transition. However, the diagram could in principle be extended downwards to include the isotropic phase, which indeed is always "flow-aligning". The birefringence angle of the isotropic phase (in a sense the analogous of the Leslie angle of nematics) starts from 45° in the linear limit, to progressively decrease with increasing $\dot{\gamma}$, eventually approaching zero in the same asymptotic way as the Leslie angle in the nematic phase.

2. Negative Normal Stress Effect

It is well known that polymeric fluids in shear flow not only exhibit tangential stresses but also normal ones. To be specific, and recalling that in incompressible fluids only normal stress "differences" are rheologically significant, two quantities related to normal stresses are relevant: (i) the first, or primary, normal stress difference $N_1 = \sigma_{xx} - \sigma_{yy}$, where σ_{xx} is the normal stress in the shear direction and σ_{yy} that in the shear gradient direction; (ii) the second normal stress difference $N_2 = \sigma_{yy} - \sigma_{zz}$, where z is the "neutral", or vorticity, direction. Both these differences can be experimentally determined as a function of the shear rate; the first, however, is more readily measurable, and is usually the most important.

For ordinary polymers, the first normal stress difference N_1 is always "positive", a positive sign implying that there exists a traction in the shear direction or, equivalently, a compression in the direction of the gradient. For example, in one of the typical geometries used for rheological characterization, that of a cone-and-plate rotational device, a positive sign of N_1 means that the material in the gap pushes the two platens apart; the normal force required to maintain the platens at a fixed distance is in fact used to measure N_1. The direction of the rotation being irrelevant in this experiment, N_1 must be an even function of the shear rate $\dot{\gamma}$; in the limit of vanishing shear rates, N_1 is thus predicted to be proportional to $\dot{\gamma}^2$. At high shear rates, deviations from this quadratic law are expected, and are found in fact; in ordinary polymers, however, N_1 remains an ever-increasing function of $\dot{\gamma}$ throughout.

Contrary to this simple behavior, N_1 was found to show quite peculiar features in LCPs. Kiss and Porter [40] first reported the behavior shown in Fig. 2.5. Starting from low shear rates, N_1 first increases with increasing $\dot{\gamma}$, then goes through a maximum and decreases to reach negative values; by further increasing the shear rate, N_1 comes up again into positive values, and keeps increasing thereafter. These "anomalous" results obtained by Kiss and Porter, with as many as two changes in sign of N_1 were confirmed in a variety of other systems, that is, in most lyotropic systems [41–45] and even in some thermotropic ones [46]. In some cases, depending on the shear-rate range which is accessible to measurements for a particular system, only some part of the complete N_1 pattern shown in Fig. 2.5 was determined.

Such a strange behavior of N_1, unexplained for some time, was eventually found to be well fitted by the predictions of the nonlinear theory of rod-like polymers [36,38,47]. Indeed, the negative normal stress effect is predicted to occur in a region of shear rates encompassing the dynamic transition from tumbling to flow-aligning (at $\dot{\gamma} = \dot{\gamma}_{crit}$) discussed in the previous section. In such a region of shear rates, the solution of

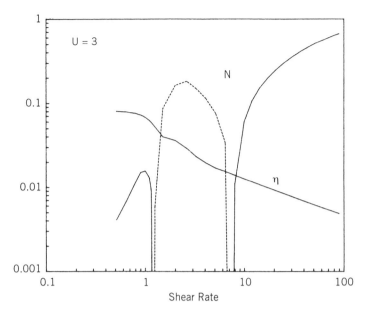

Figure 2.5   The viscosity and the first normal stress difference at large shear rates, as calculated from theory for a tumbling nematic (reprinted with permission from [47]). The predictions are in good qualitative agreement with the experimental results (see [40–46]); in particular, the central branch of normal stresses is negative.

Eq. (37) shows an orientational distribution function which departs considerably from its equilibrium "shape". As explained in [36], what generates the negative sign of N_1 is the fact that the departure is in the direction of a reduced order, that is, the spread of rod orientations is larger than at equilibrium. In contrast, for values of $\dot{\gamma}$ somewhat larger than $\dot{\gamma}_{crit}$, that is, when not only tumbling is suppressed but also the Leslie angle has become positive (see previous section), the departure from equilibrium of the orientational distribution reverts direction, that is, the orientational spread of the rods becomes smaller than at equilibrium, and progressively decreases by further increasing $\dot{\gamma}$. Correspondingly, N_1 reverts to a positive sign, and keeps increasing monotonically thereafter.

The excellent qualitative agreement between theory and experiments as regards this anomalous normal stress behavior of LCPs has an important "diagnostic" value; LCPs that show the negative normal stress effect must be tumbling nematics at low shear rates, even if their chemical structure does not strictly conform to the rod-like model, that is, even if their molecules are only semi-rigid for one reason or other. Indeed, above and beyond the detailed predictions of the theory developed so far, which might only partly apply to polymers that are not really rod-like (or that do not obey the particular form of nematogenic potential assumed in the theory), it appears sensible to link the occurrence of negative normal stresses to the tumbling nature of the nematic, and to the corresponding tendency for the orientational spread to become "larger" than at equilibrium as a nonlinear effect due to the rotational character of the shear flow. As regards the range of $\dot{\gamma}$ values where negative normal stresses are to be expected, Fig 2.5 shows that it is located soon after the beginning of the shear-thinning region of the viscosity curve, that is, when the shear rate is indeed large enough to induce significant nonlinearities. In the same shear rate range, seemingly at the transition from tumbling to wagging, the simulated viscosity curve exhibits a sort of "kink", a feature that can in fact be observed in most experimental curves, and which had been noted explicitly by Moldenaers [43].

The internal consistency of the conclusion reached so far, whereby negative normal stresses imply that the nematic is of the tumbling type at low shear rates, is confirmed by some different data, also collected by Kiss and Porter [40]. In one case, they found that the behavior of N_1 with shear rate did not conform to the complex pattern of Fig. 2.5. Rather, N_1 was found to be always positive, and kept increasing with increasing shear rate, the only anomalous feature being a "bump" in the curve at the approximate location where, by analogy with the other data, one would have expected the negative stresses. Consistently, one should infer that this particular system is not of the tumbling type. Such a conclusion can be tentatively accepted on the basis of the fact that the concentration in this system is very close to the isotropic–nematic transition; thus, the equilibrium order parameter S might well be low enough to give rise to a flow-aligning nematic throughout the entire range of shear rates (compare Figs. 2.3 and 2.4). The conjecture becomes virtually a certainty when one examines the theoretical predictions for $U < U^*$ (that is, for a nematic of the flow-aligning type) in the nonlinear range. Indeed, one finds that the N_1 curve has a shape similar to the experimental one, with a "bump" in the right place (see Fig. 2.6).

It is noted at this point that, if the above interpretation is accepted, the Maier–Saupe potential appears to be more suitable to the dynamical description of these lyotropic systems than is the Onsager potential, if only on a phenomenological basis. We have previously explained how the Onsager potential predicts that the lower value of S in the nematic phase, that is, the one at the isotropic–nematic transition, is as large as 0.8, and therefore that the "first" nematic phase is of the tumbling type. Now, although such a large value of S is close to those that are typically measured in LCPs, it remains true that, by adopting the Onsager potential in the rod-like molecular model, a curve such as that of Fig. 2.6

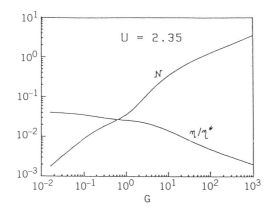

Figure 2.6   The viscosity and the first normal stress difference versus the shear rate for a nematic of the flow-aligning type, as calculated from the theory (reprinted with permission from [36]). The calculated curves in this figure are in qualitative agreement with experiments on lyotropic LCPs right after the isotropic–nematic transition (see [40, Fig. 11]).

cannot be explained. Perhaps the apparent discrepancy between large experimental values of the order parameter on the one hand, and the fact that tumbling may in some cases be suppressed on the other, can be reconciled by considering the semi-rigid nature of the actual chains. Indeed, Semenov [48] has shown that if the chain length is sufficiently larger than its persistence length tumbling is suppressed. A similar result has been recently obtained for nematic Kuhn chains, by treating their dynamics as that of elastic dumbbells [49].

We conclude this section by noting that negative normal stresses are seldom observed in thermotropic LCPs. Yet there are good reasons to believe that this relative scarcity is not due to absence of tumbling but, rather, to the fact that N_1 can only be measured up to relatively low shear rates (differently from viscosity, which is measured at higher shear rates in the capillary apparatus). We will take up this point again in a later section.

3.   The Full Three-Dimensional Results

The main results obtained with the two-dimensional version of the theory were fully confirmed by the numerical code developed by Larson [38] to cope with the three-dimensional problem. Larson used the Onsager potential instead of the Maier–Saupe one; thus, only values of the equilibrium order parameter larger than ca. 0.8 were accessible to his simulations, always corresponding to tumbling behavior at low shear rates. The features of the dynamic transitions which occur with varying shear rate, from tumbling to wagging to flow-aligning, were, however, the same as obtained in two-dimensions with the Maier–Saupe potential [36, 47]; quite similar viscosity and N_1 curves were thereby predicted. This "robustness" of the qualitative features of the theoretical results with respect to a change in nonsecondary aspects of the theory such as dimensionality and type of potential is indeed gratifying. It implies that essential aspects of the dynamical behavior of tumbling LCPs in shear flows have in fact been caught.

In three dimensions, the second normal stress difference N_2 can be added to the predictions. In a paper by Magda et al. [50], where original data are also reported (N_2 is more difficult to measure than N_1), Larson's code is used to predict N_2, and the results compare very well with the data. Also the second normal stress difference changes sign twice by varying $\dot{\gamma}$, and in the opposite way with respect to N_1. It starts being negative at low shear rates, then turns to a positive sign for intermediate values of $\dot{\gamma}$, finally becoming negative again at high ones. The values of $\dot{\gamma}$ where the sign inversions of N_2 take place fall close to, but are not quite coincident with, those of N_1. The theory appears to catch even these fine details. The three-dimensional calculations of Larson also clearly show the loss of uniaxiality of the orientational distribution function induced by the shear flow (alluded to in Section II.B). As a consequence of flow, the ellipsoid representing the order parameter tensor **S** is no longer axially symmetric about the director **n**.

We have not mentioned so far that, in either the two- or the three-dimensional simulations, whenever the regime of motion is periodic (tumbling and wagging regimes), the stresses that have been compared with the experimental data were calculated as average values over a cycle. The implications of such a procedure will be examined further on, when discussing polydomains. In the three-dimensional results of Larson briefly reported above [38,50], the cycle of tumbling (or the up-and-down oscillation of the director, called wagging) were simulated with the director in the plane of shear. Needless to say, in the two-dimensional calculations, only such a choice is available. In three dimensions, however, it is also possible to work out simulations with the director off the shear plane. This was indeed accomplished by Larson and Öttinger [25], who found a quite complex dynamic behavior. Depending on the initial value of the off-plane angle and on the shear rate, the director is either attracted to the plane of shear, where it behaves as already described, or else ends up oriented along the neutral axis (a condition called "log-rolling" by Larson and Öttinger).

A discussion on the possible relevance of these results is again deferred to the section on polydomains. Here, it may be worth emphasizing that the joint paper by Larson and Öttinger in fact reports on two separate works, insofar as each of the authors used a wholly different technique to tackle the problem. Larson solved the Smoluchowski equation of Hess [27] and Doi [28] by expanding the orientational distribution function in spherical harmonics. Öttinger adopted the technique known as "Brownian dynamics", whereby the physical process of a population of rods undergoing rotational Brownian motion, as well as interacting with each other via the mean-field potential, and being orientationally convected by the flow field, is simulated directly on the computer. With the latter technique, it is not required that the rotational diffusion coefficient D be pre-averaged, an approximation that cannot be avoided when solving the Smoluchowski equation; rather, D can be taken to be a function of $\mathbf{u}$ (following the theory of Doi and Edwards [24] for concentrated rod systems) without too much additional effort. The fact that the results obtained by Öttinger with a variable D do not differ significantly from those of Larson can be taken as a good indication that the pre-averaging approximation is indeed acceptable.

Other results, beyond those mentioned here, have been obtained by numerical simulations, in either two or three dimensions. For example, for a high enough shear rate, such that the system is flow-orienting, the approach to the steady state (starting from some arbitrary initial condition) is usually accompanied by damped oscillations of both director and stresses [47], similar to those experimentally observed in many instances. It remains true, however, that the gist of the simulations discussed in this and the previous sections was that of interpreting the strange behavior of steady-state normal stresses. Two main conclusions can be drawn by the indisputable success obtained in that regard. One is that the basic model laid down by Hess [27] and Doi [28] (i.e., treating the dynamics of a nematic phase through Brownian rods which interact via a mean-field potential) appears to catch the essentials of the behavior observed in main-chain LCPs. The second conclusion is that most, if not all, main-chain LCPs are of the tumbling type. In order to make them behave as flow-orienting nematics, a large enough value of the shear rate must be imposed. Direct optical evidence of tumbling has been obtained on carefully prepared monodomain samples of some lyotropic nematics [20, 21]. Additional evidence of tumbling, involving thermotropic LCPs as well, is discussed in the section on polydomains.

4. *Approximations and Constitutive Equations*

Although the kinetic theory of rod-like LCPs has been found successful, it would be quite unthinkable to use the Smoluchowski equation, Eq. (23) (or Brownian dynamics simulations, for that matter) to obtain predictions in situations significantly more complex than considered so far. We are referring here, for example, to flow in complex geometries as encountered in applications, where the velocity gradient is generally not uniform in space and is even unknown beforehand, the problem typically being determined by conditions imposed on the boundaries of the flow field. Furthermore, even when the geometry and the boundary conditions imply a uniform field of the velocity gradient for an ordinary polymeric liquid, there is no guarantee that the situation is a simple one for LCPs, for which the director field must also be considered. In most cases, the director field is extremely complex and essentially unknown, corresponding to the so-called polydomain texture.

Any hope of dealing with spatial inhomogeneities of one nature or other requires that some approximation be made, such that it is no longer necessary to solve the Smoluchowski equation explicitly for the orientational distribution function. Rather, we need to establish a direct link between the kinematics of motion and the relevant molecular averages providing both the order parameter tensor (hence the director) and the stress tensor. Such a direct link is usually called a "constitutive equation", although, of course, equally constitutive is the relationship between kinematics and stress which first goes through the task of determining the distribution function explicitly. A constitutive equation is known already, namely that of LE theory, which however is expected to apply to LCPs without concern only for slow flows, and in the absence of defects (see Sections I.D and I.E).

The procedure for deriving the (approximate) constitutive equation for some average quantity from a Smoluchowski equation is as follows. If, by way of example, we want to consider the average $\langle \mathbf{uu} \rangle = \mathbf{S}$, then $\mathbf{uu}$ is multiplied by all terms of the Smoluchowski equation (Eq. (23) in our case), and the integral is taken over the whole $\mathbf{u}$-space. The left-hand side of the equation then gives the time derivative $d\mathbf{S}/dt$. The terms on the right-hand side are manipulated via integration by parts, producing algebraic combinations of the same average $\langle \mathbf{uu} \rangle$, together with the velocity gradient tensor $\boldsymbol{\kappa}$, and, unfortunately, with "higher order" averages as well (such as $\langle \mathbf{uuuu} \rangle$ at the very least; it may be worse, depending on the expression used for the potential V). Were it not for these higher order terms, the resulting equation would be the exact constitutive equation for tensor $\mathbf{S}$, in the form of an ordinary time differential equation. Because of those terms, the problem remains open; nor does it close if the operation is repeated by multiplying Eq. (23) by $\mathbf{uuuu}$, because then at least the average $\langle \mathbf{uuuuuu} \rangle$ would also appear, and so forth. The problem can be closed only if some approximation is made, that is, if the higher order averages are expressed in terms of lower order ones.

Hess [27] and Doi [28] proposed the simple "decoupling" approximation $\langle \mathbf{uuuu} \rangle \approx \langle \mathbf{uu} \rangle \langle \mathbf{uu} \rangle$. The resulting constitutive equation was then used to obtain rheological predictions, some of which, the viscosity curve for example, were later found not to have suffered much from the approximation. Indeed, it was with this simple, yet powerful, tool that Doi [28] could explain the typical reduction in viscosity occurring in a lyotropic system at the isotropic–nematic transition, with increasing polymer concentration. This result of Doi marked the first success of the rod-like molecular model in explaining anomalous rheological features of LCPs. On the other hand, the decoupling approximation does influence the sign of the Leslie coefficient α_3, thus creating the artificial result that LCPs are always flow-aligning. We have noted already, in Section II.D, that the ratio α_3/α_2 (the sign of which determines the tumbling or nontumbling nature of the nematic) is very small in magnitude, and it can easily cross the zero. As explained in Section II.D, the exact calculations of Semenov [32] and Kuzuu and Doi [33] had in fact shown that tumbling should prevail in most cases, although the far-reaching consequences of that result were not seen at the time.

We have demonstrated in the previous section how tumbling is responsible for some important rheological effects, such as the negative normal stresses. We will see later that tumbling can also be invoked to explain other rheological anomalies, for example the oscillations during transients at low shear rates. Most importantly, it can be used to approach the complex problem of textures in polydomains. Thus, with today's knowledge, it is required that any approximation should retain the tumbling character of the LCPs at low shear rates, except when the equilibrium order parameter is very low, of course (see Fig. 2.3).

It seems that there are essentially two ways to accomplish this result. In one of them [51], simpler to use, the same level of the decoupling approximation is maintained, that is, the constitutive equation only involves the average $\langle \mathbf{uu} \rangle = \mathbf{S}$. Now, although the approximation for $\langle \mathbf{uuuu} \rangle$ can be chosen in such a way as to obtain tumbling, the shortcoming is that the shear flow is never capable of changing the character of the nematic, no matter how large the shear rate is. In other words, tumbling can be there all right, but the dynamic transition to flow-aligning gets lost. As we shall see, this approximation can nevertheless prove useful in some cases. In the same paper [51], although only for the two-dimensional case, it was shown that rising to the next order, that is, keeping $\langle \mathbf{uuuu} \rangle$ untouched and using an approximation for $\langle \mathbf{uuuuuu} \rangle$, might represent an acceptable solution to the closure problem, albeit somewhat more complex to use.

These kinds of approximations were explored in three dimensions by Maffettone [52], who ended up with a proposal which seems indeed more than adequate. We show here in Fig. 2.7 the predictions of his equation for the second normal stress difference N_2 in a shear flow (N_2 is probably the rheological quantity most sensitive to the quality of a prediction). His curves can hardly be distinguished from the simulations of Larson [38], and they compare with the data of Magda et al. [50] equally well. Notice that no adjustable parameter has been introduced, although the approximation itself is of course intrinsically arbitrary. As also mentioned at the end of Section I, one advantage of molecular modeling of nonlinear behavior (over phenomenological continuum mechanics methods) is that arbitrary assumptions which become necessary to simplify matters can be checked against "exact" calculations, even before checking them against experiments.

In concluding this section on constitutive equations, we remind the reader that we are still confined to a spatially homogeneous situation (monodomain). As a consequence, whenever the dynamical regime is

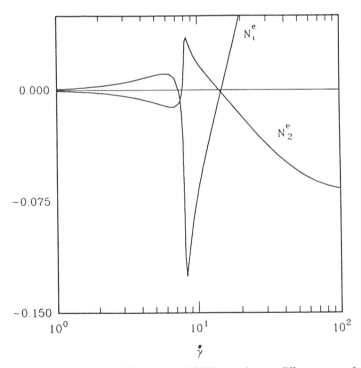

Figure 2.7 Calculated first (N_1) and second (N_2) normal stress differences as a function of shear rate (reprinted with permission from [52]). Predictions compare favorably with the experimental results (see [50]).

either tumbling or wagging, the constitutive equation predicts a periodic response. Here, as before, the quantities reported in the figures (to be compared with experimental results obtained in stationary conditions) are averages over a cycle. As discussed presently, this is a simple way of accounting for the polydomain structure of the actual sample.

F. The Linear Limit Revisited: The Puzzle of the Shear Viscosity

So far, we have shown significant features of the rheological behavior resulting from the nonlinearity of the response at high shear rates; they are especially dramatic when the nematic has a tumbling character to start with. Needless to say, these results could never be obtained from LE theory, tumbling or nontumbling, because of the intrinsically linear character of that theory.

It remains true, however, that when shear rate is decreased, the nonlinear "excitement" must vanish, and the rod-like molecular model must give results that are consistent with LE theory. To be precise, since the molecular model considered so far is spatially homogeneous, the model must approach a reduced form of LE theory, deprived of Frank elasticity, and only containing the viscous Leslie stresses. This is indeed what the model does, as shown in detail by Kuzuu and Doi [33].

Now, in such a low-shear-rate limit, the shear viscosity is obviously a constant (as linearity implies) also in the tumbling case. Tumbling generates a periodic shear stress at low shear rates as well; nevertheless, at any position in the cycle (i.e., for any orientation of the director), the shear stress is proportional to the shear rate in the linear limit. An average over the cycle (however made) therefore gives a viscosity that is independent of shear rate. Such a (dull) prediction is indeed shown in fig. 2.5, where the viscosity curve approaches its "Newtonian" plateau towards the left of the diagram.

Reality, however, is more exciting than predictions. Figure 2.8 shows the now well known viscosity curve for lyotropic LCPs, with its subdivision in three zones as proposed by Onogi and Asada [5] (and, even before, by Kulichikhin et al. [53]). Regions II and III of this curve correspond, respectively, to the Newtonian plateau, and to shear thinning resulting from the nonlinear behavior at high shear rates (as in Fig. 2.5). Conversely, Region I reveals a quite unexpected (from the point of view of theory) nonlinearity, which appears at very low shear rates.

To complete the picture of the experimental information on viscosity curves, a few more remarks are in order. In the first place, Region I, although a common feature of most (if not all) lyotropic LCPs, is not always found. For some systems, it only appears at the highest polymer concentrations [5]; it is as yet undecided whether its absence at lower

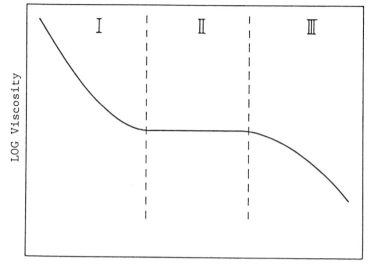

Figure 2.8 Shear viscosity versus shear rate for LCPs, showing the characteristic "three-region" curve (reprinted with permission from [5]).

concentrations (but still in the nematic phase) is real or else simply means a displacement to unmeasurably low shear rates. In some systems, Region II is not truly horizontal as depicted in Fig. 2.8; rather, the curve shows shear thinning throughout all three regions [42,45]. These remain clearly distinguishable nevertheless, because of the two inflections (in opposite directions) that the curve makes when changing slope. The latter type of curve is also found in thermotropic LCPs [54], for some of which, however, the distinction in zones disappears, a single power law (i.e., a single slope in a log–log diagram) fitting the viscosity curve throughout [55]. It should further be mentioned that, in either lyotropic or thermotropic LCPs, the negative slope in Region I, although system dependent, is always significantly less than unity, and often close to 0.5. Note also that the change in viscosity across Region I is not a minor affair, as it can reach several orders of magnitude in some systems. It is finally remarked that, in the isotropic phase of these same polymers (as well as in general), a viscosity curve with a shape similar to that in Fig. 2.8 has never been observed.

It may safely be concluded that, with few possible exceptions, both lyotropic and thermotropic LCPs behave nonlinearly in the nematic phase at very low shear rates. The problem is that no simple interpretation of this nonlinearity is at hand, as will be apparent after examining the various possibilities. Let us first consider a yield stress phenomenon, brought about by some form of intermolecular aggregation. The three-dimensional structure resulting from such aggregations forms a weak solid, which collapses (thus allowing the system to flow) when the applied shear stress rises above a yield value σ_0. In such a case, the apparent viscosity would approach the value $\sigma_0/\dot{\gamma}$ at low shear rates, i.e., the viscosity curve should exhibit a -1 slope, which is in fact never observed.

Let us next examine whether or not Frank elasticity, so far ignored, might explain the effect. In other words, could it be that the "complete" LE theory (as opposed to the reduced one, where Frank elasticity is absent) does the trick? In abstract principle, this is not impossible; we have discussed in Section 1.D that nonlinear effects can arise in LE theory due to a coupling between velocity and director fields. Quite apart from the difficulty of explaining order-of-magnitude changes due to such couplings, the main argument against the Frank elasticity possibility is that, for such an explanation to be acceptable, the measured viscosity values should depend on the experimental characteristic dimension l (compare Section 1.D), and follow the Ericksen scaling law of Eq. (15). On the contrary, the data do not show any dependence on gap size l, let alone obeying Ericksen scaling. The observed nonlinearity is in fact intrinsic to the material undergoing shear flow, and is not due to director-anchoring effects at the walls of the apparatus.

A tentative explanation, also based on Frank elasticity and LE theory, was advanced some time ago by one of the authors [56], and taken up by Wissbrun [57]. The suggestion was that, not the external dimension l should be considered in the scaling, but the "internal" one of the domains, which is itself created by the flow process. Although possibly containing a grain of truth (as we shall presently see), the weakness of the argument was its circularity. Indeed, if we call a this internal dimension of the material, a can only be determined (within the context of LE theory) by a balance between viscous and Frank elasticity forces, that is, $a^2 \approx K/(\eta\dot{\gamma})$, where K stands for the elastic constants and η for the Leslie coefficients. It follows that the "internal" Ericksen number stays constant when $\dot{\gamma}$ is changed (see Eq. (14)), whereby the measured viscosity should be equally constant (see Eq. (15)). In other words, within the context of LE theory, it is impossible to have both the characteristic dimension of the problem and the measured viscosity changing with varying the shear rate. Either the dimension changes, and the viscosity should then stay constant (which it does not in Region I), or else the dimension is imposed externally, but then the viscosity should also depend on gap size, and follow Ericksen's scaling (which it does not either).

We are forced to conclude that LCPs that show the typical Region I behavior (that is, most of them) do not obey LE theory at low shear rates. At least not strictly; there must be some missing factor. We have a clue to a possible missing factor, however. In Section I.E, we have anticipated the fact that Frank elasticity (and therefore LE theory) does not hold true close to defects, where the mathematics of the theory show divergences. The missing factor could be found, perhaps, in the behavior of the defects of the nematic phase, which are known to be particularly dense in LCPs. Indeed, the concept that Region I should be associated with the polydomain structure of LCPs has always been present in the relevant literature, and is clearly stated in the presentation of the three-region curve of the viscosity by Onogi and Asada [5]. This leads us naturally to the last part of this chapter.

III. THE POLYDOMAIN PROBLEM

We have repeatedly mentioned that, more often than not, LCP samples are polydomains, this word indicating that the anisotropy of the material is not spatially homogeneous. The director varies from one point to another, over distances that are typically of the order of several micrometers, or even less. Over similar distances, "defects" are also observed, usually in the form of lines, which may become visible under the microscope because of the light they scatter. There has been an ongoing discussion (although never particularly "hot") in the pertinent literature about differences and similarities between LCPs and low molecular weight nematics as regards defects, and the polydomain "texture" in general. There appears to be a general consensus that defects tend to be more numerous in LCPs, and more difficult to eliminate. Monodomain samples can be obtained under very special conditions [1,20,21].

Because the orientational state of the LCP polydomain often determines molecular orientations in the solid phase, which in their turn influence end-use properties of the material, good control of the polydomain structure would have great potential value. It is also well known that flow can alter the polydomain structure significantly. In spite of that, and because of inherent difficulties, the effect of flow on the polydomain structure is far from being well understood. There has been some progress, however, as we shall see. The inverse problem, that is, the effect of polydomain structure on the rheological properties of LCPs appears to be, up to a point at least, less crucial. Indeed, in previous sections we have reported successful comparison of theory with experiments although theory ignored all details of the polydomain, if not its very existence, while the experiments had certainly been made on polydomain samples.

Here, we first examine the models of LCP rheology which, although accounting for the polydomain structure, do not explicitly investigate the dynamical role played by the defects. We then report on recent work concerning both the statics and the dynamics of defects, which might help in understanding the effect of flow on LCP texture.

A. Simple Polydomain Models of LCP Rheology

Intrinsic to the polydomain is the concept of spatial variations of the director, that is, of orientational distortions, which generate Frank elasticity forces. Following the success of the molecular rod-like theory, the first approach to polydomain rheology was, however, that of neglecting Frank elasticity altogether, the underlying justification being that the elastic constants in LCPs are not much larger than in low molecular weight nematics while the viscosity is, and by several orders of magnitude; viscous forces in LCPs should then be dominant over those arising from Frank elasticity. The assumption was therefore made that, although there is a multitude of "domains" in the polydomain structure (i.e., a multitude of director orientations), each of them behaves independently of the others, and evolves in time according to the predictions of the monodomain theory.

As a first consequence of this assumption, when the monodomain theory predicted a periodic regime in shear flow (tumbling or wagging) and, correspondingly, a periodic stress, the stationary overall stress in the polydomain was calculated as an average of the monodomain stress over the period, taken to be equivalent to an average in space over domains. As mentioned in previous sections, such was the procedure adopted in both the two-dimensional [37,47,51] and three-dimensional calculations [38,50,52] to obtain overall stress values (later called "mesoscopic" averages by Larson and Doi [58]) to be compared with experiments on polydomain samples.

A second obvious consequence of the same assumption is that, when the tumbling nematic reverts to a flow-aligning behavior at high shear rates, all domains would end up oriented in the same way, that is, they would merge in a single domain with the director essentially aligned in the shear direction (the Leslie angle being very small). There exists indeed direct evidence of good alignment at high shear rates [59,60], but whether or not the domains have really merged into a single one is still an open question. The alternative is that defects somehow remain "trapped" in the aligned sample, ready to "spring up" again when the shear is removed, thus re-establishing a polydomain texture.

Still in the flow-aligning regime, it is conceivable to use the same assumption to make predictions on transient behavior of the polydomain, like the start-up of a shear flow. For a single domain, the initial condition for such a problem is that the orientational distribution of the rodlike molecules $f(\mathbf{u})$ is the equilibrium distribution about some arbitrarily assigned director. For the polydomain case, one needs to assume some initial condition for the "director distribution". The latter is a function $F(\mathbf{n})$ such that $F(\mathbf{n}) \, d\Omega$ gives the volume (or mass) fraction of material having a director oriented within the solid angle $d\Omega$ about $\mathbf{n}$. Once an initial director distribution $F_0(\mathbf{n})$ is assigned (for example a uniform distribution, corresponding to randomly oriented domains), it is a simple task to determine the director distribution in the polydomain at any subsequent time, $F(\mathbf{n}, t)$, as long as each domain evolves independently of the others. The mesoscopic stress, obtained by averaging the stress contributions of individual domains according to $F(\mathbf{n}, t)$, was predicted to exhibit damped oscillations while approaching the steady state [47], in much the same way as observed in many instances (see the next section for a discussion of experiments).

A serious problem with the simple assumption considered so far becomes apparent, however, when considering transient behavior in the tumbling (or wagging) regime. Indeed, let us consider a shear start up in the tumbling regime, and assume some arbitrary initial distribution of directors $F_0(\mathbf{n})$. Then, unless $F_0(\mathbf{n})$ is very special (and there is no physical reason for it to be), a stationary state at the mesoscopic level is never achieved. The polydomain is predicted [47] to approach a periodic response asymptotically, itself reflecting the periodic behavior of the individual domains (all of which have the same period but different phases). The situation is similar to that of a viscous suspension of non-Brownian elongated particles, such as prolate ellipsoids, in a shear flow; if each particle goes undisturbed through its Jeffery orbit, that is, if the hydrodynamic interactions among the particles are neglected, both the orientational distribution and the dissipation (hence the viscosity of the suspension) are predicted to be (for ever) periodic in the shear deformation [61].

It became apparent, therefore, that the interactions among domains should be somehow accounted for. The considerations to be reported in the rest of this section, and in the next one, are based on the assumption that the domain interactions can be described in terms of Frank elasticity, somehow coupled to the tumbling nature of LCPs. We show that, up to a point, models using these ingredients compare favorably with many experimental results. After the next section, however, we will find it necessary to move on to more elaborate theories of domain interaction.

One simple way of eliminating the indefinite oscillations of $F(\mathbf{n}, t)$ in the tumbling regime was that of representing interactions via a pseudo-diffusion term [47], corresponding to a fictitious (or effective) angular diffusivity of the director, given by $K/(a^2\eta)$, where K is the Frank elasticity constant, η is the characteristic viscosity of the nematic phase, and a is the characteristic domain size. Such a pseudo-diffusion term would damp down the oscillations in the start-up of a shear flow. It would also restore randomness of the domains, previously aligned by a flow process, after the flow has been switched off. It should be noted in this regard that $a^2\eta/K$ has the dimensions of a time, and plays the role of a relaxation time. Needless to say, such a phenomenological term does not represent any real diffusion process; the idea was that it perhaps accounted, at a mesoscopic level, for the effect of Frank elasticity in the polydomain.

It was not clear, however, what to do with the size a of the domains. Previously, the suggestion had been made [62] that, in a steady shear flow of a tumbling nematic, the group $\dot{\gamma}a^2\eta/K$ should be a constant, that is, that a^2 should be inversely proportional to shear rate. This suggestion was based on LE theory as applied to a tumbling nematic in an order-of-magnitude sense, whereby the "internal" characteristic dimension of the polydomain, that is, the domain size generated by tumbling, would result from a balance of viscous and elastic forces (compare Section II.F). Even granted that this was the case, how would the domain size behave during transients?

A phenomenological answer to this question was provided by the peculiar scaling laws which emerged from the rheological measurements under time dependent conditions of Moldenaers [43], Moldenaers and Mewis [63], Mewis and Moldenaers [64], Doppert and Picken [65], Moldenaers et al. [66], Larson and Mead [67], and from the analysis of tumbling under transient conditions made by Burghardt and Fuller [68] on the basis of LE theory. The approach to the polydomain problem, based on a combination of LE theory and of these scaling laws, culminated in the constitutive equation of Larson and Doi [58].

B. Scaling Laws for LCPs: The Larson–Doi Constitutive Equation

The peculiar scaling laws alluded to above are the following. Let us first consider a step-up experiment, which consists of jumping from a steady shear flow at a very low shear rate to some larger value of the shear rate. The shear stress will approach the new steady state through a number of damped oscillations [43]. The scaling law consists of the fact that the results of a number of experiments made with different shear rates $\dot{\gamma}$ (after the jump) essentially superimpose if the stress is normalized to the steady state value (after the jump), and the product $t\dot{\gamma}$ is used instead of time t [64–66]. A similar superposition is achieved in flow reversal experiments (consisting of switching from a shear flow with some $\dot{\gamma}$ to a shear flow in the opposite direction with the same value of $\dot{\gamma}$), where oscillations are also found [66]. Up to this point, the message is that the strain $t\dot{\gamma}$ rather than time t is the important variable. The message becomes more intriguing when considering stress relaxation experiments [43,63]. Here, if an initial fast relaxation is ignored, and only the longer time behavior is considered, it is found that the relaxation curves again superimpose by plotting against $t\dot{\gamma}$ instead of t, where $\dot{\gamma}$ is the shear rate of the flow "preceding" the relaxation process. In this case, $t\dot{\gamma}$ has no direct physical meaning since the strain is in fact zero in a relaxation experiment. Finally, in experiments of strain recovery after a shear flow [67], it is found that the total strain recovered is a constant for a given material, that is, it is independent of the value of $\dot{\gamma}$ which existed before the removal of the shear stress, and typically amounts to a few units of shear (which reveals a strong "elasticity"). The behavior in time of the recovery process again scales with $t\dot{\gamma}$. It is important to note that all these scaling laws were found to hold in the "central" region of shear rates, where the viscosity is a constant (Region II of Fig. 2.8), while deviations were observed when $\dot{\gamma}$ was increased to the point that Region III is entered. To our knowledge, no systematic transient data are available as yet for shear rates belonging to Region I.

Burghardt and Fuller [68], using LE theory, simulated the transient response of nematics in shear flow, of both the flow-aligning and tumbling types. As reported in Section I.B, steady state solutions of the equations of LE theory already existed, indicating that flow-aligning nematics develop an orientational boundary layer close to the confining walls [3], while tumbling nematics show "winding up" of the director in the shear gradient direction, that is, across the sample thickness l [11,12]. Burghardt and Fuller demonstrated that, while these windings of the director field build up in a transient situation, the stress is predicted to oscillate in much the same way as is observed in step-up and in flow reversal experiments. Conversely, nothing in any way close to the observed behavior could be produced by assuming flow-alignment. As regards strain recovery, flow-aligning nematics were predicted to show values less than unity in all cases, whereas tumbling ones could easily go up to several units, as observed in LCPs. Thus, the calculations of Burghardt and Fuller further confirmed the tumbling nature of LCPs.

Burghardt and Fuller [68] did more than that, however; they found that their simulations, which were made by assuming a sheared slab of material of thickness l, could interpret the observed dependences on shear rate only by setting $l = a$, where a obeys the equation (Er is the Ericksen number, as in Eq. (14))

$$\text{Er} = \dot{\gamma}\, \frac{\eta a^2}{K} = \text{constant} \tag{39}$$

A similar result was reached independently by Larson and Mead [67] in their analysis of strain recovery.

This result served two purposes. On the one hand, if a is identified with the domain size that prevails under steady state conditions, Eq. (39) gives phenomenological substance to the previously mentioned proposal [62]. Secondly, it opens up the way to understand the scaling laws of transient behavior in general terms.

In order to better appreciate this important point, let us here rewrite Eq. (16) in the form

$$\frac{\sigma}{\sigma_{\text{stat}}} = f(\dot{\gamma}t) \tag{40}$$

where $\sigma_{\text{stat}} = \eta\dot{\gamma}$ is the shear stress under stationary conditions, and the dependence on initial conditions is left understood. Equation (40) just as Eq. (16), holds true when Frank elasticity is neglected, while it would not be expected to be valid in general, that is, when Frank elasticity is included, because in such a case an Ericksen number (based on apparatus dimension) should also appear as an argument of the function. If, however, the apparatus dimension is irrelevant, and is replaced by an "internal" dimension (itself determined by the same material parameters, plus $\dot{\gamma}$ and t), then dimensional analysis tells us that Eq. (40) is complete even when Frank elasticity plays a role. Notice that no Frank elasticity constant can appear explicitly in Eq. (40), although the two independent ratios of elastic constants are included implicitly in the material function f (as well as all independent ratios of Leslie coefficients).

By the way it is written, Eq. (40) directly applies to the flow reversal case, where each experiment only involves a single value of $\dot{\gamma}$. The initial condition of a flow reversal experiment is a steady state, and we must assume, again consistently with dimensional analysis, that the initial distribution of directors in the polydomain is independent of shear rate. Thus Eq. (40) fully explains the scaling observed in flow reversal experiments.

For the step-up experiment, Eq. (40) should be rewritten as

$$\frac{\sigma}{\sigma_{\text{stat}}} = f\!\left(\dot{\gamma}t,\, \frac{\dot{\gamma}}{\dot{\gamma}_{\text{init}}}\right) \tag{41}$$

where $\dot{\gamma}_{\text{init}}$ is the value of $\dot{\gamma}$ before the jump, and f is a different material function of course. According to Eq. (41), that is, according to dimensional analysis, curves of different experiments should superimpose in reduced coordinates if the ratio $\dot{\gamma}/\dot{\gamma}_{init}$ is kept constant. In the actual experiments [43], $\dot{\gamma}_{init}$ itself was held constant instead of the $\dot{\gamma}$-ratio; the fact that the curves nearly superimpose all the same shows that the dependence of f on the $\dot{\gamma}$-ratio must be weak. The dependence on initial conditions is not weak in general, however. Start-up experiments, i.e., transient experiments starting from rest [43,45,65], although also showing damped oscillations, do not always obey simple scaling laws [45], and sometimes did not show good reproducibility [43]. Clearly, a rest state is not uniquely defined for these materials [45]. Further studies are in progress on this aspect [69].

Dimensional analysis requires that the relaxation experiment should be described by an equation formally identical to Eq. (40), which indeed predicts the observed scaling [63], although $\dot{\gamma}$ times t does not have any direct physical meaning in this case. For the strain recovery experiment, dimensional analysis leads to

$$\gamma = f(\dot{\gamma}t) \qquad (42)$$

where γ is the strain recovered at time t, and f is yet another nondimensional material function. Last but not least, it should be remembered (compare the discussion at the end of Section II.F) that the assumption embodied by Eq. (39) implies a Newtonian steady state viscosity. It is therefore expected that the scaling laws discussed so far only apply to Region II of the shear rate range in Fig. 2.8.

All these results arising from dimensional analysis coupled with the assumption of Eq. (39), were nicely incorporated in a constitutive equation proposed by Larson and Doi [58]. This constitutive assumption is made up of two parts. One deals with the evolution of the domain size a, which is assumed to obey the equation

$$\frac{d(a^2)}{dt} = \frac{K}{\eta} - \dot{\gamma}a^2 \qquad (43)$$

(Their original equation is written somewhat differently, but is in fact equivalent to Eq. (43)). If numerical coefficients are ignored, Eq. (43) properly reduces to Eq. (39) in a steady state; the first term on the right-hand side of Eq. (43) gives the rate of increase in the (square) domain size when the flow is switched off. Larson and Doi also derived an evolution equation for the "mesoscopic" order parameter tensor $\bar{\mathbf{S}}$; in order to solve this equation, the domain size a must first be derived from Eq. (43). Finally, once a and $\bar{\mathbf{S}}$ have been calculated, the mesoscopic stress is also obtained.

We do not report the details of the constitutive equation of Larson and Doi here because, as noted by the authors themselves, convenient assumptions were used which are in fact arbitrary. Thus, although adequate in predicting the material functions of Eqs. (40)–(42) qualitatively, the constitutive equation proposed in [58] is not expected to apply in general. An intriguing example of disagreement is offered by the recent results of Hongladarom and Burghardt [70] showing that, after a shear flow has been stopped, the director alignment with the shear direction "improves with time" (in a first part of the relaxation process, at least). This result is at variance with the prediction of the constitutive equation of Larson and Doi which, just as done in [47], is constructed in such a way that the mesoscopic orientation "diffuses" towards randomness when the flow is stopped. Larson and Doi [58] also noted that even Eq. (43) per se is not good enough; experiments show that, if instead of a constant shear flow, an oscillating shear strain of small magnitude (but arbitrary frequency) is applied, then the domain size will keep increasing in time as though the material were at rest, that is, as if the right-most term in Eq. (43) were absent.

In fact, the true merit of the constitutive equation of Larson and Doi is that of providing the correct scaling for LCPs in Region II, a scaling which indeed is common to other fluid systems lacking an intrinsic relaxation time [71]. One further important result of this scaling, not mentioned so far, is as follows. At steady state, the mesoscopic stress can only be linear in the velocity gradient; therefore, the normal stresses in a shear flow must also be proportional to $|\dot{\gamma}|$ in Region II (instead of quadratic in $\dot{\gamma}$, as is true for ordinary polymers at low shear rates). This is a remarkable prediction, in reasonable agreement with the observations [40,41,63].

C. A Closer Look Into Tumbling and Defects

It so appears that LE theory, combined with the ansatz of Eq. (39), gives good scaling predictions in Region II. Nevertheless, the situation cannot be considered entirely satisfactory as it stands. One reason is that we are left rather in the dark about details of the polydomain texture, which might be important for the applications of these materials. A second good reason is that, on closer inspection, the logical status of Eq. (39) is found to be very weak. Let us start from this theoretical aspect first.

The excuse for invoking an internal variable dimension a, determined by the flow itself, is tumbling. Were it not for tumbling, one would expect the characteristic dimension to be determined by the apparatus (it would be the sample thickness in most cases), and the nice scalings discussed in the previous section would be lost. The question then becomes how tumbling succeeds in modifying the length scale of our problem. As mentioned repeatedly, LE theory admits solutions for tumbling nematics [11,12]. It is worth looking at them more in detail.

Let us consider a nematic slab of thickness $2l$ subjected to a simple shear flow. Let us temporarily restrict the director to the plane of shear so that a single angle θ is required to specify it at any point in the material. Within LE theory, time independent solutions of this problem would satisfy Eq. (10) with $\dot\theta = 0$, that is,

$$K \frac{d^2\theta}{dy^2} = (\alpha_3 - \alpha_2)\dot\gamma \, \frac{\sin^2\theta + \lambda \cos^2\theta}{1 + \lambda} \tag{44}$$

where $\lambda = -\alpha_3/\alpha_2$ (compare Eq. (9)), and y is the coordinate in the direction of the gradient, that is, $-l < y < l$. In Eq. (44), just as in Eq. (10), the one-constant approximation is used for simplicity. This equation states that the elastic torque on the director (left-hand side of Eq. (44)) must equal the viscous one (right-hand side) at all points.

In tumbling nematics, which are characterized by the fact that λ is a positive number, the right-hand side of Eq. (44) never crosses zero; with varying y, and therefore with varying θ, the value of the right-hand side of Eq. (44) will oscillate about some nonzero mean value. If we are not interested in the details of these oscillations, and only look for the behavior of the function $\theta(y)$ satisfying Eq. (44) on a coarse scale, Eq. (44) can be simplified to become [72]

$$K \frac{d^2\theta}{dy^2} = \eta\dot\gamma \tag{45}$$

where the viscosity η can be interpreted as $(\alpha_3 - \alpha_2)$ times the mean value of the fraction in Eq. (44).

Equation (45), in its apparent simplicity, reveals a very important quality of tumbling, namely that (unless some other factor enters the picture, as discussed at the end of this section) the director windings cannot be uniformly distributed throughout the sample thickness; rather, they would become increasingly tighter as the walls are approached. Indeed, integrating Eq. (45) once, and accounting for symmetry, gives

$$\frac{d\theta}{dy} = \frac{\eta\dot\gamma}{K} \, y \tag{46}$$

which shows that the director gradient is zero at the midplane of the nematic slab, to become at the walls

$$\left| \frac{d\theta}{dy} \right|_w = \frac{\eta\dot\gamma l}{K} \tag{47}$$

With typical values of the parameters, this gradient would be as large as 10^{10} m^{-1} or even more. If we call a the distance over which the director makes one turn (which can be taken to represent the domain size), Eqs. (46) and (47) show that such a quantity varies along the thickness coordinate, to become at the walls (numerical factors are neglected):

$$a = \frac{K}{\eta \dot{\gamma} l} \tag{48}$$

This result is at variance with Eq. (39). Which of the two conflicting equations, Eq. (39) or Eq. (48), makes sense? It is immediately apparent that any scaling based on Eq. (48) would be absurd for at least two reasons. It would reintroduce the sample thickness l in the scaling laws, while the rheological experiments on LCPs clearly indicate that intrinsic, size-independent bulk properties are being measured. Secondly, with typical values of the parameters in Eq. (48), the domain size a would fall below molecular dimensions.

The question then is where is the fault in the above analysis. It cannot be in the in-plane director assumption because, even if the director has an out-of-plane component, the above result is not modified (in an order-of-magnitude sense) unless the in-plane component goes virtually to zero, that is, unless the director becomes aligned everywhere with the neutral direction, an extreme possibility that the experiments do not support. Let us next consider the steady-state assumption, that is, the fact that Eq. (44) only holds true if $\dot{\theta}$ is zero. In words, Eq. (44) expresses the concept that the Frank elasticity associated with the director distortions has effectively halted actual tumbling. Now, if in the real world the director field is not stationary in the shear flow, and actual tumbling takes place, surely the viscous torque gets decreased, and so does the counteracting torque due to Frank elasticity. This is best seen by going back to Eq. (10), which shows that indeed a negative $\dot{\theta}$ (a rotation of the director consistent with the vorticity of the shear flow) can relieve the "load" of torque balancing from the elastic term. Only two possibilities, both unsatisfactory, can exist, however. Either actual tumbling relieves that "load" completely, but then Frank elasticity disappears entirely from the picture, contrary to all the evidence discussed in the previous section which indicates a central role of such an elasticity during transients. Alternatively, if the "load" is relieved only in part, that is, if the elastic term containing the second derivative survives, then again the result of Eq. (48) is not essentially modified, and we have not made any progress yet.

Let us finally consider defects, either pre-existing in the material or generated by the tumbling itself. From the point of view of LE theory, defects are discontinuities in the director field (compare Section I.E).

Now, since discontinuities often represent a "death sentence" for mathematical developments, one might think that the scaling result of Eq. (48) gets destroyed merely by assuming, as indeed is realistic, the existence of discontinuities in the director field. Such is not the case, in fact, and in order to understand this important point, it is convenient to consider the physical meaning of the terms appearing in the equations, from Eq. (45) to Eq. (48). Equation (45) shows that the local value of the elastic torque per unit volume involves the second derivative of θ. Thus, the physical reason behind the "abnormal" growth of the first derivative of θ up to unrealistic values (as given by Eq. (47)) is that the torque per unit surface area at the wall of the slab (given by $K|d\theta/dy|_w$) must be such as to support the "integral" of the viscous torque, which is of the same sign everywhere, over the whole half-thickness of the sample. Such an integral over volume is indifferent to volumeless discontinuities.

We are stuck in a very serious contradiction. On the one hand, tumbling within the context of LE theory formally leads to the scaling of Eq. (48) which is unrealistic. On the other hand, scaling laws based on LE theory together with the ansatz of Eq. (39) (which appears to come from nowhere) are successful in Region II. There seems to be no way out of this contradiction other than exploring the possibility that defects are in fact much more "physical" objects than mere discontinuities in the director field, and that they play an "active role" in the dynamics of a tumbling nematic.

Indeed, in order to obtain a torque per unit surface area that does not grow indefinitely from midplane to wall, the only possibility appears to be one where "defects counteract the torque of the tumbling nematic". In other words, we envisage a situation where the interaction of the shear field with the defects generates a torque which is opposite to that exerted by the same shear field on the nematic. Such a "two-phase" model, by implying the possibility that the viscous torques of the two phases equilibrate each other over the length scale of the domains, would indeed be compatible with a texture that does not vary from the center to the walls of the sheared slab, that is, with a bulk response intrinsic to the material. Of course, the "defect phase" cannot obey LE theory; nor it is expected to do so, if we identify such a phase with the so-called defect cores, where distortions exceed the range of validity of Frank elasticity. We are thus motivated to start our "explorative journey" into defect cores. We hope to find a rationale for the ansatz of Eq. (39), as well as an explanation for the mysterious Region I behavior.

D. Defect Cores

In order to use mathematical tools to describe what goes on within defect cores, we must first get rid of discontinuities, such as those arising in the director field. In other words, we must find a "descriptor" of the anisotropy of the material, other than the director, which changes smoothly throughout the space occupied by a defect.

In a uniform nematic phase at equilibrium, the director **n** lies along the symmetry axis of the orientational distribution. This is also the symmetry axis of all the moments of the distribution, the simplest of which is the second-order moment **S** (see Eqs. (25) and (26)). Therefore, **S** is the first natural choice to be explored in our search for a different descriptor, and **S** proves in fact to be adequate to the task. Indeed, jumping ahead of the calculations to be described later, Fig. 3.1 shows that tensor **S**, geometrically represented by an ellipsoid, changes smoothly in a defect core. Away from the defect, the ellipsoid has the "shape" assigned to it by the value of the order parameter of the defectless nematic phase (see Eq. (26)); then, by entering the defect core, this shape changes in a continuous way up to, and including, the very "center" of the defect.

The example shown in Fig. 3.1 is particular [73]. It represents the so-called hedgehog defect, which is the single defect enjoying a spherical symmetry (the director lines radiate out of a point like the quills of the hedgehog). As a consequence of this symmetry, the ellipsoid always

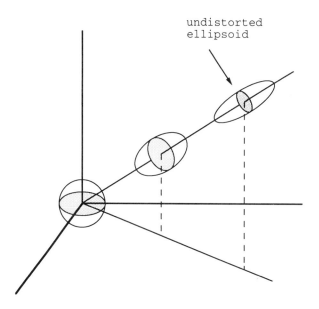

Figure 3.1 Deformation of the ellipsoid representing the molecular distribution while the center of a hedgehog defect is approached. (The defect center is at the origin of the Cartesian axes.) A reduction of the order parameter takes place, corresponding to a progressive "isotropization" of the nematic.

remains axially symmetric while it "deforms" in approaching the center of the defect; at the very center, it becomes a sphere. In all other kinds of defects, the axial symmetry of S is lost in the deformation induced by the defect; in other words, although we are talking of nematics that are intrinsically uniaxial, biaxiality of the orientational distribution and of its moments is the rule in defect cores [74–76]. Instead, the important message conveyed by Fig. 3.1 is expected to be general; namely, that the field of S (i.e., the function S(R) where R is the position vector) is continuous throughout a nematic sample, defects included; different from n(R), which is singular at the points, or lines, constituting the center of the defect cores. Therefore, S(R) can safely be used as a descriptor of anisotropy in a comprehensive mathematical treatment of defective nematics, as tumbling LCPs seem to be.

A treatment based on continuum mechanics appears unsuitable, however, even in problems of statics (not to speak of dynamics). The reason is as follows. From the continuum viewpoint, we should start from an expression of the distortion energy which, in analogy with Frank elasticity theory, could be written as an objective (frame indifferent) scalar function of S and ∇S, quadratic in ∇S. As S is a tensor instead of a vector, the general expression for this energy will contain many more terms than in the case of Frank elasticity [75]. This is not the main difficulty, however. The conceptual problem arises in considering the coefficients of the terms of such an expression which, differently from K_1, K_2, and K_3 of the theory of Frank, could not be taken as constants in this case [75]. They should be taken as scalar functions of S, that is, functions of the three scalar invariants of S (only two of which change with R, because the trace is fixed to unity, see Eq. (25)). Since these invariants describe the "shape" of the ellipsoid, they are expected to change a lot in a defect core. Now, continuum mechanics is unable to make predictions about the functional forms of these coefficients; nor are these functions accessible to measurement (differently from, say, the temperature dependence of Frank's constants). Thus, within the continuum or phenomenological approach, one is forced to make arbitrary assumptions, the validity of which is virtually impossible to assess.

In view of the general character of the consideration made above, which in fact goes beyond the specific case [22], let us discuss it further. The physical justification for a quadratic form of the distortion energy in the field gradients is clear; it corresponds to the assumption that the changes occur "gradually" in the field (where the yardstick for this graduality is some relevant molecular dimension). Although the assumption of gradual change fully justifies the quadratic form, it is not sufficient for assuming constant coefficients. These stay constant throughout the field only if the "material" itself does not change. For example, if in a nematic obeying Frank elasticity, the temperature varies with $\mathbf{R}$, the dependence of the coefficients on temperature must be accounted for; from the continuum viewpoint, a change in temperature corresponds to a change in the material. Gradual as this variation may be, the cumulative effect can be large, and ignoring the temperature dependence of the coefficients could lead to serious errors. Now, the change in shape of the ellipsoid (i.e., of the order parameters of the nematic), which occurs in defect cores, is the equivalent of a large temperature variation in the previous example, with the important difference that, contrary to the case of temperature, we have no direct way to measure the effects of this "internal" variable (the ellipsoid shape) on material properties.

The conclusion is that we better resort to some molecular model in order to predict the behavior of $\mathbf{S(R)}$ within defect cores. Simplified continuum models, such as that of Ericksen [77], could nevertheless prove useful to simulate broad qualitative features of defective nematics. One drastic assumption in Ericksen's model is that $\mathbf{S}$ is uniaxial under all conditions.

The model described here provides an extension of the Maier–Saupe expression for the mean-field orientational potential. The extension is worked out by considering, as is usually done, the interactions between a test molecule and those falling in a suitable neighborhood. In our case, however, the orientational distribution of the surrounding molecules is taken to be space-dependent, in order to account for the distortions induced by a defect (or by any other cause). We omit the details of the derivation, which can be found in the original references [73,78], and only report here the simplest possible expression for the orientational potential issuing from this approach. This is

$$V(\mathbf{u}, \mathbf{R}) = -kTU\left(\mathbf{S} + \frac{1}{2}\,\ell^2\nabla^2\mathbf{S}\right) : \mathbf{uu} \tag{49}$$

where $\nabla^2\mathbf{S}$ is the Laplacian of $\mathbf{S}$, and ℓ is an interaction distance. The dependence on $\mathbf{R}$ of both $\mathbf{S}$ and $\nabla^2\mathbf{S}$, in Eq. (49) is understood. Equation (49) can be used in a time dependent situation as well, that is, when $\mathbf{S}$ (hence V) also depends on t. Equation (49) properly reduces to Eq. (33) in the absence of spatial gradients.

The simplicity of Eq. (49) arises from several assumptions. One is the "gradual variation" hypothesis that we have discussed already. Indeed, it can be shown that terms containing up to the second derivatives of $\mathbf{S}$, as in the above Laplacian, correspond to a distortion energy that is quadratic in $\nabla\mathbf{S}$. The implication is that, for Eq. (49) to be valid, the magnitude of the second term in the parentheses must be small with respect to that of the first. The gradual variation assumption seems to be indispensable to proceed with the calculations, also when, as in [78], the problem is tackled via molecular modeling. A second, less crucial, simplifying assumption was made in obtaining Eq. (49); it corresponds, in the Frank elasticity limit, to the one-constant approximation. Without making use of this simplification, expressions for the three constants of Frank elasticity were obtained [78], which compare favorably with some data of rigid molecule nematics.

Frank elasticity is recovered from Eq. (49) when the ellipsoid representing S approaches a constant, axially symmetric shape, and only its orientation changes with R. Such is indeed the case for defectless nematics or, in any event, for regions of the nematic away from defects (and phase boundaries). Within defect cores, where the ellipsoid "deforms", Frank elasticity breaks down, and we can speak of "large" distortions. A distortion is considered large, therefore, insofar as it strongly modifies the "local material" with respect to the situation prevailing in the uniform state. This concept of large distortions is in no way contradictory with that of gradual variation, which has to do with how rapidly (in space) these big changes occur.

A self-consistency argument can be applied to the potential in Eq. (49), which generalizes that used by Maier and Saupe [31] for the spatially uniform nematic. The field equation for $S(R)$ is thus obtained in the (deceivingly) simple form

$$S = P + U\ell^2(Q - PP) : \nabla^2 S \qquad (50)$$

where P and Q (a second and a fourth rank tensor, respectively) are known functions of the product of U times S [73]. Figure 3.1 is a schematic representation of the solution of Eq. (50) for the hedgehog defect. For any other type of defect (corresponding to different boundary conditions at infinity), we expect that the solution of Eq. (50) is obtained in the form of a field of a biaxial tensor; explicit calculations have not yet been made, however.

In the case of the hedgehog defect of Fig. 3.1, the material becomes isotropic at the centerpoint. For other types of defects, this very special result, that is a complete isotropization, is not to be expected. In all cases, however, it can be surmised that the material in the defect cores becomes much less anisotropic than it is in the spatially uniform nematic phase far away from defects. As we shall see, we make use of this concept in interpreting the dynamic behavior of defective nematics.

Another general aspect of defect cores is their "size", defined as the distance from the defect centerpoint (or centerline) after which the ellipsoid has effectively recovered the shape belonging to the undistorted equilibrium nematic. This size is always small, of the order of a few "molecular" dimensions, that is, a few times ℓ in the model of Eq. (49). The calculations made in [73] show that the core size depends on the order parameter S of the undistorted nematic; the defect core is predicted to shrink for values of S approaching unity. At a value of S of about 0.8, the gradual variation assumption begins to break down (in the hedgehog defect, at least); the order parameter inside the core changes too much over too small a distance.

Although the size of defect cores is comparatively small, the overall dynamical influence of the "defect phase" can be a large one. This aspect is discussed in the next section. Before concluding here on the statics, we should mention that yet another simplifying assumption was tacitly included in Eq. (49). We have ignored the possibility that, as a consequence of the spatial inhomogeneity of the orientational order, density could vary with $\mathbf{R}$. Such density fluctuations might be particularly relevant in lyotropic systems. In view of the qualitative character of our investigation into polydomain behavior, all of these simplifying assumptions (and possibly more) will be maintained throughout.

E. Role of Defect Cores in the Dynamics of Tumbling Nematics

In molecular modeling, the starting point of a dynamical treatment is the Smoluchowski equation (such as Eq. (23) or Eq. (37)) for the pertinent distribution function. In the present problem, where spatial variations are considered, the distribution function of rod-like molecules should be defined (see [79]) as $\psi(\mathbf{u}, \mathbf{R}; t)$ such that, at any given time t, $\psi(\mathbf{u}, \mathbf{R})\, d\Omega\, dV$ gives the fractional number of all molecules in the sample that are located in the neighborhood dV of $\mathbf{R}$, and which are oriented within $d\Omega$ about $\mathbf{u}$. If, however, as mentioned at the end of the previous section, polymer density gradients are ignored, we can still use a "purely orientational" distribution function, where $\mathbf{R}$ merely plays the role of a parameter (just as t). In other words, a function $f(\mathbf{u}; \mathbf{R}, t)$ is defined such that, at any $\mathbf{R}$ and t, $f(\mathbf{u})\, d\Omega$ gives the fraction of "local" molecules oriented within $d\Omega$ about $\mathbf{u}$.

Furthermore, consistent with the above approximation, we can also avoid the explicit inclusion in the Smoluchowski equation of terms accounting for translational diffusion [80], and make the assumption that their effect on $f(\mathbf{u})$ is somehow absorbed in an "effective" rotational diffusivity.

In conclusion, an approximate Smoluchowski equation for the polydomain problem can be written that is formally identical to Eq. (23) (or to

Eq. (37) in two dimensions), except that $f(\mathbf{u}; t)$ is replaced by $f(\mathbf{u}; \mathbf{R}, t)$, and the orientational potential V must account explicitly for the existence of gradients in space, for example, through the Laplacian term of Eq. (49). The mean-field potential V is responsible for the existence of the nematic phase in the first place; it appears natural, therefore, to account for orientational distortions, in dynamics as well as in statics, mainly through their effect on V.

In a recent work [81], a solution for such a polydomain equation in two dimensions was worked out. Even in two dimensions, the mathematical problem remains a formidable one, so that, in order to proceed, an approximation of the type discussed in Section II.E.4 was used. The closure adopted was at the lowest level (i.e., it resulted in an equation for tensor $\mathbf{S}$), but it was such that the tumbling character of the nematic was maintained. It can be surmised, therefore, that the approximation preserved the essential physics prevailing at low shear rates.

Here, again, we are not reporting details of that calculation, and jump directly to the result that was obtained, and to the possible implications thereof. It was shown that the defect phase, that is, the cores of the defects, can act as an "anchoring wall" for the director of the nematic phase. More precisely, the story runs as follows. As a condition of the mathematical problem, it was assumed that the defects already existed in the system in the form of lines parallel to the shear direction, uniformly spaced one from another, and fixed in time. The boundary conditions were, however, such that the director was free to tumble everywhere in space, cores of the defects included. The calculations then revealed the following effect. If the distance between consecutive defects is taken smaller than some (shear rate dependent) critical value, the evolution in time of the director field is found to approach a stationary state. In contrast, if the assumed distance is larger than that critical value, then tumbling goes on indefinitely. In other words, if the defect lines are sufficiently close, "tumbling is suppressed".

Although the mathematics may seem somewhat weird, the physical interpretation of this result is in fact very simple. One should first remember that tumbling is a property of nematics that have a large value of the order parameter. If S falls below some limiting value ($S = 0.53$ for Maier–Saupe nematics; cf. Section II.D), then tumbling is replaced by flow-aligning behavior. On the other hand, a drastic reduction of the order parameter is in fact a common property of defect cores, as we have discussed in the previous section. It is then to be expected that the defect phase 'per se' does not want to tumble.

The intrinsic flow-aligning character of the defect cores is not generally sufficient to arrest tumbling of the entire sample, however. This will occur only if the "amount" of defect phase is above some critical proportion with respect to the nematic phase surrounding the defect cores, so that the balance of opposite viscous torques in the two phases (which are elastically linked one to another) is in favor of the first. Such a proportion depends on several factors (to be discussed in the next section), among which is the shear rate. The volume fraction of the defect phase required to arrest tumbling does not need to be very large, however, especially if the viscosity of the defect phase is larger than that of the nematic. This property of defects is to be expected, again owing to the reduced value of the order parameter in defect cores. In this regard, it should be remembered that the viscosity of LCPs undergoes a significant reduction at the transition from the isotropic phase ($S = 0$) to the nematic one. Going from the nematic to the defect phase represents a sort of transition in the opposite direction, which is driven by the distortion due to the defect topology. Furthermore, such an opposite transition takes place under conditions (a higher concentration and/or a lower temperature) for which the viscosity difference between the two phases should be even larger than it is at the thermodynamic transition.

If the viewpoint is accepted whereby defects can arrest tumbling (this idea should still be considered as a "working hypothesis", due to the numerous assumptions and approximations used in the derivation [81] referred to above), then we may envisage the following dynamic mechanism during start-up of a shear flow. Let us take an initial situation where, because of previous relaxation of the distortions towards some minimum energy, the defects in the sample are scarce. After the flow is started, tumbling of the director will occur all over the place. There is a general consensus that actual tumbling is likely to generate defects, due to growth of various instabilities; the volume fraction of the defect phase is therefore expected to increase with time. A point will be reached, however, where the defect density will be large enough to arrest tumbling. The defect generation itself will therefore stop at that point, and a steady state will have been reached, with a well defined defect density.

On the other hand, if the defect density has temporarily "overshot" the balance, that is, has reached a larger value than the minimum required to arrest tumbling, then actual tumbling will be absent of course, and healing of the defects can take place. The defect density will therefore drop in time, again approaching (from above) the value corresponding to the exact balance. Some recent optical experiments by De'Neve and Navard, showing damped oscillations of a transmitted intensity, were interpreted in this way [82].

F. Scaling Arguments Based on Defect Cores

The calculations reported in [81] showed that the "critical" distance between consecutive defect lines, that is, the minimum distance that is necessary to stop tumbling, decreased with increasing shear rate. More in detail, this critical distance appeared to attain a constant value at very low shear rates; it then started decreasing with increasing $\dot{\gamma}$ to approach a power law dependence, possibly with a $-1/2$ exponent, at (relatively) high shear rates. Here, however, we want to derive scaling relationships in a much simpler way. Indeed, we only appeal to LE theory, although our argument uses the concept that the material in defect cores is intrinsically nontumbling. We also allow for a viscosity of the defect phase that is probably (if not necessarily so) much larger than that of the nematic.

Here, as in [81], let us imagine that defects mostly consist of lines parallel to the shear direction. (This cannot be strictly true, of course, because defect lines must form closed loops [3]. Light scattering experiments [83] and some optical experiments [84] do show, however, defects elongated in the shear direction; recent electron micrographs of a thermotropic LCP, quenched after a shear flow, have revealed a dense population of defect loops elongated in the shear direction [85].)

Let us assume that the distance between neighboring defects indeed corresponds to the critical distance that has arrested tumbling; the director field in the nematic phase is then stationary in time, "hanging", as it were, from the defect cores, to which it is anchored. For the stationary nematic, we can write Eq. (44) once more as

$$K \frac{d^2\theta}{dy^2} = \eta\dot{\gamma}(\sin^2\theta + \lambda \cos^2\theta) \tag{51}$$

where we have called η the group of constants $(\alpha_3 - \alpha_2)/(1 + \lambda)$. Differently from the use of this equation in Section III.C, we are interested here in local details; thus, we keep the θ dependence on the right-hand side. Equation (51) applies only for a director in the plane of shear, θ being the angle that the director makes with the shear direction. It also holds true, however, in an order of magnitude sense, if the director has an out-of-plane component. In such a case, θ should be interpreted by using the projection of the director onto the shear plane instead of the director itself.

In LCPs $\lambda = -\alpha_3/\alpha_2$ is a very small number [1,33], and this fact is often referred to in terms of a large anisotropy of the viscous behavior of the polymeric nematic. Such being the case, the right-hand side of Eq. (51) is very sensitive to θ. In physical terms, if θ is small, the viscous torque is also small, that is, of order $\eta\dot{\gamma}\lambda$, rising up to order $\eta\dot{\gamma}$ otherwise. In the range of small values of θ, Eq. (51) can be written as

$$K \frac{d^2\theta}{dy^2} = \eta\dot{\gamma}(\theta^2 + \lambda) \tag{52}$$

showing that the second derivative of θ stays virtually constant in the neighborhood of $\theta = 0$, to jump up suddenly to much larger values when θ becomes of order $\sqrt{\lambda}$ and larger. We can surmise that the rapid increase in θ, and in viscous torque, that takes place when θ has passed this critical angle generates another defect. In other words, we can perhaps assume that the director field "hanging" from neighboring defects under stationary conditions shows a maximum value of θ of order $\sqrt{\lambda}$. Then, if a is the distance between neighboring defects, we get from Eq. (52) the scaling law,

$$\frac{\eta\dot{\gamma}a^2}{K} = \frac{1}{\sqrt{\lambda}} \tag{53}$$

Equation (53) compares favorably with the ansatz of Eq. (39), that is, with the idea that the domain size a decreases with increasing shear rate in such a way that $\dot{\gamma}a^2$ is a constant. Equation (53) also offers a possible interpretation to a constitutive value of the characteristic Ericksen number in terms of the anisotropy ratio λ.

An important implication of the assumption we have made is that the director is everywhere (defects aside, of course) virtually parallel to the plane formed by the shear and the neutral directions. With respect to that plane, the director forms a small angle, which attains its maximum value (in the negative direction) halfway between defects. One should in fact remember that the director would like to tumble, and is resisted from doing so by the elastic torque that ultimately finds its anchoring port in the nearby defects. The model does not make predictions for what concerns orientations "within" that plane. These are probably determined by the type of defects, and by the symmetry of the problem, which should not distinguish left from right with respect to the shear direction.

The scaling of Eq. (53) cannot be maintained indefinitely when $\dot{\gamma}$ is decreased, however, as the corresponding indefinite growth of the distance a would eventually violate another physical limitation imposed by defect cores. So far, we have in fact ignored the requirement that the viscous torque within defect cores must not reach the point where it is overcome by that in the nematic. This event would occur notwithstanding the fact that the director of the nematic always remains in the plane formed by the shear and the neutral directions, just as a consequence of the fact that the volume of the nematic anchored to a defect becomes too large.

If we call d the size of the defect core, the balance equation of the two opposing torques is written as

$$\eta_d \dot{\gamma} d^2 = \eta \dot{\gamma} \lambda a^2 \tag{54}$$

where η_d is the relevant viscosity of the defect phase. The left-hand side of Eq. (54) is the viscous torque per unit length of defect line, $\eta_d \dot{\gamma}$ being the torque per unit volume of the defect phase. The right-hand side of Eq. (54) gives the overall torque in the volume of the nematic phase which is anchored to that defect (per unit length of the defect line); the torque per unit volume in the nematic is written in accordance with Eq. (52). We have not introduced a "λ-factor" of the defect phase because the defect phase is probably much less anisotropic than the nematic proper; the λ-factor of the defect phase is therefore of the order of unity.

The shear rate $\dot{\gamma}$ cancels in Eq. (54). Thus, Eq. (54) provides a shear-rate independent upper limit for the distance a, given by

$$a_{max}^2 = \frac{\eta_d}{\eta} \frac{1}{\lambda} d^2 \tag{55}$$

Equation (55) shows that a_{max} is itself a constitutive quantity, in no way related to the sample dimensions.

The $\dot{\gamma}$-dependent domain size a of Eq. (53) will be smaller than a_{max} if and only if $\dot{\gamma}$ obeys the following inequality:

$$\dot{\gamma} > \frac{K\sqrt{\lambda}}{\eta_d d^2} \tag{56}$$

We tentatively suggest that the inequality of Eq. (56) defines the lower boundary of Region II of the viscosity curve. Remembering that the upper boundary of Region II occurs at the onset of nonlinear behavior in the nematic itself, that is, when $\dot{\gamma} \approx D$, we can define the complete range of Region II in nondimensional form through the inequalities

$$\frac{K\sqrt{\lambda}}{D\eta_d d^2} < \frac{\dot{\gamma}}{D} < 1 \tag{57}$$

There remains to be shown that such a range indeed exists. In other words, we must still prove that the predicted value of the left-most group in Eq. (57), called G, is smaller than unity. To this end, we rewrite G as

$$G = \frac{K}{D\eta\ell^2} \frac{\eta}{\eta_d} \frac{\ell^2}{d^2} \sqrt{\lambda} \qquad (58)$$

where ℓ is the molecular (or interaction) characteristic length, such as that appearing in the potential of Eq. (49). The nondimensional group $K/(D\eta\ell^2)$ only contains properties of the nematic phase, and is of order unity. This can be seen by considering that $K \approx ckT\ell^2$, where c is the number density of molecules [78], and $\eta \approx ckT/D$ [6,33]. Hence, G reduces to the expression

$$G = \frac{\eta}{\eta_d} \left(\frac{\ell}{d}\right)^2 \sqrt{\lambda} \qquad (59)$$

Equation (59) shows that G is made up of three factors, each of which is smaller (possibly much smaller) than unity.

Since we do not have a prediction for the viscosity in Region I, we cannot be sure that the G group really marks the transition between Region I and Region II. More to the point, we can perhaps state that the inequalities

$$G < \frac{\dot{\gamma}}{D} < 1 \qquad (60)$$

(with G given by Eq. (59)) define the range of shear rates where the polydomain should obey the scaling laws discussed in Section III.B.

The dependence of domain size a on $\dot{\gamma}$ in a steady state shear flow is summarized in Fig. 3.2. In this figure, together with the behavior predicted by Eq. (55) (Region I), and by Eq. (53) (Region II), the transition to Region III is also indicated. The latter transition is marked by an upturn in the curve, corresponding to the onset of flow-aligning behavior in the nematic phase.

The dependence of G on polymer concentration in lyotropic systems is ambiguous. Since an increase in concentration moves the order parameter towards unity, molecular theory [32,33] predicts that λ either increases or decreases, depending on S itself (see Fig. 2.3), that is, depending on whether the nematic is close to the dynamic transition, or else close to a "perfect" order ($S = 1$), a condition that is difficult to ascertain with the actual LCPs which are in fact semi-rigid. A decrease is a sensible expectation for η/η_d, although perhaps to a lesser extent than it might superficially appear, because η in the present context is in fact essentially

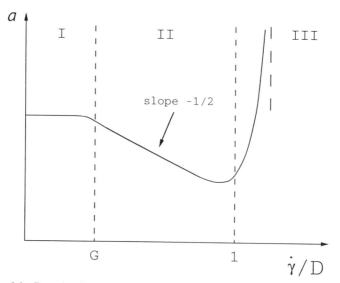

Figure 3.2 Domain size as a function of shear rate throughout the three regions reported in fig. 2.8. The group G (see Eq. (59)) should mark the transition from Region I to Region II. The size a diverges at the dynamic transition from tumbling to flow-aligning behavior (see Fig. 2.4).

$|\alpha_2|$ (cf. Eqs. (44) and (51)). On the other hand, ℓ^2/d^2 changes in the opposite direction, because an increase in the order parameter of the nematic decreases the core size d (see previous section). If G indeed marks the transition from Region I to Region II, then the experiments indicate that G grows with increasing polymer concentration, perhaps through a prevalence of the effect on core size. From the theoretical viewpoint, further progress should consider the actual topology, and the internal structure, of these defects.

Recently, significant experimental progress has been made in this regard. As briefly mentioned previously, De'Neve [85] was able to obtain electron micrographs of the defects that form in a tumbling thermotropic polymer in a shear flow. These micrographs show that the defect loops lay in the plane of shear, and are mostly formed by twist lines. The micrographs also indicate that the director orientation inside the loops is along the shear axis, while that outside the loops points in the neutral direction. De'Neve associates these orientations to the dynamic "attractors" examined by Larson and Öttinger [25]. This may well be true, but perhaps the observed orientations might also be explained directly through the topology of the twist defects, together with the symmetry of the shear flow. An explanation of the unexpected results of Hongladarom and Burghardt [70] could perhaps also be sought in the relaxation behavior of these loops.

Again considering the results of De'Neve [85] for a thermotropic LCP, it is worth emphasizing that, although the defect evidence indicates tumbling behavior, no negative normal stress effect was detected. The explanation offered by De'Neve, with which we agree, is that a negative N_1 is not observed simply because the accessible shear rates for normal stress measurement are too low for this system. Probably, a similar situation is encountered in other thermotropic systems, too. Yet another direct evidence of tumbling for a thermotropic LCP can be found in [86].

We note finally that the scaling argument developed in this section is in fact insensitive to whether or not the defect lines actually run parallel to the shear direction. The assumption was thought useful to better visualize the situation, and is probably a sensible scheme of the real defects well inside Region II. However, the argument survives after lifting that assumption, that is, it applies equally well to the case where the defect loops have some more complex structure. The three-dimensional configurations of the defects, and their change with increasing shear rate, might perhaps determine the shape of the viscosity curve in Region I. We are tempted by the idea that these highly viscous defect cores act as suspended fibers in the nematic medium.

IV. SUMMARY AND CONCLUSIONS

The excursus we have made through our present understanding of LCP rheology can be summarized in a few basic points. In the first place, one should remember that, almost invariably, LCPs are tumbling nematics, lyotropic and thermotropic polymers alike. This is the main reason why their rheology is so complex. From the point of view of applications, the tumbling nature must be accounted for, as it is prone to create a very complex pattern of molecular orientations. The tumbling nature of LCPs is of the utmost importance in shear flows, much less so in elongational deformations such as those encountered in a spinning line. In elongational flows, the LCP molecules behave very much like any other anisotropic particle; in fact, LCPs orient in the elongation direction better than ordinary polymers, because of the cooperative nature of a nematic phase. It is for this reason that the focus of this review was placed on shear flows only. On the other hand, shear flows are ubiquitous. A shear flow is present in the spinneret that feeds the spinning line, and a shear flow takes place when a sample is squeezed in between the optical glasses for microscopic observation.

Molecular modeling has been very successful in describing the nonlinear effects that take place at high shear rates. Starting from the early works of Hess [27] and Doi [28], a complex, yet fully convincing picture has gradually emerged which appears to fit, qualitatively at least, most observations in the nonlinear range. Further progress in the description might perhaps arise by accounting for semi-rigidity of the actual chains. This seems to be a very difficult task, however. At low shear rates, molecular modeling has proved useful in two ways. Through the work of Semenov [32] and Kuzuu and Doi [33], it has provided indications on how the Leslie coefficients vary with the order parameter, and how can it be explained that LCPs are so anisotropic in their viscous behavior. The second contribution of molecular modeling at low shear rates has to do with defect cores, and with their possible role in tumbling dynamics.

Another important point to keep in mind is that the Leslie–Ericksen theory never really applies to LCPs in a strict sense (except for specially designed experiments), yet almost always it can be applied to them (at low shear rates) in some slightly modified way. The modifications, if apparently minor ones, make all the difference, however. In this way, the scaling laws observed in LCPs during transient situations can be explained.

There remains a lot to be done in the way of understanding LCP flow behavior; it appears as though so far we have barely "scratched the surface". Indeed, in our review we have not done justice to a large number of experimental contributions, some of them highly enticing. The reason for not mentioning them at all, or maybe only too briefly, is that we do not understand the implications. For example, there are several works reporting data on LCP response in oscillatory shear, which are expected to reveal structural details. The behavior shown is in fact complex and system dependent. One can understand that changes are occurring, but what these changes really are, and why they occur, is not yet understood. Similarly obscure is the mechanism of band formation during relaxation of a shear flow; this phenomenon has also received much attention, but a convincing explanation is still lacking. And the examples might continue; we must work towards a more complete solution of the polydomain problem.

ACKNOWLEDGMENTS

Our work on LCP rheology has been funded at various stages by the Ministry for University (MURST), by the National Research Council (CNR), and by the European Community Commission under BRITE/EURAM contract No. BREU-0125-C(A). Our friends and coworkers, N. Grizzuti, S. Guido and P.L. Maffettone, who directly contributed to the process of unraveling the intricacies of LCP rheology, are gratefully acknowledged for their constant support during the preparation of this review.

APPENDIX A

A function of the unit vector **u** can be seen as a mapping on the unit sphere. The gradient $\nabla_u f$ of the scalar function $f(\mathbf{u})$ for any given **u** can therefore be visualized in the following way. First consider the plane tangent to the sphere at the "point" **u** (see Fig. A.1). The vector $\nabla_u f$ belongs to that plane, points towards the direction of the steepest ascent of the function, and its magnitude is determined by such maximum "slope".

In order to calculate $\nabla_u f$, it is expedient to proceed as follows. The function $f(\mathbf{u})$ is first extended to the case where **u** is no longer restricted to be a vector of unit magnitude; in other words, the three Cartesian components of **u** are looked upon as independent variables. Such a continuation of $f(\mathbf{u})$ is not unique; but any particular choice will do, that is, the nonuniqueness is irrelevant for what follows.

We then calculate the gradient of the continued function in the ordinary three-dimensional Cartesian space R^3; the components are then just the partial derivatives $\partial f/\partial u_x$, $\partial f/\partial u_y$, $\partial f/\partial u_z$. The vector representing this gradient is not tangent to the unit sphere; rather, it points towards some other direction, as indicated in Fig. A.1. Its projection onto the tangent plane, however, gives us the required quantity $\nabla_u f$. In order to accomplish the projection, we can make use of cross multiplications to the unit vector **u** itself. Thus, if the symbol $\partial f/\partial \mathbf{u}$ is used to indicate the gradient in R^3 of the continued function, we get

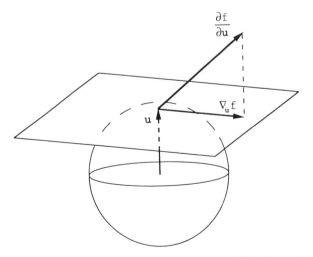

Figure A.1 Geometrical meaning of $\nabla_u f$. For any given **u**, $\nabla_u f$ is the projection of $\partial f/\partial \mathbf{u}$ on the tangent plane to the unit sphere.

$$\nabla_{\mathbf{u}} f = \mathbf{u} \times \frac{\partial f}{\partial \mathbf{u}} \times \mathbf{u}$$

For the sake of completeness, it is mentioned that the rotational operator used by Doi [6,28] differs from $\nabla_{\mathbf{u}} f$ only by one of the cross multiplications indicated above, that is, it can also be visualized as a vector lying on the tangent plane, pointing to the orthogonal direction to that of the steepest ascent of the function.

REFERENCES

1. A. Ciferri, Ed., *Liquid Crystallinity in Polymers*, VCH, New York, 1991.

2. A. M. Donald and A. H. Windle, *Liquid Crystalline Polymers*, Cambridge University Press, Cambridge, 1992.

3. P. G. deGennes, *The Physics of Liquid Crystals*, Clarendon Press, Oxford, 1974.

4. R. M. Kannan, J. A. Kornfield, N. Schwenk and C. Boeffel, *Macromolecules*, **26**, 2050 (1993).

5. S. Onogi and T. Asada, in *Rheology*, Vol. I, G. Astarita, G. Marrucci and L. Nicolais, Eds., Plenum Press, New York, 1980, pp. 127–147.

6. M. Doi and S. F. Edwards, *The Theory of Polymer Dynamics*, Clarendon Press, Oxford, 1986.

7. F. C. Frank, *Discuss. Faraday Soc.* **25**, 19 (1958).

8. J. L. Ericksen, *A.R.M.A.* **4**, 231 (1960).

9. F. M. Leslie, *A.R.M.A.* **28**, 265 (1968).

10. O. Parodi, *J. Phys.* **31**, 581 (1970).

11. P. Manneville, *Mol. Cryst. Liq. Cryst.* **70**, 223 (1981).

12. T. Carlsson, *Mol. Cryst. Liq. Cryst.* **104**, 307 (1984).

13. I. Zuniga and F. M. Leslie, *Liq. Cryst.* **5**, 725 (1989).

14. W. H. Han and A. D. Rey, in *Theoretical and Applied Rheology*, Vol. II, P. Moldenaers and R. Keunings, Eds., Elsevier, Amsterdam, 1992, pp. 531–533.

15. P. E. Cladis and S. Torza, *Phys. Rev. Lett.* **35**, 1283 (1975).

16. J. L. Ericksen, *Trans. Soc. Rheol.*, **13**, 9 (1969).

17. J. Fisher and A. G. Fredrickson, *Mol. Cryst. Liq. Cryst.* **8**, 267 (1969).

18. S. Chandrasekhar and G. S. Ranganath, *Adv. Phys.* **35**, 507 (1986).

19. N. J. Alderman and M. R. Mackley, *Faraday Discuss. Chem. Soc.* **79**, 149 (1985).

20. M. Srinivasarao, and G. C. Berry, *J. Rheol.* **35**, 379 (1991).

21. W. R. Burghardt and G. G. Fuller, *Macromolecules* **24**, 2546 (1991).

22. G. Marrucci, in *Theoretical and Applied Rheology*, Vol. I, P. Moldenaers and R. Keunings, Eds., Elsevier, Amsterdam, 1992, pp. 3–8.

23. M. Doi, *J. Phys.* **36**, 607 (1975).

24. M. Doi and S. F. Edwards, *J. Chem. Soc. Faraday Trans. 2* **74**, 560 (1978).

25. R. G. Larson and H. C. Öttinger, *Macromolecules* **24**, 6270 (1991).

26. J. G. Kirkwood and P. L. Auer, *J. Chem. Phys.* **19**, 281 (1951).

27. S. Hess, *Z. Naturforsch.* **31A**, 1034 (1976).

28. M. Doi, *J. Polym. Sci. Polym. Phys.* **19**, 229 (1981).

29. L. Onsager, *Ann. N. Y. Acad. Sci.* **51**, 627 (1949).

30. P. J. Flory, *Proc. R. Soc. London Ser. A* **234**, 73 (1956).

31. W. Maier and A. Saupe, *Z. Naturforsch.* **13A**, 564 (1958); **14A**, 882 (1959); **15A**, 287 (1960).

32. A. N. Semenov, *Sov. Phys. J.E.T.P.* **58**, 321 (1983).

33. N. Kuzuu and M. Doi, *J. Phys. Soc. Jpn.* **52**, 3486 (1983); **53**, 1031 (1984).

34. G. Marrucci, *Mol. Cryst. Liq. Cryst.* **72L**, 153 (1982).

35. M. Doi, *Faraday Symp. Chem. Soc.* **18**, 49 (1983).

36. G. Marrucci and P. L. Maffettone, *Macromolecules* **22**, 4076 (1989).

37. F. Cocchini, C. Aratari and G. Marrucci, *Macromolecules* **23**, 4446 (1990).

38. R. G. Larson, *Macromolecules* **23**, 3983 (1990).

39. G. Marrucci, *Rheol. Acta* **29**, 523 (1990).

40. G. Kiss and R. S. Porter, *J. Polym. Sci. Polym. Symp.* **65**, 193 (1978).

41. G. Kiss and R. S. Porter, *J. Polym. Sci. Polym. Phys.* **18**, 361 (1980).

42. P. Navard, *J. Polym. Sci. Polym. Phys.* **24**, 435 (1986).

43. P. Moldenaers, Ph.D. Thesis, Katholiecke Universiteit Leuven, 1987.

44. G. C. Berry, *Mol. Cryst. Liq. Cryst.* **165**, 333 (1988).

45. N. Grizzuti, S. Cavella, and P. Cicarelli, *J. Rheol.* **34**, 1293 (1990).

46. A. D. Gotsis and D. G. Baird, *Rheol. Acta* **25**, 275 (1986).

47. G. Marrucci and P. L. Maffettone, J. Rheol. **34**, 1217 (1990); **34**, 1231 (1990).

48. A. N. Semenov, *Sov. Phys. J.E.T.P.* **66**, 712 (1987).

49. P. L. Maffettone and G. Marrucci, *J. Rheol.* **36**, 1547 (1992).

50. J. J. Magda, S. G. Baek, L. de Vries and R. G. Larson, *Macromolecules* **24**, 4460 (1991).

51. P. L. Maffettone and G. Marrucci, *J. Non-Newtonian Fluid Mech.* **38**, 273 (1991).

52. P. L. Maffettone, *J. Non-Newtonian Fluid Mech.* **45**, 339 (1992).

53. S. P. Papkov, V. G. Kulichikhin, V. D. Kalmykova, and A. Ya. Malkin, *J. Polym. Sci. Polym. Phys.* **12**, 1753 (1974).

54. K. F. Wissbrun, *J. Rheol.* **25**, 619 (1981).

55. D. S. Kalika, D. W. Giles, and M. M. Denn, *J. Rheol.* **34**, 139 (1990).

56. G. Marrucci, in *Advances in Rheology*, Vol. 1, B. Mena, A. Garcia-Rejon and C. Rangel-Nafaile, Eds., Universidad Nacional Autonoma de Mexico, 1984, pp. 441–448.

57. K. F. Wissbrun, *Faraday Discuss. Chem. Soc.* **79**, 161 (1985).

58. R. G. Larson and M. Doi, *J. Rheol.* **35**, 539 (1991).

59. K. Hongladarom, W. R. Burghardt, S. G. Baek, S. Cementwala, and J. J. Magda, *Macromolecules* **26**, 772 (1993).

60. T. De'Neve, P. Navard, and M. Kleman, *J. Rheol.*, **37**, 515 (1993).

61. J. Happel and H. Brenner, *Low Reynolds Number Hydrodynamics*, Noordhoff, Leyden, 1973.

62. G. Marrucci, *Pure Appl. Chem.* **57**, 1545 (1985).

63. P. Moldenaers and J. Mewis, *J. Rheol.* **30**, 567 (1986).

64. J. Mewis and P. Moldenaers, *Mol. Cryst. Liq. Cryst.* **153**, 291 (1987).

65. H. L. Doppert and S. J. Picken, *Mol. Cryst. Liq. Cryst.* **153**, 109 (1987).

66. P. Moldenaers, G. G. Fuller, and J. Mewis, *Macromolecules* **22**, 960 (1989).

67. R. G. Larson and D. W. Mead, *J. Rheol.* **33**, 1251 (1989).

68. W. R. Burghardt and G. G. Fuller, *J. Rheol.* **34**, 959 (1990).

69. N. Grizzuti and P. Moldenaers, private communication.

70. K. Hongladarom and W. R. Burghardt, *Macromolecules* **26**, 785 (1993).

71. M. Doi and T. Ohta, *J. Chem. Phys.* **95**, 1242 (1991).

72. G. Marrucci, *Macromolecules* **24**, 4176 (1991).

73. F. Greco and G. Marrucci, *Mol. Cryst. Liq. Cryst.* **210**, 129 (1992).

74. T. C. Lubensky, *Phys. Rev.* **2A**, 2497 (1970).

75. A. Poniewierski and T. J. Sluckin, *Mol. Phys.* **55**, 1113 (1985).

76. N. Schopohl and T. J. Sluckin, *Phys. Rev. Lett.* **59**, 2582 (1987).

77. J. L. Ericksen, *IMA Preprint Series* N. 559 (1989).

78. G. Marrucci and F. Greco *Mol. Cryst. Liq. Cryst.* **206**, 17 (1991).

79. M. Doi, T. Shimada, and K. Okano, *J. Chem. Phys.* **88**, 4070 (1988).

80. F. Greco and G. Marrucci, *Mol. Cryst. Liq. Cryst.* **212**, 125 (1992).

81. G. Marrucci and F. Greco, *J. Non-Newtonian Fluid Mech.* **44**, 1 (1992).

82. T. De'Neve and P. Navard, in *Theoretical and Applied Rheology*, Vol. II, P. Moldenaers and R. Keunings, Eds., Elsevier, Amsterdam, 1992, pp. 522–523.

83. B. Ernst and P. Navard, *Macromolecules* **22**, 1419 (1989).

84. R. G. Larson and D. W. Mead, *Liq. Cryst.* **12**, 751 (1992).

85. T. De'Neve, Ph.D. Thesis, Ecole Nationale Superieure de Mines de Paris, 1993.

86. M. Srinivasarao, O. Garay, H. H. Winter, and R. S. Stein, *Mol. Cryst. Liq. Cryst.*, **223**, 29 (1992).

AUTHOR INDEX

Numbers in parentheses are reference numbers and indicate that the author's work is referred to although his name is not mentioned in the text. Numbers in *italic* show the pages on which the complete references are listed.

405

SUBJECT INDEX